ICY BODIES
OF THE SOLAR SYSTEM

IAU SYMPOSIUM No. 263

INTERNATIONAL ASTRONOMICAL UNION

UNION ASTRONOMIQUE INTERNATIONALE

ICY BODIES
OF THE SOLAR SYSTEM

PROCEEDINGS OF THE 263th SYMPOSIUM OF THE INTERNATIONAL ASTRONOMICAL UNION HELD IN RIO DE JANEIRO, RIO DE JANEIRO, BRAZIL AUGUST 3–7, 2006

Edited by

JULIO A. FERNÁNDEZ
Departamento de Astronomía, Faculdad de Ciencias, Montevideo, Uruguay

DANIELA LAZZARO
Observatório Nacional, COAA, Rio de Janeiro, Brasil

DINA PRIALNIK
Depto. of Geophysics & Planetary Sciences, Tel Aviv University, Tel Aviv, Israel

and

RITA SCHULZ
ESA Research abd Scientific Support Department, Noordwijk, The Netherlands

CAMBRIDGE UNIVERSITY PRESS
The Edinburgh Building, Cambridge CB2 8RU, United Kingdom
32 Avenue of the Americas, New York, NY 10013-2473, USA
477 Williamstown Road, Port Melbourne, VIC 3207, Australia
Ruiz de Alarcón 13, 28014 Madrid, Spain
Dock house, The Waterfront, Cape Town 8001, South Africa

First published 2010

Printed in the United Kingdom at the University Press, Cambridge

Typeset in System LATEX 2_ε

A catalogue record for this book is available from the British Library

Library of Congress Cataloguing in Publication data

This book has been printed on FSC-certified paper and cover board. FSC is an independent,
non-governmental, not-for-profit organization established to promote the responsible
management of the world's forests. Please see www.fsc.org for information.

ISBN 9780521764889 hardback
ISSN 1743–9213

Table of Contents

Part I. Overview

Part II. The Icy Planetesimals and Accretion Processes in the Protoplanetary Disk

Part III. Dynamical Aspects of Icy Bodies. The Oort Cloud

Part IV. Icy Satellites of the Outer Planets

Part V. Icy Dwarf Planets and TNOs

Part VI. Transition Objects

Preface

The study of the different populations of solar system small bodies is very important for understanding the accretion process in the protoplanetary disk, the different materials that condensed, and even the origin and transport of life among different worlds. This is particularly true for the ice-rich planetesimals formed beyond the snowline. Space missions, like Stardust, Deep Impact and Cassini, are bringing to us new insight about icy bodies like comets and satellites of the outer planets. Particularly interesting is the possibility that liquid water might have been present in the interiors of large icy bodies in the past, or even at present. Large surveys like Catalina, LINEAR and the upcoming Pan-STARRS or the Large Synoptic Survey Telescope (LSST) will allow us to improve our knowledge about the size and space distribution of populations of icy bodies like comets, Centaurs and TNOs.

The first ideas for this Symposium were raised during the 2006 IAU General Assembly, in Prague. At that time we felt that the great volume of new information about the different Solar System "icy bodies" would justify the proposal of a dedicated symposium, to be held in conjunction with the next IAU GA. Since its approval as part of the scientific program of the Rio de Janeiro IAU GA, we then aimed to attract planetary scientists from the different sub-areas and from a broad geographical distribution. To acheive this we took advantage of the fact that the IAU offers for this purpose a generous allotment of travel grants to assist colleagues with financial difficulties.

We are quite happy with the result: we received about 190 registrations, from which about 130 participants finally attended the symposium from around 20 different countries. The program was divided into 15 scientific sessions with 11 invited speakers, 48 oral contributions and 72 poster contributions, these discussed in three dedicated poster sessions. One key general review was also presented, as for the other symposia during the GA. The topics addressed in the symposium covered different aspects of icy bodies going from formation conditions in the protoplanetary disk, reservoirs and dynamical transport within the solar system, physics, space missions, and transition objects comet-asteroid, the latter a hot topic given the observation of activity in some main-belt asteroids. Last but not least, the relevance of icy bodies for life on Earth and elsewhere in the solar system was also addressed, in particular given the possibility that some large icy satellites of the Jovian planets might contain subsurface oceans. We present here part of the contributions to the symposium organized following the corresponding scientific program.

We are grateful to the IAU EC and Division Presidents for having selected this symposium to be held in conjunction with the XXVIIth General Assembly, an occasion when astronomers from all the fields of astronomy are gathered together. It is important to notice that this was the only symposium fully devoted to planetary sciences, an area that has had a great development in the last few decades. It is a great pleasure to acknowledge the members of the SOC, which ensured a very interesting scientific program, as well as the support from the National Organizing Committee of the XXVIIth IAU General Assembly. The confortable venue and the wonderful city of Rio de Janeiro resulted in the perfect setting for a memorable meeting.

Julio A. Fernández, Daniela Lazzaro, Dina Prialnik, Rita Schulz, Editors
Montevideo, Rio de Janeiro, Tel Aviv, Noordwijk, November 30, 2009

THE ORGANIZING COMMITTEE

Scientific

M.A. Barucci (France)
J.A. Fernández (co-chair, Uruguay)
M.P. Haynes (IAU, ex-officio)
K. Meech (USA)
D. Prialnik (Israel)
R. Schulz (co-chair, The Netherlands)
G. Valsecchi (Italy)

H. Campins (USA)
S. Ferraz-Mello (co-chair, Brazil)
Z. Knežević (Serbia)
K. Noll (USA)
H. Rickman (Sweden)
I. Toth (Hungary)
J.-I. Watanabe (Japan)

NOC of the XXVIIth IAU General Assembly

D. Lazzaro (chair)
A. Bruch
F.X. de Araújo
E. Janot-Pacheco
S. Lorenz-Martins
M.G. Pastoriza
L.P.R. Vaz

B. Barbuy (co-chair)
L. da Silva
J.R. de Medeiros
S.O. Kepler
W.J. Maciel
A. Silva-Valio
T. Villela

Acknowledgements

The symposium is sponsored and supported by the IAU Division III (Planetary Systems Sciences) and by the IAU Commissions No. 7 (Celestial Mechanics & Dynamical Astronomy), No. 15 (Physical Studies of Comets & Minor Planets), No. 20 (Positions & Motions of Minor Planets, Comets & Satellites), No. 22 (Meteors, Meteorites & Interplanetary Dust) and No. 51 (Bio-Astronomy).

The National Organizing Committee gratefully acknowledge the founding by the
International Astronomical Union,
Ministério da Ciência e Tecnologia – MCT,
Coordenação de Aperfeiçoamento de Pessoal de Nível Superior – CAPES/MEC,
Conselho Nacional de Desenvolvimento Científico e Tecnológico – CNPq/MCT,
Observatório Nacional – ON/MCT,
Laboratório Nacional de Astrofísica – LNA/MCT
Astronomy & Astrophysics
Fundação de Amparo à Pesquisa do Estado do Rio de Janeiro – FAPERJ
Fundação de Amparo à Pesquisa do Estado de São Paulo – FAPESP
and
Prefeitura do Rio de Janeiro.

Part I

Overview

Icy Bodies of the Solar System
Proceedings IAU Symposium No. 263, 2009   © International Astronomical Union 2010
J. Fernández, D. Lazzaro, D. Prialnik & R. Schulz, eds.   doi:10.1017/S1743921310001420

Icy Bodies in the New Solar System

David Jewitt

Dept. Earth and Space Sciences and Institute for Geophysics and Planetary Physics, UCLA,
3713 Geology Building, 595 Charles Young Drive East, Los Angeles, CA 90095-1567

Abstract. This brief paper summarizes a "key general review" with the same title given at
the IAU meeting in Rio de Janeiro. The intent of the review talk was to give a broad and
well-illustrated overview of recent work on the icy middle and outer Solar system, in a style
interesting for those astronomers whose gaze is otherwise drawn to more distant realms. The
intent of this written review is the same.

Keywords. Comets, Oort Cloud, Kuiper Belt, Main-belt comets.

1. Introduction

The last 20 years have seen an incredible burst of research on the small bodies of
the Solar system, particularly addressing the icy objects in its middle (from Jupiter
to Neptune) and outer (beyond Neptune) parts. This burst has been driven largely by
ground-based telescopic surveys, revealing previously unknown populations in regions
formerly thought to be empty. Through physical observations with the world's largest
telescopes, the characters of many known icy bodies have also been more firmly established. Separately, dynamicists have greatly added to our appreciation of the complexity
of the Solar system through their clever exploitation of ever-faster and cheaper computers
and the need to fit new observations. Lastly, several small bodies (comets, asteroids and
planetary satellites) have been approached by spacecraft, providing invaluable close-up
data on objects previously quite beyond human reach.

As a result of all this activity, many of us have come to realize that the small bodies,
although they contain a negligible fraction of the total mass in the Solar system, in fact
carry a disproportionately large fraction of the scientifically useful information. They
are macroscopic analogs of the radioactive elements (which, although they are mass-wise
insignificant play a central scientific role as our best cosmic chronometers). This is especially true concerning information about the origin and evolution of the Solar system,
for two reasons. Firstly, many of the small bodies have escaped substantial thermal alteration since the formation epoch. Their chemical and molecular constitutions therefore
approach the initial conditions, or at least approach them much more closely than do
large bodies like the Earth (in which the gravitational binding energy is so large as to
have caused melting at accretion). Secondly, the small bodies are so numerous, and their
orbits so accurate, that they can be used to map dynamical parameter space and to trace
processes occurring in the protoplanetary disk and after.

Figure 1 emphasizes the resulting change in perception of the Solar system in a visual
way. It illustrates the old view of the Solar system, in which there are nine important
objects outside the Sun, and a host of smaller, relatively unimportant bodies that have
been unceremoniously thrown into a bin called "Other". The irony of the new Solar
system is that it is precisely these throw-away "Other" objects which have provided
much of the scientific excitement and sense of renewal in our field. They are the objects
of this review.

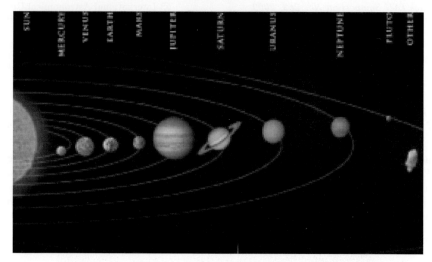

Figure 1. Your grandmother's Solar system: outside the Sun, 9 objects worthy of names and a grab-bag full of insignificant things called "Other". Recent research into "Other" has yielded incredible scientific treasure out of all proportion to the tiny mass of the objects therein. Figure courtesy NASA.

In this review, I will focus on the main reservoirs of icy objects in orbit about the Sun. These are, in the order of their discovery, the Oort cloud, the Kuiper belt and the outer asteroid belt. First, though, it is worth having a look at some of the objects of our attention.

1.1. *Photo-interlude*

Figure 2 shows in-situ images of the nucleus of Jupiter family comet comet 81P/Wild 2, taken from two different perspectives by NASA's Stardust mission. This is a ∼4 km scale icy object whose orbit is consistent with a source in the Kuiper belt. Although it is icy, the albedo of the nucleus is very small (∼0.06), probably because the ices in this

Figure 2. Nucleus of comet P/Wild 2, a recently escaped Kuiper belt object. The density of this ∼4 km diameter nucleus is not known, but believed to be less than 1000 kg m^{-3}, consistent with an ice-rich, porous make-up. Craters in the surface are probably not impact craters, but may result from mass-loss driven by sublimation. Image courtesy NASA's Stardust mission.

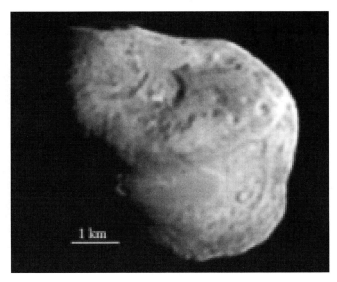

Figure 3. Nucleus of comet P/Tempel 1. The nucleus is irregular, with ∼6 km effective diameter, while the bulk density is uncertain, but lies in the range 200 to 1000 kg m^{-3}. Note the large tongue-shaped smooth region in the lower half of the nucleus (a ground-hugging flow on an object whose escape velocity is ∼1 m s^{-1}?) and the relative lack of large craters compared to Figure 2. Image courtesy NASA's Deep Impact mission.

bodies are buried in a refractory, perhaps carbon-rich mantle. The numerous depressions in the surface of P/Wild 2 at first glance resemble impact craters, but closer examination suggests that they are more likely to be related to the outgassing phenomenon. They are, frankly, not well understood.

A different Jupiter family comet (Kuiper belt source) nucleus is shown in Figure 3. While similar in size to P/Wild 2, this nucleus is less densely cratered and a large swath of the surface is occupied by a smooth, lobate flow-like feature. Again, the surface geology is a mystery for which several solutions have been proposed. The bigger mystery is why these two JFC nuclei should look so different.

Figure 4 shows Saturn's large irregular satellite, Phoebe. With diameter ∼220 km, it is nearly two orders of magnitude larger than the cometary nuclei shown in the previous figures. The density is about 1630 ± 45 kg m^{-3} (Porco *et al.* 2005), requiring a rock-ice composition but admitting the possibility of considerable internal porosity. The heavily cratered surface much more closely resembles the highly cratered Lunar highlands, suggesting that the influence of near-surface ice (and its sublimation) is much less important on this body (as expected for its more distant location; Saturn is at 9.5 AU from the Sun and far too cold for water ice to sublimate). Ices are identified in spectra of the surface. The significance of Phoebe is that it, and other irregular satellites, might have been captured by the planets following formation in the Kuiper belt.

The final Figure 5 shows Saturn's regular satellite Hyperion. This large (∼400 km scale) object is notable for several reasons. It has an aspherical shape that allows planetary torques to imbue this satellite with a chaotic rotation. It has a densely cratered ice-rich surface that is unlike any other so-far imaged. Most of the camera-facing side of Hyperion in Figure 5 is occupied by an impact crater with ∼20 km deep walls: the kinetic energy of the projectile that formed this crater must surely have been comparable to the energy needed to disrupt the satellite. And the density of Hyperion is a remarkably low 540 ± 50 kg m^{-3}, indicating substantial porosity. As a regular satellite, Hyperion

Figure 4. Saturnian satellite Phoebe, ∼220 km in diameter. Image courtesy
NASA's Cassini mission.

Figure 5. The 400 km scale Saturnian satellite Hyperion, an extraordinarily porous body
with a density only half that of water. Image courtesy NASA's Cassini mission.

probably formed in Saturn's accretion disk and has no likely connection with the major
Sun-orbiting ice reservoirs. Nevertheless, in its low density and highly battered surface
it is probably representative of many objects in the outer Solar system that have yet to
be imaged close-up by spacecraft.

2. The Three Ice Reservoirs

The three heliocentric ice reservoirs, the Oort cloud, the Kuiper belt and the main-belt,
are the main topics of this review. Other repositories of ice are also worthy subjects of
intensive scientific study, including the nuclei of comets, dead (or defunct) comets, the

Damocloids (probably dead Halley-type comets), the planetary satellites and the Trojans, both those of Jupiter and of Neptune. Limited space forces me to suppress them in this article.

The three reservoirs supply comets in three dynamical classes, the long-period comets (Oort cloud), the short-period (specifically "Jupiter family") comets (Kuiper belt) and the main-belt comets (outer asteroid belt). A simple way to think about the different comet types is in terms of their Tisserand parameters measured with respect to Jupiter (Vaghi 1973). The parameter is

$$T_J = \frac{a_j}{a} + 2\left[(1-e^2)\frac{a}{a_j}\right]^{1/2}\cos(i) \tag{2.1}$$

in which a_j and a are the semimajor axes of Jupiter and of the object and e and i are the eccentricity and inclination of the object's orbit (Jupiter is assumed to have $e_j = i_j = 0$ in this problem). This parameter, T_J, is a constant of the motion in the restricted three-body problem. In practice, it is related to the close-approach velocity relative to Jupiter. Jupiter has $T_J = 3$, comets from the Oort cloud have $T_J < 2$, comets from the Kuiper belt have $2 \leqslant T_J \leqslant 3$ and main-belt comets have $T_J > 3$. The parameter is an ambiguous discriminant when applied too closely (e.g. is an object with $T_J = 2.99$ really different from one with $T_J = 3.00$?) because the restricted three-body problem is only an approximation, there are important perturbations from many planets, and so on. But a majority of comets can be meaningfully linked to their source regions through this single number, T_J.

3. Oort Cloud

The principal evidence for the existence of the Oort Cloud lies in the distribution of the binding energies of long-period comets (Oort 1950). The binding energies are proportional to the reciprocal semimajor axes, as plotted in Figure 6. The distribution is peaked to small, positive (i.e. gravitationally bound) values of $1/a$, corresponding to the large semimajor axes of comets falling into the Solar system from a distant, bound source. On average, the energy of the orbit of an infalling comet will be slightly modified by interactions with the planets (note that the barycenter of the system lies near the photosphere of the Sun, rather than at its center, mostly as a result of the mass and 5 AU semimajor axis of Jupiter). The key observation made by Oort is that the width of the peak in Figure 6 is small compared to the expected change in orbital energy from planetary scattering, meaning that comets in the peak have not previously passed through the Solar system many times. The comets in the peak must include a large fraction of first-arrivals. Over time, scattering in successive revolutions deflects comets out of the peak, either to be lost back to interstellar space or captured into more tightly bound orbits (larger $1/a$ in Figure 6. The shape of the Oort cloud is inferred to be spherical based on the (within observational limits) isotropic distribution of the arrival directions of long-period comets.

Comets with $1/a < 0$ in Figure 6 include objects that have been scattered into unbound trajectories by planetary perturbations during previous orbits, and those whose orbits have been unbound by non-gravitational forces from anisotropic mass loss. There are no known examples of the strongly hyperbolic comets that would be expected if comets entered the Solar system from interstellar space (Moro-Martin *et al.* 2009).

The estimated size and mass of the Oort cloud have been revised over the past half Century. The cloud has no sharp edge, but the scale is 50,000 AU to 100,000 AU based

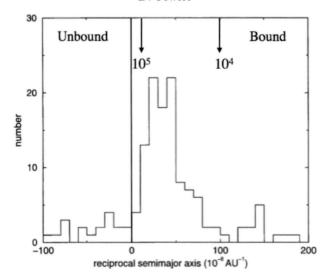

Figure 6. Distribution of the reciprocal semimajor axes of comets. Adapted from Marsden and Williams 1999.

on the best available comet orbit determinations. The number of comets in the cloud is estimated by comparing the rate of arrival of new long period comets with the rate of erosion expected from external perturbations due to the combined action of passing stars and the torque from our own galaxy (Higuchi *et al.* 2007). The estimate lies in the range 10^{11} to 10^{12} (Oort 1950, Francis 2002). The mass in the Oort cloud is less certain still, since the size distribution of the long-period comet nuclei is very poorly known. If all comets in the Oort cloud have radii of 1 km and densities equal to that of water, their combined mass would be in the range 0.1 $M_\oplus$ to 1 $M_\oplus$.

The emplacement of the comets into the Oort cloud requires a two (or more) step process. First, strong scattering by a growing planet leads to the ejection of comets from the protoplanetary disk of the Sun. Those ejected at substantially less than the local escape speed fall back into the disk to be scattered again (or to collide with a growing planet). Those ejected substantially faster than the local escape speed are lost forever into the interstellar medium. Comets in a narrow range of ejection speeds are susceptible to the action of external perturbations from passing stars and from the galactic tide. Their perihelia can be lifted out of the planetary domain and their orbits can be circularized to occupy the Oort cloud. Estimates of the efficiency of emplacement into the Oort cloud vary. Massive Jupiter (escape speed $\sim$60 km s^{-1} compared with local escape speed from the Solar system $\sim$20 km s^{-1}) tends to launch comets into the interstellar medium. Less massive Uranus and Neptune (escape speed $\sim$20 km s^{-1}, local Solar system escape speed $\sim$8 km s^{-1}) emplace comets into the Oort cloud more efficiently. Overall, the efficiency for emplacement is in the range 1% to 10%. If the mass of the Oort cloud is 0.1 to 1 $M_\oplus$, this means that $1M_\oplus$ to $100M_\oplus$ of comets must have been ejected during the process of planet formation.

3.1. *Current Issues*

• To fit the distribution of reciprocal semimajor axes (Figure 6) Oort was forced to invoke a "fading parameter", whose purpose is to decrease the number of returning comets relative to the number observed. In other words, dynamics alone would produce a flatter histogram for $1/a > 0$ than is observed. Oort suggested that freshly arriving comets might burn off a layer of "interstellar frosting" and so become fainter than expected on subsequent returns, thereby decreasing their number in any magnitude-limited plot. The "fading parameter" is really a "fudge parameter" needed to make the model fit the data, and the physical nature of the fading remains unspecified, although suggestions abound (Levison *et al.* 2002, Dones *et al.* 2004). It is tempting to speculate that the need for the fading parameter might be removed by some change in the dynamical part of the Oort cloud model but a recent careful study has failed to identify any such change (Wiegert and Tremaine, 1999, Dones *et al.* 2004). Still, if the incoming long-period comets really fade, the important question remains "why?".

• In Oort's model, the long-period comets arrive primarily from distant locations because the larger orbits are more susceptible to modification by tides from the Galaxy and from passing stars. It has been suggested that the Oort cloud might contain a more tightly bound component at smaller mean distances (e.g. 5,000 AU instead of 50,000 AU), known as the Inner Oort Cloud (IOC). These objects, if they exist, could provide a source from which to replenish the classical or Outer Oort Cloud (OOC). It has been suggested that Kuiper belt object Sedna is an IOC body but, with a perihelion of only 76 AU and a modest inclination and eccentricity, it seems more likely to have a dynamical connection to the Kuiper belt. It has also been suggested that the IOC is the *dominant* source of the long period comets and that, by implication, Oort's inferred much larger comet cloud is unimportant (Kaib and Quinn 2009).

Two big issues regarding the IOC remain unresolved. First, and most importantly, does the IOC exist? Precisely because this region is difficult to study observationally, we possess few serious constraints on its contents. We have to admit that, observationally, large numbers of objects with considerable combined mass could exist in the IOC without violating any observational constraints. But this does not mean that the objects are there. Second, how was this region populated (if it was)? The perturbations from passing stars and from the Galactic tide are too small to be responsible. Instead, a densely populated IOC might be explained if the Sun formed in a dense star cluster, where the mean distance between stars was much smaller than at present and the magnitude and rate of perturbations were both much larger. Formation in a cluster seems not unlikely: only a minority of stars form in isolation.

• Can we empirically constrain the mass that was ejected from the forming planetary system into the interstellar medium? The orbit of a planet launching material from the Solar system should be greatly affected if the ejected mass rivals that of the planet, which sets a loose upper bound. However, massive Jupiter (310 M$_\oplus$) anchors the system, and potentially very large masses could have been ejected.

4. Kuiper Belt

The Kuiper belt is the region beyond Neptune ($a_j = 30$ AU) in which vast numbers of small bodies have recently been discovered. At first thought to be a simple, unprocessed remnant of the Sun's accretion disk, the Kuiper belt has been shaped in numerous ways by previously unsuspected evolutionary processes. Significantly, the current mass is ~ 0.1 M$_\oplus$, Jewitt *et al.* 1996, Trujillo *et al.* 2001), which is far too small for the known Kuiper belt objects to have grown by binary accretion on any reasonable timescale (Kenyon and Luu 1999). The most likely explanation is that the modern-day Belt is the remnant of a (100× to 1000×) more massive initial ring. This primordial Kuiper belt could have shed its mass through some combination of dynamical processes (Gomes 2009) and collisional grinding.

The velocity dispersion in the belt is $\Delta V \sim 1$ to 2 km s^{-1} (Jewitt *et al.* 1996, Trujillo *et al.* 2001), comparable to the gravitational escape velocity of the largest objects. With

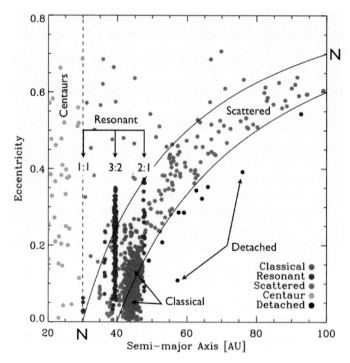

Figure 7. Semimajor axis vs. orbital eccentricity for the outer Solar system. The important dynamical sub-types of Kuiper belt objects are marked on the plot. Only three of the many populated resonances are indicated. The locus of orbits having perihelia $q = 30$ AU, equal to Neptune's semimajor axis, is shown as the curve NN. The lower but unlabeled curve shows the locus of orbits with $q = 40$ AU.

this ΔV, collisions in the belt are primarily erosive, not agglomerative, leading to the idea that the Kuiper belt is eroding away. It is reasonable to think of the Kuiper belt as the Sun's own "debris disk" (Wyatt 2008); a ring of parent bodies which collide to produce dust. In fact, Kuiper belt dust has already been detected by the Voyager spacecraft (Gurnett *et al.* 1997, although the investigators apparently did not realize the source), leading to an estimate of the normal optical depth in dust $\tau_{kb} = 10^{-7}$ (Jewitt and Luu 2000). For comparison, the corresponding optical depths of the best-known (brightest) nearby debris disks are $\tau \sim 10^{-3}$ to 10^{-4} (Wyatt 2008). If displaced to a nearby star, our own Kuiper belt would fall far below the thresholds of observational detection.

   The orbits of Kuiper belt objects (KBOs) are empirically grouped into different sub-types (see Figure 7). Most classical KBOs orbit between about 42 AU and 48 AU and have small eccentricities ($e \lesssim 0.2$). They are dynamically stable on Gyr timescales basically because their perihelia never approach the dominant local perturber, Neptune. Resonant KBOs can cross Neptune's orbit, but are phase-protected from close encounters by their resonant locations and hence can survive on Gyr timescales. The most famous resonant KBO is 134340 Pluto. Resonant KBOs provide strong evidence for past planetary migration (Fernandez and Ip 1984, Malhotra 1995). The Scattered KBOs have higher eccentricities ($e \sim 0.5$, see Figure 7) and perihelia in the $30 \leqslant q \leqslant 40$ AU range. Numerical integrations show that the orbits are unstable on Gyr timescales owing to perihelic interactions with Neptune. These interactions are responsible for kicking up the eccentricities, sending the Scattered KBOs out to large distances (the current record aphelion is held by 2000 OO67, at $Q = 1300$ AU). The long-term instability of this population means that it is a remnant that has been dynamically depleted over the age of the Solar system, perhaps by factors of a few to ten (Volk and Malhotra 2008). The Detached KBOs are like the Scattered objects except that their perihelia lie beyond 40 AU, where integrations suggest that Neptune perturbations are unimportant over the age of the Solar system. The orbits of these objects suggest that another force, perhaps perturbations from an unseen planet (Gladman and Chan 2006, Lykawka and Mukai 2008), from a passing star (Ida *et al.* 2000), or resonant effects (Gomes *et al.* 2005) must have lifted the perihelia out of the planetary domain. The Centaurs, while strictly not part of the Kuiper belt, are thought to be recent escapees from it. Various definitions exist: the simplest is that they have perihelia *and* semimajor axes between Jupiter (5 AU) and Neptune (30 AU). The Centaurs interact strongly with the giant planets and have correspondingly short lifetimes, with a median of order 10 Myr (Tiscareno and Malhotra 2003, Horner *et al.* 2004).

4.1. *Current Issues*

- What was the initial mass and radial extent of the Kuiper belt?
- How was that mass lost?
- How were the Detached KBOs emplaced?
- From where in the Kuiper belt do the Jupiter family comets originate?
- Why does the Classical KBO orbit distribution have a sharp outer edge near 48 AU?
- Did the young Kuiper belt more closely resemble the debris disks of other stars?

5. Main-Belt Comets

The Oort Cloud and Kuiper Belt comet reservoirs are relatively well known but the comets of the main-belt represent a still new and relatively unfamiliar population. The main-belt comets (MBCs) are distinguished from other comets by having asteroid-like $(T_J > 3, a < a_j)$ orbits (Hsieh and Jewitt 2006). In fact, except for their distinctive physical appearances (including comae and tails) these objects would remain indistinguishable from countless main-belt asteroids in similar orbits, as is evident from Figure 8.

The spatially resolved comae and tails of the MBCs are caused by sunlight scattered from $\sim$10 μm-sized dust particles. Several processes might eject grains from the MBCs, but the only one that seems plausible is gas drag from sublimated ice. For example, a rapidly rotating asteroid might lose its regolith centripetally. Electrostatic charging of the surface by Solar photons and/or the Solar wind might cause potential differences that could launch particles, as has been observed on the Moon. But neither of these mechanisms can explain why mass loss from the best-observed MBC (133P/Elst-Pizarro) is episodic and recurrent, being concentrated in the quarter of the orbit after perihelion. As in many short-period comets at the same distances, gas has not yet been detected; limits to the gas production based on spectroscopic data are of order 1 kg s^{-1} (Jewitt *et al.* 2009), some 2 to 3 orders of magnitude smaller than in typical near-Earth comets.

It is surprising that ice has survived for billions of years in this high temperature $(T \sim 150$ K at $a = 3$AU) location. Stifling of sublimation by a surface refractory mantle of meter-scale thickness is part of the answer (Schorghofer 2008, c.f. Prialnik and Rosenberg 2009). Sublimation (at $\sim$1 m yr^{-1}) should lead to the production of mantles on cosmically very short timescales (100 yr?). Activation of mass loss then requires a trigger; we suspect that impacts by boulders are responsible (Figure 9), but have no proof. The interval between sufficient impacts is likely $\sim$10^4 yr, giving a duty cycle of $\sim$1%.

Evidence for the past presence of water in outer-belt asteroids is strong. Solid-state spectroscopic features that are indicative of hydrated minerals have been found in 10%

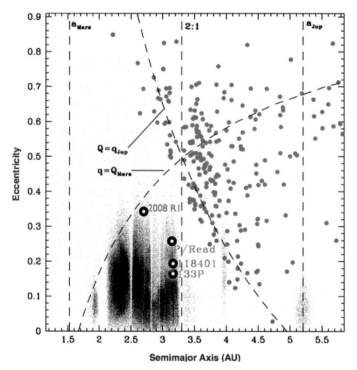

Figure 8. Semimajor axis vs. orbital eccentricity for the inner Solar system. The classical comets (mostly short-period comets derived from the Kuiper belt) are shown as grey circles in the upper right of the diagram. Main-belt asteroids are represented by single-pixel black points, mostly in the lower left. The four known MBCs are indicated with large, hollow circles. Clearly, they occupy the same region of the diagram as is occupied by the main-belt asteroids. Vertical dashed lines show the semimajor axes of Mars and Jupiter and the location of the 2:1 mean-motion resonance with Jupiter. Dashed arcs show the locus of orbits that are Mars-crossing (labeled $q = Q_{Mars}$) and Jupiter-crossing ($Q = q_{Jup}$).

or more of asteroids beyond 3 AU. More directly, meteorites from the outer-belt contain hydrated minerals such as serpentine, carbonates and halite. To the author's eyes, the most incredible example is shown in Figure 10, in which is pictured a chondrite meteorite that was observed to fall near the town of Monahans in Texas (Zolensky *et al.* 1999). The meteorite fell in a dry place and was collected and immediately handed to scientists for study, precluding the possibility of Terrestrial contamination. Inside the meteorite are found, among other clear diagnostics for the past presence of *liquid* water, millimeter-scale cubic halite (NaCl) crystals. The age of the halite determined from Rb-Sr radioactive decay is 4.7 ± 0.2 Gyr, essentially equal to the age of the Solar system. Some of these salt crystals further contain brine pockets (10 μm scale) and, in Figure 10, a gas bubble floating in a brine pocket is shown (Zolensky *et al.* 1999). This remarkable specimen leaves no room for doubt about the role of water in the asteroid belt. The asteroid parent of Monahans contained ice that was melted, probably in the first few Myr after the formation of the Solar system. The liquid water reacted chemically with adjacent minerals forming hydrated species. Evaporation lead to increasing salinity and the formation of brines in which salt crystals grew. Then, as ^{26}Al and other short-lived elements decayed away, the asteroid cooled and any remaining water froze. A subsequent collision blasted off fragments, one of which struck Texas and appears in Figure 10.

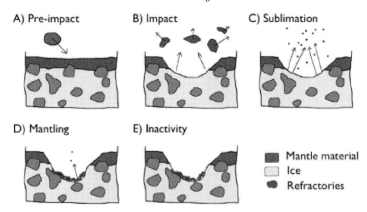

Figure 9. Conceptual model of the trigger process for main-belt comets. A) The pre-impact surface is mantled in refractory matter (dark layer) beneath which ice (light) and refractory blocks (grey) coexist. B) Impact of a meter-sized interplanetary boulder blows off the mantle, exposing ice. C) The ice sublimates, propelling entrained dust into the coma and tail of the comet. D) Refractory blocks too large to be ejected settle in the bottom of the crater eventually, E), stifling further mass loss. The process is repetitive with the interval between impacts of meter-sized boulders being perhaps 5,000 to 10,000 yrs. The stifling of sublimation should be much more rapid, perhaps taking only ∼100 yrs of exposure at 3 AU.

Watery asteroids are especially interesting in the context of the origin of the Earth's oceans and other Terrestrial planet volatiles. It is likely that the Earth accreted at temperatures too high for water to be held by adsorption (indeed, the Earth took 50 to 100 Myr to attain final mass and was molten for most of this time). Instead, volatiles may have been retained only after the Earth had cooled to form a surface crust. Impacts of comets and water-containing asteroids are the most likely sources of these volatiles. Recent discussion has moved against delivery of water by comets, based on the difference between the mean deuterium to hydrogen ratio measured in comets ($D/H \sim 3{\times}10^{-4}$) and the ratio in standard mean ocean water (SMOW $D/H = 1.6{\times}10^{-4}$) (Meier and Own 1999). The cometary measurements are, however, of low significance and limited to a few comets, with recent evidence suggesting that a wider range might be present. It has been asserted that the asteroid belt represents a more efficient source of water, in the sense that impact probabilities from sources in the asteroid belt are higher than from corresponding sources in the Kuiper belt (Morbidelli *et al.* 2000). Most likely, water is a mix from different sources, but the outer asteroid belt is almost certain to be one of these.

Where did the MBCs form? Occam's Razor suggests that they formed in-place, a possibility that is consistent with current models and data in which the snow-line in the protoplanetary disk spent some time closer to the Sun than the current asteroid belt (i.e. asteroids could have trapped ice at formation). No convincing dynamical pathway from the Kuiper belt has been identified, given the current layout of the Solar system. However, if the layout were substantially different in the past (e.g. because of planetary migration) then it is possible that some objects from the outer regions could have been trapped in the asteroid belt.

5.1. *Current Issues*

- How many MBCs are there? Do *all* outer-belt asteroids contain ice?
- Is the triggered sublimation model of MBC activity (c.f. Figure 9) correct?

Figure 10. The incredible H5 chondrite Monahans, in which are found the clear products of aqueous alteration, including cubic salt (NaCl) crystals (upper left). Inside the salt are several liquid brine pockets (upper right) and, inside some of the brine pockets, are gas bubbles. The halite is radioactively dated to 4.7 ± 0.2 Gyr. Monahans is most likely a piece of an outer asteroid belt object in which much of the ice melted soon after formation, and boiled away. It is perhaps closely related to the MBCs. Figure adapted from Zolensky *et al.* (1999).

- How is the ice in MBCs isotopically or compositionally distinct from the ice in comets from the Oort or Kuiper reservoirs?
- Did they form locally (as seems most likely) or were they captured from elsewhere?

6. The Unknown

The clear and very uplifting lesson of all this is that we remain only dimly aware even of our immediate astronomical backyard. The Kuiper belt with its unexpected dynamical subdivisions and structures, the main-belt comets, the (perhaps) densely populated inner Oort cloud, were all completely unknown when, for example, most of the readers of this article were born. The obvious next question is "What else is out there?".

To gain some idea of the vast swaths of the Solar system in which substantial objects could exist (even in large numbers) and yet remain undetected, see Table 1. The Table shows that Earth, for example, would appear at red magnitude 24 if displaced to ∼600 AU and would certainly be undetected in any existing all-sky survey. Even ice giants (Neptune in Table 1) and gas giants (Jupiter) would be undetected given the present state of our all-sky surveys. It is widely assumed that unseen massive planets would be detected by their gravitational perturbations of the known planets (or comets). Table 1 shows that this is not the case, and optical detection (if taken to magnitude 24 over the whole sky) is the more sensitive method. While pointing to our extreme state of ignorance concerning the contents of the Sun's potential well, the Table also can be read as a sign of good news. There is lots of work to do and all-sky surveys to 24th magnitude (and beyond) are both technologically feasible and planned. All we need now is money.

Table 1. What Else is Out There?[1]

Object	$V(1,1,0)$[2]	R_{24}[AU][3]	R_{grav}[AU][4]
Pluto	−1.0	320	−
Earth	−3.9	620	50
Neptune	−6.9	1230	130
Jupiter	−9.3	2140	340

Notes:
[1] Modified from Jewitt (2003).
[2] The absolute magnitude, equal to the apparent magnitude when corrected to unit heliocentric and geocentric distances and to zero phase angle.
[3] The distance at which the apparent red magnitude falls to $m_R = 24$, the approximate limit of the planned Large Synoptic Survey Telescope.
[4] The distance beyond which the gravitational influence of the object on the major planets would go undetected. Pluto has no entry because its gravitational influence is undetectable at any relevant distance. Calculations are from Hogg *et al.* 1991 (see also Zakamska and Tremaine 2005).

Acknowledgements

I thank Michal Drahus and Aurelie Guilbert for comments, and NASA's Planetary Astronomy program for support.

References

Dones, L., Weissman, P. R., Levison, H. F., & Duncan, M. J. 2004, in: M. C. Festou, H. U. Keller & H. A. Weaver (eds.), *Comets II*, (Tucson: Univ. Arizona Press), p. 153

Fernandez, J. A. & Ip, W.-H. 1984, *Icarus*, 58, 109

Francis, P. J. 2005, *Ap.J.*, 635, 1348

Gladman, B. & Chan, C. 2006, *Ap.J. Lett.*, 643, L135

Gomes, R. D. S. 2009, *CeMDA*, 104, 39

Gomes, R. S., Gallardo, T., Fernández, J. A., & Brunini, A. 2005, *CeMDA*, 91, 109

Gurnett, D. A., Ansher, J. A., Kurth, W. S., & Granroth, L. J. 1997, *Geophys. Res. Lett.*, 24, 3125

Higuchi, A., Kokubo, E., Kinoshita, H., & Mukai, T. 2007, *AJ*, 134, 1693

Hogg, D. W., Quinlan, G. D., & Tremaine, S. 1991, *AJ*, 101, 2274

Horner, J., Evans, N. W., & Bailey, M. E. 2004, *MNRAS*, 354, 798

Hsieh, H. H. & Jewitt, D. 2006, *Science*, 312, 561

Ida, S., Larwood, J., & Burkert, A. 2000, *ApJ*, 528, 351

Jewitt, D., Luu, J., & Chen, J. 1996, *AJ*, 112, 1225

Jewitt, D. C. & Luu, J. X. 2000, in: V. Mannings, A. P. bOss & S. S. Russel (eds.), *Protostars and Planets IV*, (Tucson: Univ. Arizona Press), p. 1201

Jewitt, D. 2003, *Earth Moon and Planets*, 92, 465

Jewitt, D., Yang, B., & Haghighipour, N. 2009, *AJ*, 137, 4313

Kaib, N. A. & Quinn, T. 2009, *Science*, 325, 1234

Kenyon, S. J. & Luu, J. X. 1999, *ApJ*, 526, 465

Levison, H. F., Morbidelli, A., Dones, L., Jedicke, R., Wiegert, P. A., & Bottke, W. F. 2002, *Science*, 296, 2212

Lykawka, P. S. & Mukai, T. 2008, *AJ*, 135, 1161

Malhotra, R. 1995, *AJ*, 110, 420

Marsden, B. G. & Williams, G. V. 1999, in: B. G. Marsden & G. V. Williams (eds.) *Catalogue of cometary orbits, 13th ed.*(Cambridge: Central Bureau for Astronomical Telegrams and Minor Planet Center)

Morbidelli, A., Chambers, J., Lunine, J. I., Petit, J. M., Robert, F., Valsecchi, G. B., & Cyr, K. E. 2000, *Meteorit. and Planet. Sci.*, 35, 1309

Moro-Martin, A., Turner, E. L., & Loeb, A. 2009, *ApJ*, 704, 733

Oort, J. H. 1950, *Bull. Astron. Inst. Neth*, 11, 91

Porco, C. C., *et al.* 2005, *Science*, 307, 1237

Prialnik, D. & Rosenberg, E. D. 2009, *MNRAS*, 399, L79

Schorghofer, N. 2008, *ApJ*, 682, 697

Tiscareno, M. S. & Malhotra, R. 2003, *AJ*, 126, 3122

Trujillo, C. A., Jewitt, D. C., & Luu, J. X. 2001, *AJ*, 122, 457

Vaghi, S. 1973, *Astron. Ap.*, 24, 107

Volk, K. & Malhotra, R. 2008, *ApJ*, 687, 714

Wiegert, P. & Tremaine, S. 1999, *Icarus*, 137, 84

Wyatt, M. C. 2008, *Ann. Rev. Astron. Ap.*, 46, 339

Zakamska, N. L. & Tremaine, S. 2005, *AJ*, 130, 1939

Zolensky, M. E., Bodnar, R. J., Gibson, E. K., Jr., Nyquist, L. E., Reese, Y., Shih, C.-Y., &
    Wiesmann, H. 1999, *Science*, 285, 1377

Part II

The Icy Planetesimals and Accretion Processes in the Protoplanetary Disk

Icy Bodies of the Solar System
Proceedings IAU Symposium No. 263, 2009
J.A. Fernández, D. Lazzaro, D. Prialnik & R. Schulz, eds.

The Location of the Snow Line in Protostellar Disks

Morris Podolak

Dept. of Geophysics and Planetary Sciences, Tel Aviv University,
Tel Aviv, Israel
email: morris@post.tau.ac.il

Abstract. The snow line in a gas disk is defined as the distance from the star beyond which the water ice is stable against evaporation. Since oxygen is the most abundant element after hydrogen and helium, the presence of ice grains can have important consequences for disk evolution. However, determining the position of the snow line is not simple. I discuss some of the important processes that affect the position of the snow line.

Keywords. Planetary systems: protoplanetary disks, dust, extinction, solar system: formation

1. Introduction

Oxygen is the most abundant element after hydrogen and helium, and water is one of the most abundant oxygen-bearing molecules. When water is a vapor, the mass fraction of solids in a solar composition disk is $\sim 4.2 \times 10^{-3}$. After water condenses, the solid mass fraction jumps to $\sim 1.3 \times 10^{-2}$ (Anders & Grevasse 1989). The properties of a protostellar disk will therefore change significantly when we cross the boundary between water vapor and ice grains. These changes can have important consequences for other processes taking place in the disk.

One consequence is the change in opacity that accompanies the transition from a region where water is a vapor to a region where it is a solid. This is illustrated in Fig. 1 which is based on the Rosseland mean opacities of Pollack *et al.* (1985). At temperatures above around 170 K, the opacity is due to mainly rocky grains, but just below this temperature the opacity jumps by nearly a factor of 5 because of the sudden condensation of water vapor to ice grains. At still lower temperatures the opacity again drops because, in taking the Rosseland mean for low temperatures, the longer wavelengths are given more weight. At these longer wavelengths the ice grains are inefficient scatterers and absorbers, so the opacity decreases. Such changes in opacity can have an important effect on the evolution of the gas disk.

Another example of the possible significance of the snow line is the mechanism suggested by Stevenson and Lunine (1988) for forming Jupiter. Sunward of the snow line the disk temperature is high enough so that all the water is in the vapor phase. Once you cross the snow line and water begins to condense, the mass fraction of water in the vapor phase decreases. This causes a composition gradient across the snow line which drives diffusion of water vapor into the colder region and causes this region to become enhanced in solids. Stevenson and Lunine (1988) argued that this enhancement could lead to the formation of Jupiter. The problem with this scenario is that this mechanism forms only one giant planet. Some other mechanism would be required for Saturn. But

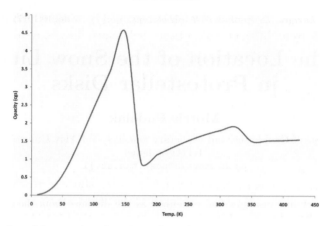

Figure 1. Opacity of dust grains thought to have been present in the early solar nebula as a function of temperature. The opacities are taken from Pollack *et al.* (1985).

if it is correct, then the snow line had to have been somewhere in the vicinity of Jupiter, i.e. at around 5 AU.

Unfortunately, Jupiter appears to be sending us contradictory messages. An analysis of Jupiter's composition by Lodders (2004) finds that Jupiter is anomalously low in water, and concludes that the snow line was much further away from the sun; perhaps in the neighborhood of 10 AU.

A third reason for trying to determine the position of the snow line is that it tells us where we can expect to see icy bodies. This has particular significance in view of the recent discovery of main-belt comets (Hsieh and Jewitt 2008). These bodies have orbits that are dynamically similar to those of the main belt asteroids (Jewitt *et al.* 2009) yet display cometary activity. If main-belt comets were indeed formed in the asteroid belt, the snow line must have been closer to the sun than around 3 AU.

2. Theoretical Calculation of the Snow Line

What can theory tell us? One early attempt to compute the position of the snow line was by Hayashi (1981). He argued that the stellar flux should decrease as r^{-2} where r is the distance from the star, so that the temperature at some point in the disk, $T(r)$ should be given by

$$\sigma T(r)^4 = \frac{1}{4}\sigma T_*^4 \left(\frac{R_*}{r}\right)^2 \tag{2.1}$$

Here R_* is the stellar radius, T_* is the stellar surface temperature, and σ is the Stefan-Boltzmann constant. For a solar composition gas at a pressure typical for protostellar disks the temperature at which water starts to condense is roughly 170 K. Setting $T(r) = 170\,\mathrm{K}$ and using the solar radius and temperature in eq. (2.1) gives $r = 2.7\,\mathrm{AU}$. This is very close to the value needed to explain the main-belt comets.

The problem is that Hayashi's model is too simple. More modern views of the proto-stellar disk see it as physically thin, and optically thick. In the case of a *passive disk* with no internal heat sources, the temperature is determined by integrating the flux falling on a point in the disk from different parts of the stellar surface. The resulting temperature distribution is [see, e.g. Armitage (2009)]

$$\sigma T_{rad}^4(r) = \frac{1}{\pi}\sigma T_*^4 \left[\sin^{-1}\left(\frac{R_*}{r}\right) - \left(\frac{R_*}{r}\right)\sqrt{1 - \left(\frac{R_*}{r}\right)^2} \right] \approx \frac{2}{3\pi}\sigma T_*^4 \left(\frac{R_*}{r}\right)^3 \quad (2.2)$$

for $r \gg R_*$. Since this temperature is due to radiative heating I have denoted it by $T_{rad}(r)$

This expression too must be modified to allow for the fact that the disk is flared, and the outer regions present a more direct face to the star and therefore absorb more light. As a result the flaring is even greater, and the coefficient $2/(3\pi)$ in eq. 2.2 must be replaced by some function of r. Finally, if the central star is accreting material from the disk, there will be a flux of material moving through the disk which will provide a source of viscous heating. This provides an additional flux that can be written as (Armitage 2009)

$$\sigma T_{visc}^4(r) = \frac{3GM_*\dot{M}}{8\pi r^3}\left(1 - \sqrt{\frac{R_*}{r}}\right) \quad (2.3)$$

where M_* is the mass of the central star, $\dot{M}$ is the mass accretion rate, and G is Newton's constant. The actual temperature profile in the disk will be given by the sum of the radiative and viscous fluxes

$$\sigma T^4(r) = \sigma T_{rad}^4(r) + \sigma T_{visc}^4(r) \quad (2.4)$$

Sasselov & Lecar (2000) solved eq. (2.4) numerically, and found that the snow line fell between 0.7 and 1.5 AU, depending on the strength of the accretional heating. While this allows *in situ* formation of the main-belt comets, it also implies that *all* the main-belt asteroids should be icy bodies. This would appear to put the snow line too close to the sun.

3. Grain Thermal Balance

Although the model of Sasselov & Lecar (2000) does a much better job of calculating the temperature of the disk, is still assumes that the temperature at the snow line is 170 K. The actual condensation temperature of water vapor will depend on the partial pressure of water vapor in the disk, and this will vary from place to place. In addition, the temperature of an ice grain will not necessarily be the same as the temperature of the ambient gas. Rather, it will depend on the details of energy balance with the surrounding medium. This energy balance is encompassed in the three heating and two cooling terms given below [see Mekler & Podolak (1994) and Podolak & Mekler (1997) for further details]. In all cases the heating and cooling are per unit area of the grain.

<u>Radiative Heating</u> - The heating of the grain by radiation from the star (for the optically thin case) or the surrounding medium (for the optically thick case). Here a is the grain radius, λ is the wavelength of the radiation, and $S(\lambda)$ is the wavelength dependent

radiation field. Q_{abs} is the efficiency of absorption per unit area and is a function of a, λ, and the grain material. It can be computed from Mie theory (van de Hulst 1957).

$$E^h_{rad} = \frac{1}{4r^2} \int_0^\infty Q_{abs}(a, \lambda)S(\lambda)d\lambda \qquad (3.1)$$

Gas Heating - The heating of the grain by contact with the gas. Here n_{H_2} is the number density of hydrogen molecules in the gas, m_{H_2} is the mass of a hydrogen molecule, T_{gas} and T_{grain} are the gas and grain temperatures, respectively, j is the number of thermodynamic degrees of freedom of the hydrogen molecule, and k is Boltzmann's constant.

$$E^h_{gas} = \frac{n_{H_2}}{4} \sqrt{\frac{8kT_{gas}}{\pi m_{H_2}}} \frac{jk(T_{gas} - T_{grain})}{2} \qquad (3.2)$$

Note that if the grain is warmer than the gas, this term is negative.

Condensation Heating - The heating of the grain due to water condensing on it. Here n_{H_2O} is the number density of water molecules in the gas, m_{H_2O} is the mass of a water molecule, and q is the latent heat released when a water molecule condenses onto the ice grain.

$$E^h_{cond} = \frac{n_{H_2O}}{4} q \sqrt{\frac{8kT_{gas}}{\pi m_{H_2O}}} \qquad (3.3)$$

Radiative Cooling - The cooling of the grain by radiation to space. Here $Q_{emis} = Q_{abs}$ is the efficiency factor for emission of radiation, and F is the Planck function at the temperature of the grain.

$$E^c_{rad} = \int_0^\infty Q_{emis}(a, \lambda)F(\lambda, T_{grain})d\lambda \qquad (3.4)$$

Evaporative Cooling - The cooling of the grain by water sublimation. Here P_{vap} is the vapor pressure of water.

$$E^c_{evap} = q \frac{P_{vap}(T_{grain})}{\sqrt{\pi m_{H_2O} kT_{grain}}} \qquad (3.5)$$

In steady state the heating and cooling terms must be equal, and we can use the above equations to find T_{grain}. The fluxes due to the different terms are shown in Fig. 2 for the case of a $10\,\mu$m pure ice grain in an optically thick disk.

The disk model used in these calculations is taken from the work of Bell et al. (1997) and corresponds to the model parameters $\alpha = 10^{-4}$ and $\dot{M} = 10^{-8}$. We can see that near the star the major heating term is contact with the gas, and this is offset by evaporative cooling. The grains are kept much cooler than the gas, and they evaporate in a very short time. Further from the star the radiative heating becomes important, but evaporation is still the major cooling term. Since evaporative cooling is strongly dependent on grain temperature, it drops quickly with distance from the star, and by around 2.5 AU radiative cooling becomes the dominant cooling mechanism. At around 3.5 AU the evaporative flux equals the condensation flux, and the snow line is reached.

Because the temperatures near the snow line are low, the Planck peak of the radiation is in the tens of microns range, and is large compared to the radius of the grain. As a result grains of $10\,\mu$m and smaller are small compared to the wavelength and are inefficient both at emission and absorption. In addition, since most of the radiation is in the far IR, where water has strong absorption features, pure ice grains will behave pretty much like grains composed of a mixture of water and some darker material ("dirt"). As

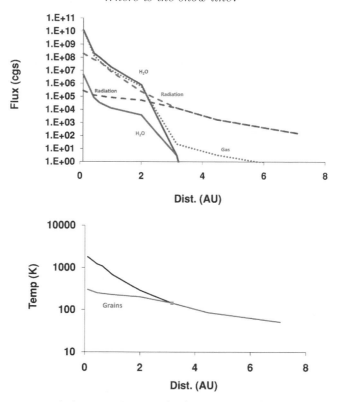

Figure 2. The upper panel shows the heating (red) and cooling (blue) terms for a 10 μm grain in an optically thick gas disk as a function of distance from the star. Water condensation and evaporation (solid curves), radiative heating and cooling (dashed curves) and energy transfer by contact with the gas (dotted curve) are shown. The lower panel shows the temperatures of the gas (black) and grains (red). The green square indicates the position of the snow line.

a result, the position of the snow line is quite insensitive to the composition or size of the grains. However, because the flux of condensing vapor depends on the pressure of the background gas, the temperature of the snow line will vary as a function of temperature as well. Depending on the density, the snow line temperature can be between 150 and 180 K. Based on these arguments, Lecar *et al.* 2006 recomputed a disk model and found that the snow line lies in the neighborhood of 2 AU, although it can be moved out as far as 3 AU if the opacity is increased by a factor of 5.

4. The Photospheric Snow Line

There is an additional way to determine the position of the snow line, and that is by observations of disks around other stars. Such work is just beginning, and recently ice was detected in a disk around HD 142527 (Honda *et al.* 2009). Here energy balance considerations are even more important. Observations see the grains near the surface of the disk, but because these grains are near the surface, they are exposed to the direct radiation from the star.

Unlike the optically thick case discussed above, here the size and composition of the grains are much more important. In particular, 0.1 μm grains will have a circumference

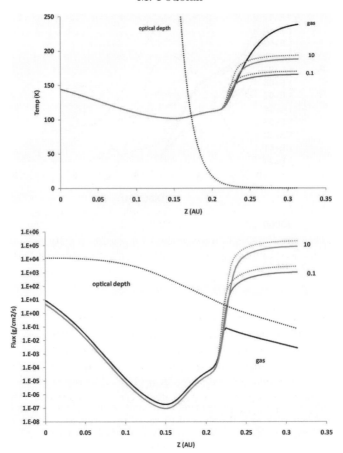

Figure 3. The upper panel shows the temperature of $10\,\mu$m (blue) and $0.1\,\mu$m (red) grains as a function of height above midplane at a radial distance of 1.57 AU from the star. The solid curves are for pure ice grains and the dotted curves are for grains with 10% by mass of generic dark material. The solid black curve is the gas temperature. The dotted black curve is the optical depth to the star in the visible. The disk model is from Jang-Condell (2008). The lower panel shows the flux of water evaporating off the grains. The solid black curve is the condensation flux onto the grains.

roughly equal to the wavelength of the impinging stellar radiation, and will therefore interact with it efficiently. However, at the relevant temperatures the wavelength of the emitted radiation will be orders of magnitude larger and will not be emitted efficiently. In addition, since water is transparent in the visible, the absorption cross section of the grains to visible radiation will differ significantly between pure ice grains and dirty ice grains.

These effects can be seen clearly in Fig. 3. For both large and small grains composed of either pure or dirty ice, the grain temperature follows the gas temperature closely so long as the optical depth to the star is high. Once the optical depth drops to order one, the grain temperatures differ from the gas temperatures, and the grain size and composition make a difference. In fact, the grains are sometimes warmer and sometimes cooler than the surrounding gas. There are several things worth noting here: First, once the visible

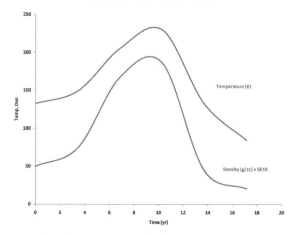

Figure 4. Temperature (blue) and density (red) profile in a density wave as a function of time. Based on the work of Mayer (personal communication).

optical depth is low enough, the evaporative flux always exceeds the condensation flux. This means that there is no snow line *per se* in the photosphere. Second, although ice grains are nowhere stable, the evaporative flux may be so low that the lifetime of the grains will exceed the lifetime of the disk. In such cases, the usual definition of snow line needs to be modified. Finally, the radial distance from the star at which ice is stable will vary with height above midplane. In such a case too, the concept of "snow line" should be replaced by "ice stability region".

5. Effects of a Grain Size Distribution

In addition to the effects of energy balance, there can also be effects due to exchange of energy (in the form of water vapor) between grains of different sizes. When only one grain size is involved, a steady state with the background gas is reached when the rate of condensation from the background onto the grain is equal to the rate of evaporation off the grain. If there is more than one grain size involved, it is possible, even in an optically thick region, for the grains to be at slightly different temperatures. In such a case there will migration of ice from the hotter grain to the colder grain, and this migration will be moderated by the background vapor pressure as determined by the gas temperature. This can lead to an interesting hysteresis phenomenon as demonstrated by the following example:

Gas disks can develop density waves (see, e.g. Meyer *et al.* 2002). These waves cause a change in the gas pressure and temperature with time. The changes are different in different parts of the wave, but a typical case is shown in Fig. 4 based on the work of Mayer (personal communication). Here the gas is assumed to have the solar ratio of water to silicon. The silicon is in the form of grains with a power law size distribution. In this particular case the number of grains was taken to vary with grain size, a, as $a^{-3.5}$, which is typical of such distributions. The gas is originally cold and the water vapor is highly supersaturated. As the density wave develops, the gas temperature rises from $\sim 135\,\mathrm{K}$ to $\sim 220\,\mathrm{K}$ and then returns to its original low value (Meyer, personal communication).

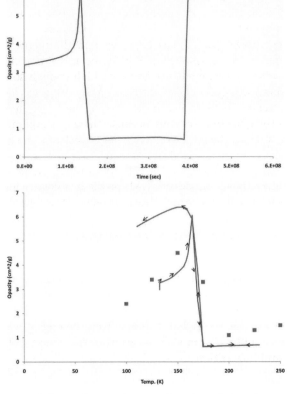

Figure 5. The upper panel shows the opacity as a function of time for a grain size distribution as a density wave develops. The lower panel shows the opacity as a function of temperature. The arrows indicate the progression as a function of time. The squares are the grain opacities of Pollack *et al.* (1985).

The upper panel shows the Rosseland mean grain opacity for this size distribution. At $t = 0$ only the silicate cores contribute, but immediately the supersaturated vapor condenses onto the cores and forms an ice shell, causing an almost instantaneous rise in the opacity. Since the flux is mainly from the gas onto the grain, the flux is nearly independent of the size and temperature of the grain, and each grain grows an ice shell of nearly the same thickness. However, since the grain temperature is determined by the interplay of several processes of heating and cooling, and some of these processes depend on the grain size, the different grains sizes will have slightly different temperatures. Differences in grain composition can also lead to such temperature differences. Since the vapor pressure is very sensitive to temperature, even slight differences in temperature can lead to significant differences in evaporation rates. The details depend on the ambient conditions, but for the case shown in the figure, the larger grains in the distribution are slightly cooler than the smaller ones. Because of this, the rate of evaporation from the smaller grains is slightly higher than the rate of evaporation from the larger grains. Since the pressure of water vapor in the disk is fixed by the temperature of the disk gas, the rate of condensation onto the grains is independent of grain size. The result is a migration of ice from the smaller grains to the larger ones, with the rate of this migration being moderated by the temperature of the background gas. At low temperatures ($T \lesssim 135\,\mathrm{K}$) the migration is so slow that it can be safely ignored, but at higher temperatures it quickly becomes important.

As the density wave develops, the gas heats up and the migration of ice speeds up. As can be seen from the upper planel in Fig. 5, the rate of migration increases with temperature (time) until the temperature becomes high enough so that the ice shells of even the coolest grains are lost. At this point the opacity drops to that of the pure silicate cores. When the gas temperature drops again, there is once more a rapid recondensation of water vapor onto the silicate cores, but because the gas temperature is still relatively high, the migration of ice to the larger grains is very quick. As the gas continues to cool, the new ice distribution remains fixed. The result is a sort of hysteresis effect, where although the temperature of the gas returns to its original value, the opacity changes. This is shown more explicitly in the bottom panel of Fig. 5, where the opacity as a function of temperature is shown. The arrows show the direction of time flow in this simulation, and the red squares show the opacities of Pollack *et al.* (1985). As can be seen, this hysteresis effect can increase the opacity at a given temperature by a factor of two in some cases.

6. Conclusions

The snow line is an important concept that influences our understanding both of the composition of solar system bodies and the processes that lead to their formation. In addition, the position of the snow line has consequences for the opacity of the disk and its evolution. However, the position of the snow line is not merely a function of the temperature of the disk gas. It also depends on the density of water vapor in the disk, and on the details of the radiation field, as well as on the size and composition of the ice grains. As a result, it is not proper to talk about a "snow line" *per se*, but of an "ice stability region" (*ISR*). The borders of this region will depend not only on distance from the central star, but also on the height above midplane. The size distribution of ice grains in this *ISR* may vary from place to place, and may even show a time dependence through the hysteresis effect described above. Theoretical predictions of where the borders of this *ISR* are located, as well as interpretations of observations of ice grains will need to base themselves on detailed modeling of the physics of such grains.

References

Armitage, P. J. 2009, *astro-ph/0701485v2*

Anders, E. & Grevasse, N. 1989, *Geochim. Cosmochim. Acta*, 53, 197

Bell, K. R., Cassen, P. M., Klahr, H. H., & Henning, Th. 1997, *ApJ*, 486, 372

Hayashi, C. 1981 *Prog. Theor. Phys. Suppl.*, 70, 35

Honda, M., Inoue, A. K., Fukagawa, M., Oka, A., Nakamoto, T., Ishii, M., Terada, H., Takato, N., Kawakita, H., Okamoto, Y. K., Shibai, H., Tamura, M., Kudo, T., & Itoh, Y. 2009, *ApJ Lett.*, 690, L110

Hsieh, H. H. & Jewitt, D. 2008, *Science*, 312, 561

Jang-Condell, H. 2008, *ApJ*, 679, 797

Jewitt, D., Yang, B., & Haghighipour, N. 2009, *AJ*, 137, 4313

Lecar, M., Podolak, M., Sasselov, D., & Chiang, E. 2006, *ApJ*, 640, 1115

Lodders, K. 2004, *ApJ*, 611, 587

Mekler, Y. & Podolak, M. 1994, *Plan. & Space Sci.*, 42, 865

Podolak, M. & Mekler, Y. 1997, *Plan. & Space Sci.*, 45, 1401

Pollack, J. B., McKay, C. P., & Christofferson, B. M. 1985, *Icarus*, 64, 471

Sasselov, D. & Lecar, M. 2000, *ApJ*, 528, 995

Stevenson, D. J. & Lunine, J. I. 1988, *Icarus*, 75, 146

van de Hulst, H. C. 1957, *Light Scattering by Small Particles*, (New York: John Wiley & Sons)

Icy Bodies of the Solar System
Proceedings IAU Symposium No. 263, 2009
J.A. Fernández, D. Lazzaro, D. Prialnik & R. Schulz, eds.

© International Astronomical Union 2010
doi:10.1017/S1743921310001444

Heavy ion irradiation of astrophysical ice analogs

Eduardo Seperuelo Duarte[1,2,3], Alicja Domaracka[1], Philippe Boduch[1], Hermann Rothard[1], Emmanuel Balanzat[1], Emmanuel Dartois[4], Sergio Pilling[2,3], Lucio Farenzena[6], and Enio Frota da Silveira[2]

[1]Centre de Recherche sur les Ions, les Matériaux et la Photonique (CEA /CNRS /ENSICAEN /Universit de Caen-Basse Normandie), CIMAP-CIRIL- Ganil, Boulevard Henri Becquerel, BP 5133, F-14070 Caen Cedex 05, France

[2]Physics Department, Pontifícia Universidade Católica, Rua Marquês de S. Vicente 225, 22453-900 Rio de Janeiro, Brazil

[3]Grupo de Física e Astronomia - CEFET/Química de Nilópolis, R. Lúcio Tavares, 1045, Centro, 26530-060, Nilópolis, Brazil

[4]Institut d'Astrophysique Spatiale, Astrochimie Expérimentale, UMR-8617 Université Paris-Sud, bâtiment 121, F-91405 Orsay, France

[5]Universidade do Vale do Paraíba (UNIVAP/IP&D), 12244-000, So José dos Campos, SP, Brazil

[6]Physics Department, Universidade Federal de Santa Catarina, Florianópolis, SC, Brazil

Abstract. Icy grain mantles consist of small molecules containing hydrogen, carbon, oxygen and nitrogen atoms (e.g. H_2O, CO, CO_2, NH_3). Such ices, present in different astrophysical environments (giant planets satellites, comets, dense clouds, and protoplanetary disks), are subjected to irradiation of different energetic particles: UV radiation, ion bombardment (solar and stellar wind as well as galactic cosmic rays), and secondary electrons due to cosmic ray ionization of H_2. The interaction of these particles with astrophysical ice analogs has been the object of research over the last decades. However, there is a lack of information on the effects induced by the heavy ion component of cosmic rays in the electronic energy loss regime. The aim of the present work is to simulate of the astrophysical environment where ice mantles are exposed to the heavy ion cosmic ray irradiation.

Sample ice films at $13\,K$ were irradiated by nickel ions with energies in the 1-10 MeV/u range and analyzed by means of FTIR spectrometry. Nickel ions were used because their energy deposition is similar to that deposited by iron ions, which are particularly abundant cosmic rays amongst the heaviest ones.

In this work the effects caused by nickel ions on condensed gases are studied (destruction and production of molecules as well as associated cross sections, sputtering yields) and compared with respective values for light ions and UV photons.

Keywords. astrochemistry, methods: laboratory

1. Introduction

Astrophysical ices are exposed to many types of irradiation either with high and low energy; they are: energetic particles from galactic cosmic rays and stellar or solar wind (in the case of the solar system), UV radiation and the particles accelerated by the giant planets magnetosphere (Gerakines *et al.* 2001; Dartois 2005; Pilling *et al.* 2009). The irradiation produces a number of effects whose knowledge is relevant to understanding the evolution of these objects. One can distinguish two major effects: (i) sputtering (the material is eroded at the surface) and (ii) physical-chemical modifications in the bulk, including the formation of different molecules.

Most of the experiments on laboratory ices have been done with protons and helium ions using infrared spectroscopy as a tool to follow and describe the evolution. Some experiments using heavy ions beams (mostly Oxygen and Argon) were performed at low energy and using mass spectrometry or RBS to analyze the ices. In the current work, the ices were irradiated by heavy ions in the electronic energy loss regime and analyzed with IR spectroscopy.

Nickel is one of the solar wind and cosmic rays components. Recently, its abundance in the solar wind was measured Karrer *et al.*(2007). As thestopping powers of Ni and Fe are close one to the other, the physical and chemistry processes induced by them are similar.

The objectives of this work are: (i) to understand the mechanisms of swift heavy ion interactions with molecular solids (ices); (ii) to identify the new species formed; (iii) to determine the destruction and formation cross sections for molecules; (iv) to determine the sputtering yields induced by heavy ions.

2. Experimental procedure

The experiments were performed in the CIMAP-GANIL laboratory situated at Caen, France, where beams in a wide range of ions (up to lead) and energies (up to GeV) are available. To simulate the astrophysical environment, the ices were formed inside a vacuum chamber by the condensation of gases on a cold substrate. The experimental device used, called CASIMIR, allows to regulate the sample temperature and measure the infrared spectra. The cryostat works by closed-cycle of helium gas. The ices were prepared by gas condensation onto a CsI substrate at 13 K. The CsI is a small disk with 13 mm diameter and 2 mm thick. It is positioned at the extremity of the cryostat. To reduce the effects of thermal radiation, a shield was installed around the substrate. Its temperature is maintained at 70 K. The ensemble is positioned inside the spectroscopy cuve. More details about the experimental procedure can be found in Seperuelo Duarte *et al.* (2009).

3. Results

Figure 1 presents the IR spectrum of CO ice before and after heavy ion irradiation. Many infrared bands were identified by means of their wavenumber positions (Trottier *et al.* 2004; Jamieson *et al.* 2006; Palumbo *et al.* 2008).

The decrease of the ice column density is related to the formation of other species and to sputtering induced by heavy ions. To analyze theseeffects, the data were fitted by an equation taking into account both the destruction cross section and the sputtering mechanism (Seperuelo Duarte *et al.* 2009)

$$N = N_0 \exp(-\sigma_d F) - Y/\sigma_d(1 - \exp(-\sigma_d F)) \qquad (3.1)$$

where N_0 is the initial column density, σ_d – the destruction cross section, and Y – the sputtering yield. Table 1 presents the cross sections and a sputtering yields measured for the ices studied in the current experiment. In the case of water ice and ammonia, a condensation from the residual gas prevented the measure of the sputtering yield.

4. Conclusions

Ice films were irradiated by MeV Ni ions. The new species produced in the samples are essentially the same as those found after proton, photon, and electron irradiation.

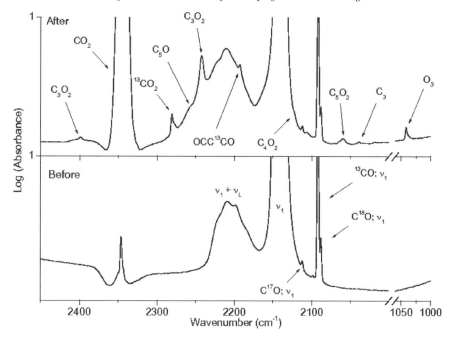

Figure 1. Infrared spectrum of CO ice before and after 46 MeV ^{58}Ni^{11+} irradiation with fluence of 1.0×10^{12} cm^{-2}. Many infrared bands were identified by means of their wavenumber positions.

Table 1. Destruction cross section and sputtering yield values of ices studied in this work.

Species	σ_d (10^{-13} cm^2)	Y (10^4 molecules/impact)
H_2O	1.1	-
CO	1.4	9
CO_2	1.7	4
*NH_3	1.4	-

Notes: *Value measured in a mixture of H_2O:CO:NH_3 (1:0.4:0.6).

Destruction and formation cross sections, as well as sputtering yields were determined. The measured sputtering yields scale with the squared electronic stopping power values, extending for higher S_e the results of Brown *et al.* (1984). Desorption induced by heavy ion sputtering is proposed to be one of the dominant processes leading to the presence of gas phase CO molecules for grains deep inside dense clouds and protoplanetary disks.

References

Brown, W. L., Augustyniak, W. M., Marcantonio, K. J., Simmons, E. H., Boring, J. W., Johnson, R. E., & Reimann, C. T. 1984, *Nucl. Instr. Meth. B*, 1, 307

Dartois, E. 2005, *Space Sci. Rev.*, 119, 293

Farenzena, L. S., Martinez, R., Iza, P., Ponciano, C. R., *et al.* 2006, *Int. J. Mass Spectrom.*, 251, 1

Gerakines, P. A. & Moore, M. H. 2001, *Icarus*, 154, 372

Jamieson, C. S., Mebel, A. M., & Kaiser, R. I. 2006, *ApJSS*, 163, 184

Karrer, R., Bochsler, P., Giammanco, C., *et al.* 2007, *Space Science Reviews*, 130, 317

Palumbo, M. E., Leto, P., Siringo, C., & Trigilio C. 2008, *ApJ*, 685, 1033

Pilling, S., Seperuelo Duarte, E., da Silveira, E. F., Balanzat, E., Rothard, H., Domaracka, A., & Boduch, P. 2009, *A&A*, submitted

Ponciano, C. R., Martinez, R., Farenzena, L. S., Iza, P., Homem, M. G. P., Naves de Brito, A., da Silveira, E. F., & Wien, K. 2006, *J. Am. Mass Spectrom.*, 17, 1120

Seperuelo Duarte, E., Boduch, P., Rothard, H., *et al.* 2009, *A&A*, 502, 599

Trottier, A. & Robert, L. B. 2004, *ApJ*, 612, 1214

Icy bodies of the solar system
Proceedings IAU Symposium No. 263, 2009
J.A. Fernández, D. Lazzaro, D. Prialnik & R. Schulz, eds.
© International Astronomical Union 2010
doi:10.1017/S1743921310001456

Low temperature CH_4 and CO_2 clathrate hydrate near to mid-IR spectra

E. Dartois[1], B. Schmitt[2], D. Deboffle[1], and M. Bouzit[1]

[1] Institut d'Astrophysique Spatiale, UMR-8617, Université Paris-Sud, bâtiment 121, 91405 Orsay, France
email: emmanuel.dartois@ias.u-psud.fr

[2] Laboratoire de Planétologie de Grenoble, Université J. Fourier - CNRS, Bâtiment D de Physique, Domaine Universitaire B.P. 53 38041 Grenoble Cedex 9, France

Abstract. The physical behaviour of methane and carbon dioxide clathrate hydrates, specific crystallographic ice crystals are of major importance for the earth and may control the stability of gases in many astrophysical bodies such as the planets, comets and possibly interstellar grains. Such models claim they provide an alternative trapping mechanism modifying the absolute and relative composition of icy bodies and can be at the source of late time injection of gaseous species in planetary atmospheres. However, there is a clear need to detect them directly. We provide in this study the laboratory recorded signatures of clathrate hydrates in the near to mid-infrared for astrophysical remote detection. These laboratory experiments will in a near future allow to follow the kinetic formation by diffusion in dedicated experiments, another important step to implement, to understand and model their possible presence in space.

Keywords. Solar system ices, clathrate hydrates, lines and bands identification, infrared spectra.

1. Introduction

Clathrate hydrates (CLH) are crystalline solid with properties close to ice, trapping inside small host molecules, and thus retaining gases into water ice crystal cages. These inclusion compounds may be important for the stability of gases in many astrophysical bodies (planets, comets, interstellar grains) as they provide a trapping mechanism playing a role in the preservation in the solid state of these molecules at temperatures higher than expected, avoiding their early escape. Their occurence would thus modify the absolute and relative composition of astrophysical (icy) bodies as well as increase preservation timescales, or e.g. provide late time (re-)injection of gaseous species in planetary atmospheres.

2. Clathrates from the laboratory to astrophysics

Many laboratory studies examined clathrate hydrates thermodynamic or kinetic behaviour (e.g. Lunine & Stevenson 1985; Fray *et al.* 2009 and references therein), but probably the best way to confirm their presence in astrophysical bodies will come from remote infrared spectroscopy observations by telescopes or space probes. Methane and carbon dioxide clathrate crystals were thus produced in our laboratory, and the specific fingerprints betraying these clathrate hydrate presence recorded by infrared spectroscopy (although infrared spectra were recorded for some species, many previous experiments focused on e.g. Raman spectroscopy or neutron diffraction studies, e.g. Klug & Whalley 1973; Davidson *et al.* 1977; Bertie & Jacobs 1982; Richardson *et al.* 1985; Fleyfel &

Devlin 1991; Sum *et al.* 1997; Gutt *et al.* 2002; Kleinberg *et al.* 2003; Nakayama *et al.* 2003; Prager & Press 2006 and upward citations).

To produce and record CLH infrared spectra a specific high pressure evacuable enclosed cell was built and placed in an high vacuum evacuated cryostat (P < 10^{-7} mbar). A stainless steel injection tube for gases entrance or evacuation is sealed at the bottom of the cell. The formation of the CLH follow a procedure that satisfies to both its nucleation and prevents the sample from becoming optically thick to the infrared beam. Water vapour is first injected into the evacuated cell and condense on the pre-cooled infrared transmitting windows, immediately followed by the guest (here CH_4 or CO_2) injection at moderately high pressures. The cell is maintained in this state, typically during two days. The remaining non enclathrated gas is then evacuated while lowering the cell to cryogenic temperatures taking care to follow a path in the P-T phase diagram close to, but just below, the vaporisation and sublimation curves for the pure guest, to stay above the expected clathrate hydrate dissociation curve. The FTIR spectra are recorded in the 5K to 150K range (depending on the CLH) with a Bruker IFS 66v at an adapted resolution of 0.15 to 0.5 cm^{-1}, with a globar IR source, KBr beamsplitter, and an HgCdTe detector cooled at LN$_2$.

3. Methane and carbon dioxide spectra

CH$_4$

The temperature dependent methane clathrate hydrate infrared spectra in the CH stretching mode region is displayed in Fig. 1. The striking feature is the observation of sub-structures as compared to typical methane ice bands, ressembling gaseous methane rovibrationnal lines, but shifted to the red. The methane is quasi free to rotate at low temperature, as already observed from neutron diffraction data (Prager & Press 2006). As the temperature increases this ability is partially quenched by collisions with the water ice cage potential surface, and then remains two broad bands assigned to methane trapped in the small and large water cages of the type I clathrate type. Details on the assignments can be found in Dartois & Deboffle (2008) and near infrared combination modes analyses will soon be available.

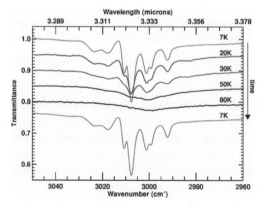

Figure 1. Low temperature methane clathrate hydrate infrared spectra in the CH stretching mode region.

CO₂

Carbon dioxide clathrate hydrate is different from methane in many respect as shown in Fig. 2 (right panel). First, the guest size is larger, the molecule linear, and the rotational ability is hindered. One sees in the antisymmetric mode only two broad bands corresponding to the two unequal cages in which CO_2 is trapped. The principal isotope main vibrational band had been recorded previously by Fleyfel & Devlin (1991) and we saturated the known bands to access the isotopes transitions as well as the combinations modes, in Fermi resonance (Fig. 2. left panel). More details are given in Dartois & Schmitt (2009).

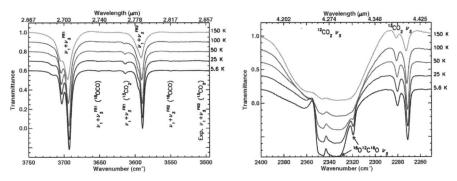

Figure 2. Temperature dependent carbon dioxide clathrate hydrate infrared spectra. Right panel : Antisymmetric stretching mode region. Left panel : Fermi resonance region ($\nu_1 + \nu_3$).

4. Discussion

Methane and carbon dioxide are both abundant molecules found in the solar system. In the martian context, first CO_2 and more recently CH_4 clathrate hydrates have been invoked, based on thermodynamical models to participate significantly to its cycle through surface and atmosphere interactions, and thus providing for CO_2 and CH_4 a sink during winter or a reservoir, respectively, to replenish the atmosphere (e.g. Genov & Kuhs 2003; Longhi 2006; Chastain & Chevrier 2007; Chassefiere 2009). In the outer solar system planets, satellites and comets, it could play a role in the icy bodies retention of species and modification of the relative abundances especially for light guest molecules (e.g. Delsemme & Swings 1952; Iro *et al.* 2003; Hersant *et al.* 2004; Mousis & Schmitt 2008 and references therein).

To allow a comparison of experimental work with astrophysical observations, some fundamental (and difficult) questions have to be adressed. In particular the formation kinetic under realistic astrophysical conditions is essential and will be conditioned by the labile water ice network interaction/reconstruction. As many physical interactions are possible with ice, and not necessarily involving the formation of a crystallographic system such as clathrate hydrates, it is of importance to be able to constrain their abundances in astrophysical media, to understand if they represent once observed an epiphenomenon for a widespread dominant crystal in astrophysics or a local state in a few objects.

Clathrate hydrate direct detection is of importance. We have shown through the infrared spectroscopic studies presented here that remote spectroscopy is one important possibility to identify CLH via their specific signatures. In particular the trapped methane molecules in the clathrate hydrate display a gaseous-like behaviour at low temperature in the water cages. Even at higher temperatures (50-150K), the vibrational spectra recorded are unique to methane and carbon dioxide clathrate hydrates, they represent a crucial identification pattern for low-temperature astrophysical icy bodies, such as planets, comets and/or interstellar grains. Methane and carbon dioxide were chosen as starting molecule to study, both because of their relevance for cometary or interstellar grains possible detection, but also because of their expected spectroscopic characteristics.

References

Bertie, J. E. & Jacobs, S. M. 1982, *J. Chem. Phys.*, 77, 3230

Chasefière, 2009, *Icarus*, in press

Chastain, B. K. & Chevrier, V. 2007, *Planet. & Space Sci.*, 55, 1246

Dartois, E. & Deboffle, D. 2008, *A&A*, 490, L19

Dartois, E. & Schmitt, B. 2009, *A&A*, 504, 869

Davidson, D. W., Garg, S. K., Gough, S. R., Hawkins, R. E., & Ripmeester, J. A. 1977, *Can. J. Chem.*, 55(20), 3641.

Delsemme, A. H. & Swings, P. 1952, *Annales d'Astrophysique*, 15, 1

Fleyfel, F. & Devlin, J.-P. 1991, *J. Phys Chem.*, 95, 3811

Fray, N., Marboeuf, U., & Schmitt, B., *Planet. & Space Sci.*, in preparation

Genov, G. & Kuhs, W. F. 2003, *Third International Conference on Mars Polar Science and Exploration*, 8011

Gutt, C., Press, W., Huller, A., & Tse, J. S. 2002, *Appl. Phys. A*, 74, 1299.

Hersant, F., Gautier, D., & Lunine, J. I. 2004, *Planet. & Space Sci.*, 52, 623

Iro, N., Gautier, D., Hersant, F., Bockelée-Morvan, D., & Lunine, J. I. 2003, *Icarus*, 161, 511

Kleinberg, R. L., Flaum, C., Griffin, D. D., Brewer, P. G., Malby, G. E., Peltzer, E. T., & Yesinowski, J. P. 2003, *J. Geophys. Res.*, 108(B10), 2508

Klug, D. D. & Whalley, E. 1973, *Can. J. Chem.*, 51(24), 4062

Longhi, J. 2006, *Journal of Geophysical Research (Planets)*, 111, 6011

Lunine, J. I. & Stevenson D. J. 1985, *ApJ Suppl. Series*, 58, 493

Mousis, O. & Schmitt, B. 2008, *ApJ Letters*, 677, L67

Nakayama, H., Klug, D. D., Ratcliffe, C. I., & Ripmeester, J. A. 2003, *Chemistry - A European Journal*, 9(13), 2969

Prager, M. & Press, W. 2006, *J. Chem. Phys.*, 125, 214703

Richardson, H. H., Wooldridge, P. J., & Devlin, J. P. 1985, *J. Chem. Phys.*, 83, 4387

Sum, A. K., Burruss, R. C., & Sloan, E. D. 1997, *J. Phys. Chem. B Mater. Surf. Interfaces Biophys.*, 101, 7371

Icy Bodies of the Solar System
Proceedings IAU Symposium No. 263, 2009
J.A. Fernández, D. Lazzaro, D. Prialnik & R. Schulz, eds.

© International Astronomical Union 2010
doi:10.1017/S1743921310001468

Angular momentum of two collided rarefied preplanetesimals and formation of binaries

Sergei I. Ipatov[1,2]

[1]Catholic University of America
Washington DC, USA
email: siipatov@hotmail.com

[2]Space Research Institute, Moscow, Russia

Abstract. The mean angular momentum associated with the collision of two celestial objects moving in almost circular heliocentric orbits was studied. The results of these studies were used to develop models of the formation of binaries at the stage of rarefied preplanetesimals. The models can explain a greater fraction of binaries formed at greater distances from the Sun. Sometimes there could be two centers of contraction inside the rotating preplanetesimal formed as the result of a collision between two rarefied preplanetesimals. Such formation of binaries could result in binaries with almost the same masses of components separated by a large distance. Formation of a disk around the primary could result because the angular momentum that was obtained by a rarefied preplanetesimal formed by collision was greater than the critical angular momentum for a solid body. One or several satellites of the primary could be formed from the disk.

Keywords. Minor planets, asteroids; Kuiper Belt; solar system: formation

1. Introduction

In recent years, new arguments in favor of the model of rarefied preplanetesimals – clumps were made (e.g., Cuzzi *et al.* 2008, Johansen *et al.* 2007, Lyra *et al.* 2008). Even before new arguments in favor of formation of planetesimals from rarefied preplanetesimals were developed, Ipatov (2001, 2004) considered that some trans-Neptunian objects (TNOs), planetesimals, and asteroids with diameter $d > 100$ km could be formed directly by the compression of large rarefied preplanetesimals, but not by the accretion of smaller solid planetesimals. Some smaller objects (TNOs, planetesimals, asteroids) could be debris from larger objects, and other smaller objects could be formed directly by compression of preplanetesimals. There are several hypotheses of formation of binaries for a model of solid bodies (e.g., Petit *et al.* 2008, Richardson & Walsh 2006, Walsh *et al.* 2008). Ipatov (2004) supposed that a considerable fraction of trans-Neptunian binaries could be formed at the stage of compression of rarefied preplanetesimals moving in almost circular orbits. Based on analysis of the angular momentum of two collided rarefied preplanetesimals, Ipatov (2009a-b) studied models of the formation of binaries at the stage of the preplanetesimals.

2. Angular momentum of two collided rarefied preplanetesimals

Previous papers devoted to the formation of axial rotation of forming objects considered mainly a model of solid-body accumulation. Besides such model, Ipatov (1981a-b, 2000, 2009b) also studied the formation of axial rotation for a model of rarefied preplanetesimals. He presented the formulas for the angular momentum of two collided rarefied

Table 1. Angular momenta of several binaries.

binary	Pluto	(90842) Orcus	2000 CF$_{105}$	2001 QW$_{322}$	(90) Antiope
a, AU	39.48	39.3	43.8	43.94	3.156
d_p, km	2340	950	170	108?	88
d_s, km	1212	260	120	108?	84
m_p, kg	1.3×10^{22}	7.5×10^{20}	2.6×10^{18}?	6.5×10^{17}?	4.5×10^{17}
m_s, kg	1.52×10^{21}	1.4×10^{19} for $\rho = 1.5$	9×10^{17}?	6.5×10^{17}?	3.8×10^{17}
L, km	19,750	8700	23,000	120,000	171
L/r_H	0.0025	0.0029	0.04	0.3	0.007
T_{sp}, h	153.3	10			16.5
K_{scm}, kg km^2 s^{-1}	6×10^{24}	9×10^{21}	5×10^{19}	3.3×10^{19}	6.4×10^{17}
K_{spin}, kg km^2 s^{-1}	10^{23}	10^{22}	1.6×10^{18} at $T_s = 8$ h	2×10^{17} at $T_s = 8$ h	3.6×10^{16}
K_{s06ps}, kg km^2 s^{-1}	8.4×10^{25}	9×10^{22}	1.5×10^{20}	5.2×10^{19}	6.6×10^{18}
K_{s06eq}, kg km^2 s^{-1}	2.8×10^{26}	2×10^{24}	2.7×10^{20}	5.2×10^{19}	6.6×10^{18}
$(K_{scm} + K_{spin})/K_{s06ps}$	0.07	0.2	0.3	0.63	0.1
$(K_{scm} + K_{spin})/K_{s06eq}$	0.02	0.01	0.2	0.63	0.1

preplanetesimals – Hill spheres (with radii r_1 and r_2 and masses m_1 and m_2) moved in circular heliocentric orbits. At a difference in their semimajor axes a equaled to $\Theta(r_1 + r_2)$, the angular momentum is $K_s = k_\Theta (G \cdot M_S)^{1/2} (r_1 + r_2)^2 m_1 m_2 (m_1 + m_2)^{-1} a^{-3/2}$, where G is the gravitational constant, and M_S is the mass of the Sun. At $r_a = (r_1 + r_2)/a \ll \Theta$, one can obtain $k_\Theta \approx (1 - 1.5\Theta^2)$. The mean value of k_Θ equals to 0.6. Mean positive values of k_Θ and mean negative values of k_Θ are equal to 2/3 and -0.24, respectively. The values of K_s are positive at $0 < \Theta < 0.8165$ and are negative at $0.8165 < \Theta < 1$.

For homogeneous spheres at $k_\Theta = 0.6$, $a = 1$ AU, and $m_1 = m_2$, the period of axial rotation $T_s \approx 9 \cdot 10^3$ hours for the rarefied preplanetesimal formed as a result of the collision of two preplanetesimals – Hill spheres, and $T_s \approx 0.5$ h for the planetesimal of density $\rho = 1$ g cm^{-3} formed from the preplanetesimal. For greater a, the values of T_s are smaller (are proportional to $a^{-1/2}$). Such small periods of axial rotations cannot exist, especially if we consider bodies obtained by contraction of rotating rarefied preplanetesimals, which can lose material easier than solid bodies. For $\rho = 1$ g cm^{-3}, the velocity of a particle on a surface of a rotating spherical object at the equator is equal to the circular and the escape velocities at 3.3 and 2.3 h, respectively.

For five binaries, the angular momentum K_{scm} of the present primary and secondary components (with diameters d_p and d_s and masses m_p and m_s), the momentum K_{s06ps} of two collided preplanetesimals with masses of the binary components moved in circular heliocentric orbits at $k_\Theta = 0.6$, and the momentum K_{s06eq} of two identical collided preplanetesimals with masses equal to a half of the total mass of the binary components at $k_\Theta = 0.6$ are presented in the Table. All these three momenta are considered relative to the center of mass of the system. K_{spin} is the spin momentum of the primary. L is the distance between the primary and the secondary, r_H is the radius of the Hill sphere, and T_{sp} is the period of spin rotation of the primary.

3. Models of formation of binaries

For circular heliocentric orbits, two objects that entered inside the Hill sphere could move there for a longer time than those entered the sphere from eccentric heliocentric orbits. The diameters of preplanetesimals were greater than the diameters of solid planetesimals of the same masses. Therefore, the models of binary formation due to the gravitational interactions or collisions of future binary components with an object (or objects) that were inside their Hill sphere, which were studied by several authors for solid objects, could be more effective for rarefied preplanetesimals.

We suppose that formation of some binaries could be caused by that the angular momentum that they obtained at the stage of rarefied preplanetesimals was greater than that could exist for solid bodies. During contraction of a rotating rarefied preplanetesimal, some material with velocity greater than the circular velocity could have formed a cloud (that transformed into a disk) of material that moved around the primary. One or several satellites of the primary could be formed from this cloud. Some material could leave the Hill sphere of a rotating contracting planetesimal, and the mass of an initial rotating preplanetesimal could exceed the mass of a corresponding present binary system. Due to tidal interactions, the distance between binary components could increase with time, and their spin rotation could become slower. For the discussed model of formation of binaries, the vector of the original spin momentum of the primary was approximately perpendicular to the plane where the secondary component (and all other satellites of the primary) moved. It is not necessary that this plane was close to the ecliptic if the difference between the distances from centers of masses of collided preplanetesimals to the middle plane of the disk of preplanetesimals was comparable with sizes of preplanetesimals. Eccentricities of orbits of satellites of the primary formed in such a way are usually small. As it was shown by Ipatov (2009b), the critical angular momentum could be attained as a result of a collision of two identical asteroids of any radii (<6000 km). At the same eccentricities of heliocentric orbits and $m_1/m_2 = $ const, the probability to attain the critical momentum at a collision is greater for smaller values of m_1 ($m_1 \geqslant m_2$) and a.

Some collided rarefied preplanetesimals had a greater density at distances closer to their centers. It might be possible that sometimes there were two centers of contraction inside the rotating preplanetesimal formed as a result of a collision of two rarefied preplanetesimals. Such formation of binaries could result in binaries with almost the same masses of components separated by a large distance. It could be also possible that the primary had partly contracted when a smaller object (objects) entered into the Hill sphere, and then the object was captured due to collisions with the material of the outer part of the contracted primary. For such a scenario, a satellite can be formed at any distance (inside the Hill sphere) from the primary. The eccentricity of the mutual orbit of components can be any (small or large) for the model of two centers of contraction.

For the binaries presented in the Table, the ratio $r_K = (K_{scm} + K_{spin})/K_{s06eq}$ is smaller than 1. Small values of r_K for most discovered binaries can be due to that preplanetesimals already had been partly compressed at the moment of collision.

At $K_s = $ const, T_s is proportional to $a^{-1/2}\rho^{-2/3}$. Therefore, for greater a, more material of a contracting rotating preplanetesimal was not able to contract into a primary and could form a cloud surrounding the primary (or there were more chances that there were two centers of contraction). This can explain why binaries are more frequent among TNOs than among large main-belt asteroids, and why the typical mass ratio of the secondary to the primary is greater for TNOs than for asteroids. Longer time of contraction of rotating preplanetesimals at greater a (for dust condensations, this was shown by several authors, e.g. by Safronov) could also testify in favor of the above conclusion. Spin and form of an object could change during evolution of the Solar System.

Ipatov (2009b) discussed the possibility of a merger of two rarefied preplanetesimals and the formation of highly elongated small bodies by the merger of two (or several) partly compressed components.

4. Conclusions

Some trans-Neptunian objects could have acquired their primordial axial momenta and/or satellites at the stage when they were rarefied preplanetesimals. Most rarefied preasteroids could have become solid asteroids before they collided with other preasteroids. Some collided rarefied preplanetesimals could have greater densities at locations that are closer to their centers. In this case, there sometimes could be two centers of contraction inside the rotating preplanetesimal formed as a result of the collision of two rarefied preplanetesimals. Such contraction could result in binaries with similar masses separated by any distance inside the Hill sphere and with any value of the eccentricity of the orbit of the secondary component relative to the primary component. The observed separation distance can characterize the radius of a greater encountered preplanetesimal.

The formation of some binaries could have resulted because the angular momentum of a binary that was obtained at the stage of rarefied preplanetesimals was greater than the angular momentum that can exist for solid bodies. Material that left a contracted preplanetesimal formed as a result of a collision of two preplanetesimals could form a disk around the primary. One or more satellites of the primary could be grown in the disk at any distance from the primary inside the Hill sphere, but typical separation distance is much smaller than the radius of the sphere. The satellites moved mainly in low eccentric orbits. Both of the above scenarios could have taken place at the same time. In this case, it is possible that, besides massive primary and secondary components, smaller satellites could be moving around the primary and/or the secondary.

For discovered trans-Neptunian binaries, the angular momentum is usually considerably smaller than the typical angular momentum of two identical rarefied preplanetesimals having the same total mass and encountering up to the Hill sphere from circular heliocentric orbits. This conclusion is also true for preplanetesimals with masses of components of considered trans-Neptunian binaries. The above difference in momenta and the separation distances, which usually are much smaller than the radii of Hill spheres, support the hypothesis that most preplanetesimals already had been partly compressed at the moment of collision, i.e. were smaller than their Hill spheres and/or were denser at distances closer to the center of a preplanetesimal. The contraction of preplanetesimals could be slower farther from the Sun, which can explain the greater fraction of binaries formed at greater distances from the Sun.

References

Cuzzi, J. N., Hogan, R. C., & Shariff, K. 2008, *ApJ*, 687, 1432
Ipatov, S. I. 1981a, *Inst. of Applied Mathematics Preprint* N 101, Moscow, 28 P, in Russian
Ipatov, S. I. 1981b, *Inst. of Applied Mathematics Preprint* N 102, Moscow, 28 P, in Russian
Ipatov, S. I. 2000, *Migration of celestial bodies in the Solar System*, URSS, Moscow, 320 P, in Russian
Ipatov, S. I. 2001, *LPS XXXII*, Abstract #1165
Ipatov, S. I. 2004, *AIP Conf. Proc.*, 713, 277;
        also *http://planetquest1.jpl.nasa.gov/TPFDarwinConf/proceedings/posters/p045.pdf*
Ipatov, S. I. 2009a, *LPS XL*, Abstract #1021
Ipatov, S. I. 2009b, *MNRAS*, submitted, *http://arxiv.org/abs/0904.3529*
Johansen, A., Oishi, J. S., Mac Low, M.-M., Klahr, H., Henning, T., & Youdin, A. 2007, *Nature*, 448, 1022
Lyra, W., Johansen, A., Klahr, H., & Piskunov, N. 2008, *A&A*, 491, L41
Petit, J.-M. *et al.* 2008, *Science*, 322, 432
Richardson, D. R. & Walsh, K. J. 2006, *Annu. Rev. Earth Planet. Sci.*, 34, 47
Walsh, K. J., Richardson, D. R., & Michel, P. 2008, *Nature*, 454, 188

Icy Bodies of the Solar System
Proceedings IAU Symposium No. 263, 2009
J.A. Fernández, D. Lazzaro, D. Prialnik & R. Schulz, eds.

© International Astronomical Union 2010
doi:10.1017/S174392131000147X

Collision probabilities of migrating small bodies and dust particles with planets

Sergei I. Ipatov[1,2]

[1]Catholic University of America
Washington DC, USA
email: siipatov@hotmail.com

[2]Space Research Institute, Moscow, Russia

Abstract. Probabilities of collisions of migrating small bodies and dust particles produced by these bodies with planets were studied. Various Jupiter-family comets, Halley-type comets, long-period comets, trans-Neptunian objects, and asteroids were considered. The total probability of collisions of any considered body or particle with all planets did not exceed 0.2. The amount of water delivered from outside of Jupiter's orbit to the Earth during the formation of the giant planets could exceed the amount of water in Earth's oceans. The ratio of the mass of water delivered to a planet by Jupiter-family comets or Halley-type comets to the mass of the planet can be greater for Mars, Venus, and Mercury, than that for Earth.

Keywords. Minor planets, asteroids; comets: general; solar system: general

1. Model of migration of small bodies and dust particles

In the present paper we discuss the probabilities of collisions of migrating small bodies and dust particles produced by these bodies with planets. Ipatov & Mather (2003, 2004a-b, 2006, 2007), Ipatov *et al.* (2004), and several other authors cited in the above papers studied the probabilities with Earth, Venus, and Mars. In our above papers (which can be found on astro-ph and http://faculty.cua.edu/ipatov/list-publications.htm), the probabilities of the collisions were calculated based on the time variations of orbits of bodies and particles during their dynamical lifetimes (until all bodies or particles reached 2000 AU from the Sun or collided with the Sun). Using the same time variations, below we consider also the probabilities of collisions with the giant planets and Mercury.

The orbital evolution of >30,000 bodies with initial orbits close to those of Jupiter-family comets (JFCs), Halley-type comets, long-period comets, trans-Neptunian objects, and asteroids in the resonances 3/1 and 5/2 with Jupiter, and also of >20,000 dust particles produced by these small bodies was integrated during their dynamical lifetimes. We considered the gravitational influence of planets, but omitted the influence of Mercury (exclusive for Comet 2P/Encke). In about a half of calculations of migration of bodies, we used the method by Bulirsh-Stoer (1966) (BULSTO code), and in other runs we used a symplectic method (RMVS3 code). The integration package of Levison & Duncan (1994) was used. For dust particles, only the BULSTO code was used, and the gravitational influence of all planets, the Poynting-Robertson drag, radiation pressure, and solar wind drag were taken into account. The ratio β between the radiation pressure force and the gravitational force varied from $\leqslant 0.0004$ to 0.4. For silicates, such values of β correspond to particle diameters d between $\geqslant 1000$ and 1 microns; d is proportional to $1/\beta$.

In our calculations, planets were considered as material points, so literal collisions did not occur. However, using the algorithm suggested by Ipatov (1988) with the correction that takes into account a different velocity at different parts of the orbit (Ipatov & Mather 2003), and based on all orbital elements sampled with a 10-500 yr step, we calculated the mean probability P of collisions of migrating objects with a planet. The step could be different for different typical dynamical lifetimes of particles. We define P as P_Σ/N, where P_Σ is the probability of a collision of all N objects with a planet. The probabilities of collisions of bodies and particles at different β with planets are presented in Fig. 1. These probabilities do not take into account the destruction of particles in collisions and sublimation, which can be more important for small particles. Our runs were made mainly for direct modelling of collisions with the Sun, but Ipatov & Mather (2007) obtained that the mean probabilities of collisions of considered bodies with planets, lifetimes of the bodies that spent millions of years in Earth-crossing orbits, and other obtained results were practically the same if we consider that bodies disappear when perihelion distance becomes less than the radius of the Sun or even several such radii.

2. Probabilities of collisions of migrating small bodies and dust particles with planets and the Sun

The probability P_E of a collision of a JFC with the Earth exceeded $4 \cdot 10^{-6}$ if initial orbits of bodies were close to those of several tens of JFCs, even excluding a few bodies for which the probability of a collision of one body with the Earth could be greater than the sum of probabilities for thousands of other bodies. The Bulirsh-Stoer method of integration and the symplectic method gave similar results. The ratios of probabilities of collisions of JFCs with Venus, Mars, and Mercury to the mass of a planet usually were not smaller than those for Earth. For most considered bodies, the probabilities P_{Me} of collisions of most bodies with Mercury (exclusive for Comet 2P/Encke, for which $P_{Me} \sim P_E$) were smaller by an order of magnitude than those with Earth or Venus.

For dust particles produced by comets and asteroids, P_E was found to have a maximum (~ 0.001-0.02) at $0.002 \leqslant \beta \leqslant 0.01$, i.e., at $d \sim 100$ μm (this value of d is in accordance with observational data). These maximum values of P_E were usually (exclusive for Comet 2P/Encke) greater at least by an order of magnitude than the values for parent comets. Probabilities of collisions of most considered particles with Venus did not differ much from those with Earth, and those with Mars were about an order of magnitude smaller. For particles produced by Halley-type comets, P was greater for Mercury than for Mars.

Using $P_E = 4 \cdot 10^{-6}$ and assuming that the total mass of planetesimals that ever crossed Jupiter's orbit was about $100m_E$ (Ipatov 1987, 1993), where m_E is the mass of the Earth, Ipatov & Mather (2003, 2004a-b, 2007) concluded that the total mass of water delivered from the feeding zone of the giant planets to the Earth could be about the total mass of water in Earth's oceans. (Similar conclusion was made by Ipatov (2001) based on other calculations.) We supposed that the fraction of water in planetesimals equaled 0.5. The ratio of the mass of water delivered to a planet by Jupiter-family comets and Halley-type comets to the mass of the planet can be greater for Mars, Venus, and Mercury, than that for Earth. This larger mass fraction would result in relatively large ancient oceans on Mars and Venus. The larger value of P for Earth we have calculated compared to those argued by Morbidelli et al. (2000) ($P_E \sim (1-3) \cdot 10^{-6}$) and Levison et al. (2001) ($P_E \sim 4 \cdot 10^{-7}$) is caused by the fact that Levison et al. (2001) did not take into account the gravitational influence of the terrestrial planets, and Morbidelli et al. (2000) considered low-eccentric initial orbits beyond Jupiter's orbit. Besides, we considered a larger number of bodies. The detailed discussion on delivery of water and the comparison of our results with the results by other authors were made by Ipatov & Mather (2007). Marov & Ipatov (2005) discussed the delivery of volatiles to the terrestrial planets.

At the present time, most authors (e.g., Lunine et al. 2003, Morbidelli et al. 2000, and Petit et al. 2001) consider that the outer asteroid belt was the main source of the delivery

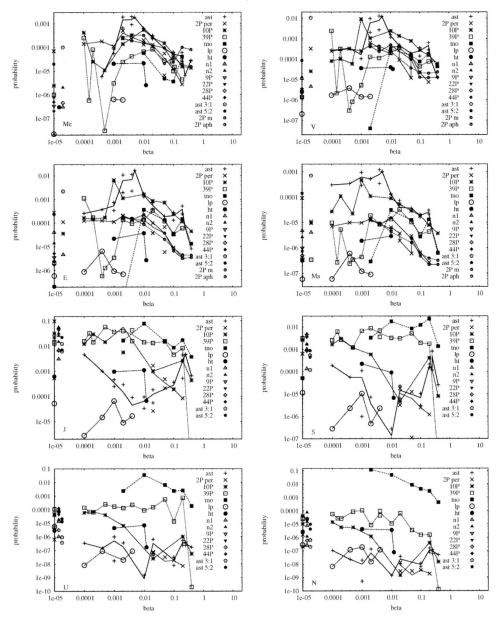

Figure 1. The probability of collisions of dust particles and bodies (during their dynamical lifetimes) with Mercury (subfigure Me), Venus (V), Earth (E), Mars (Ma), Jupiter (J), Saturn (S), Uranus (U), and Neptune (N) versus β (the ratio of the radiation to gravitational forces) for particles launched from asteroids (ast), trans-Neptunian objects (tno), Comet 2P/Encke at perihelion (2P per), Comet 2P/Encke at aphelion (2P aph), Comet 2P/Encke in the middle between perihelion and aphelion (2P m), Comet 10P/Tempel 2 (10P), Comet 39P/Oterma (39P), long-period comets (lp) at eccentricity $e = 0.995$ and perihelion distance $q = 0.9$ AU, and Halley-type comets (ht) at $e = 0.975$ and $q = 0.5$ AU (for lp and ht runs, initial inclinations were from 0 to 180°, and particles were launched near perihelia). If there are two points for the same β, then a plot is drawn via their mean value. Probabilities presented at $\beta \sim 10^{-5}$ are for small bodies ($\beta = 0$). Probabilities presented only for bodies were calculated for initial orbits close to orbits of Comets 9P/Tempel 1 (9P), 22P/Kopff (22P), 28P/Neujmin (28P), 44P/Reinmuth 2 (44P), and test asteroids from resonances 3:1 and 5:2 with Jupiter at $e = 0.15$ and $i = 10°$ ('ast 3:1' and 'ast 5:2'). For series n1 and n2, initial orbits of bodies were close to 10-20 different Jupiter-family comets (Ipatov & Mather 2004b).

of water to the terrestrial planets. Drake & Campins (2006) noted that the key argument against an asteroidal source of Earth's water is that the O's isotopic composition of Earth's primitive upper mantle matches that of anhydrous ordinary chondrites, not hydrous carbonaceous chondrites. To the discussion of the deuterium/hydrogen paradox of the Earth's oceans presented by Ipatov & Mather (2007), we can add that Genda & Ikoma (2008) showed that D/H in the Earth's ocean increased by a factor of 2-9.

Probabilities of collisions of JFCs with Saturn typically were smaller by an order of magnitude than those with Jupiter, and collision probabilities with Uranus and Neptune typically were smaller by three orders of magnitude than those with Jupiter. As only a small fraction of comets collided with all planets during dynamical lifetimes of comets, the orbital evolution of comets for the considered model of material points was close to that for the model when comets collided with a planet are removed from integrations.

Probabilities of collisions of considered particles and bodies with Jupiter during their dynamical lifetimes are smaller than 0.1. They can reach 0.01-0.1 for bodies and particles initially moved beyond Jupiter's orbit or in Encke-type orbits. For bodies and particles initially moved inside Jupiter's orbit, the probabilities are usually smaller than the above range and can be zero. Probabilities of collisions of migrating particles (exclusive for trans-Neptunian particles) with other giant planets were usually smaller than those with Jupiter. The total probability of collisions of any considered body or particle with all planets did not exceed 0.2.

Collisions of planetesimals with a star can cause variations in observed brightness and spectrum of the star. In our calculations, the fraction P_{Sun} of comets collided with the Sun during their dynamical lifetimes was about a few percent. For most JFCs, dynamical lifetimes are less than 10 Myr, and on average $P_{Sun} \sim 0.02$. For dust particles, P_{Sun} depends on β and can be considerably greater than for their parent bodies. For example, for Comet 10P/Tempel 2, $P_{Sun} \approx 0.01$, and almost all particles produced by this comet collide with the Sun (Ipatov & Mather 2006).

References

Drake, M. & Campins, H. 2006, in: D. Lazzaro, S. Ferraz-Mello & J. A. Fernandez (eds), *Asteroids, Comets, & Meteors*, IAU Symp. 229 (Cambridge: Cambridge Univ. Press), p. 381

Genda, H. & Icoma, M. 2008, *Icarus*, 194, 42

Ipatov, S. I. 1987, *Earth, Moon, & Planets*, 39, 101

Ipatov, S. I. 1988, *Sov. Astron.*, 65, 1075

Ipatov, S. I. 1993, *Solar System Res.*, 27, 65

Ipatov, S. I. 2001, *Adv. Space Res.*, 28, 1107

Ipatov, S. I. & Mather, J. C. 2003, *Earth, Moon, & Planets*, 92, 89

Ipatov, S. I. & Mather, J. C. 2004a, *Ann. New York Acad. Sci.*, 1017, 46

Ipatov, S. I. & Mather, J. C. 2004b, *Adv. Space Res.*, 33, 1524

Ipatov, S. I. & Mather, J. C. 2006, *Adv. Space Res.*, 37, 126

Ipatov, S. I. & Mather, J. C. 2007, in: IAUS 236 *Near-Earth Objects, Our Celestial Neighbors: Opportunity and Risk*, p. 55

Ipatov, S. I., Mather, J. C., & Taylor, P. A. 2004, *Ann. New York Acad. Sci.*, 1017, 66

Levison, H. F. & Duncan, M. J. 1994, *Icarus*, 108, 18

Levison, H. F., et al. 2001, *Icarus*, 151, 286

Lunine, J. I., Chambers, J., Morbidelli, A., & Leshin, L. A. 2003, *Icarus*, 165, 1

Marov, M. Ya. & Ipatov, S. I. 2006, *Solar Syst. Res.*, 39, 374

Morbidelli, A., Chambers, J., Lunine, J. I., Petit, J. M., Robert, F., Valsecchi, G. B., & Cyr, K. E. 2000, *Meteoritics & Planet. Sci.*, 35, 1309

Petit, J.-M., Morbidelli, A., & Chambers, J. 2001, *Icarus*, 153, 338

Icy Bodies of the Solar System
Proceedings IAU Symposium No. 263, 2009
J.A. Fernández, D. Lazzaro, D. Prialnik & R. Schulz, eds.

© International Astronomical Union 2010
doi:10.1017/S1743921310001481

Simulations for Terrestrial Planets Formation

Jianghui Ji[1] and Niu Zhang[1,2],

[1] Purple Mountain Observatory, Chinese Academy of Sciences, Nanjing 210008, China
email: jijh@pmo.ac.cn

[2] Graduate School of Chinese Academy of Science, Beijing 100049

Abstract. We investigate the formation of terrestrial planets in the late stage of planetary formation using two-planet model. At that time, the protostar has formed for about 3 Myr and the gas disk has dissipated. In the model, the perturbations from Jupiter and Saturn are considered. We also consider variations of the mass of outer planet, and the initial eccentricities and inclinations of embryos and planetesimals. Our results show that, terrestrial planets are formed in 50 Myr, and the accretion rate is about 60% - 80%. In each simulation, 3 - 4 terrestrial planets are formed inside "Jupiter" with masses of $0.15 - 3.6 M_\oplus$. In the 0.5 - 4AU, when the eccentricities of planetesimals are excited, planetesimals are able to accrete material from wide radial direction. The plenty of water material of the terrestrial planet in the Habitable Zone may be transferred from the farther places by this mechanism. Accretion may also happen a few times between two giant planets only if the outer planet has a moderate mass and the small terrestrial planet could survive at some resonances over time scale of 10^8 yr.

Keywords. methods:n-body simulations-planetary systems-planetary formation

1. Introduction

The discovery of the extrasolar planets (Mayor & Queloz (1995), Lee & Peale (2002), Ji *et al.* (2003)) around solar-type stars indeed provides substantial clues for the formation and origin of our own solar system. According to standard theory Safronov (1969), Wetherill (1990), Lissauer (1993), it is generally believed that planet formation may experience such several stages: in the early stage, the dust grains condense to grow km-sized planetesimals; in the middle stage, Moon-to-Mars sized embryos are created by accretion of planetesimals. When the embryos grow up to a core of $\sim 10 M_\oplus$, runaway accretion may take place. With more gases accreted onto the solid core, the embryos become more massive and eventually collapse to produce giant Jovian planets (Ida & Lin (2004)). At the end of the stage, it is around that the protostar has formed for about 3 Myr, the gas disk has dissipated. A few larger bodies with low e and i are in crowds of planetesimals with certain eccentricities and inclinations. In the late stage, the terrestrial embryos are excited to high eccentricity orbits by mutual gravitational perturbation. Next, the orbital crossings make planets obtain material in wider radial area. In this sense, solid residue is either scattered out of the planetary system or accreted by the massive planet, even being captured (Nagasawa & Ida (2000)) at the resonance position of the giant planets.

Chambers (2001) made a study of terrestrial planet formation in the late stage by numerical simulations, who set $150 - 160$ Moon-to-Mars size planetary embryos in the area of $0.3 - 2.0$ AU under mutual interactions from Jupiter and Saturn. He also examined two initial mass distributions: approximately uniform masses, and a bimodal mass distribution. The results show that $2 - 4$ planets are formed within 50 Myr, and finally survive over 200 Myr timescale, and the final planets usually have eccentric orbits with higher eccentricities and inclinations . Raymond, Quinn & Lunine (2004), Raymond, Quinn & Lunine (2006) also investigated the formation of terrestrial planets. In the simulations, they simply took into account Jupiter's gravitational perturbation, and the distribution of material are in $0.5 - 4.5$ AU. Their results confirm a leading hypothesis for the origin of Earth's water: they may come from the material in the outer area by impacts in the late stage of planet formation. Raymond, Mandell, & Sigurdsson (2006) explored the planet formation under planetary migration of the giant. In the simulations, super Hot Earth form interior to the migrating giant planet, and water-rich, Earth-size terrestrial planet are present in the Habitable Zone $(0.8 - 1.5$ AU) and can survive over 10^8 yr timescale.

In our work, we consider two-planet model, in which Jupiter and Saturn are supposed to be already formed, with two swarms of planetesimals distributed in the region among $0.5 - 4.2$ AU and $6.2 - 9.6$ AU respectively. The initial eccentricities and inclinations of planetesimals are considered. We also vary the mass of Saturn to examine how the small bodies evolve. The simulations are performed on longer timescale 400 Myr in order to check the stability and the dynamical structure evolution of the system. In the following, we briefly summarize our numerical setup and results.

2. Numerical Setup

The timescale for formation of Jupiter-like planets is usually considered to be less than 10 Myr Briceño *et al.* (2001), the formation scenario of planet embryos is related to their heliocentric distances and the initial mass of the star nebular. If we adopt the model of 1.5 MMSN (Minimum Mass Solar Nebular), the upper bound of the timescale for Jupiter-like planet formation corresponds to the timescale for embryo formation at 2.5 AU Kokubo & Ida (2002), which is just at $3 : 1$ resonance location of Jupiter. In the region $2.5 - 4.2$ AU, embryos will be cleared off by strong perturbation from Jupiter. There should be some much smaller solid residue among Jupiter and Saturn, even though the clearing effect may throw out most of the material in this area. We set embryos simply in the region $0.5 - 2.5$ AU and planetesimals at $0.5 - 4.2$ AU and $6.2 - 9.6$ AU.

We adopt the surface density profile as follows (Raymond, Quinn & Lunine (2004)):

$$\Sigma(r) = \begin{cases} \Sigma_1 r^{-3/2}, & r < snow\ line, \\ \Sigma_{snow}(\frac{r}{5AU})^{-3/2}, & r > snow\ line. \end{cases} \tag{2.1}$$

In (2.1), $\Sigma_{snow} = 4\ g/cm^2$ is the surface density at snowline, where the snowline is at 2.5 AU with $\Sigma_1 = 10\ g/cm^2$. The mass of planetary embryos is proportional to the width of the feeding zone, which is associated with Hill Radius, R_H, so the mass of an embryo increases as

$$M_{embryo} \propto r\Sigma(r)R_H \tag{2.2}$$

The embryos in the $0.5 - 2.5$ AU are spaced by Λ (Λ varying randomly between 2 and 5) mutual Hill Radii, $R_{H,m}$, which is defined as

$$R_{H,m} = (\frac{a_1 + a_2}{2})(\frac{m_1 + m_2}{3M_\odot})^{1/3} \tag{2.3}$$

where $a_{1,2}$ and $m_{1,2}$ are the semi-major axes and masses of the embryos respectively. Replacing R_H in (2.2) with $R_{H,m}$, and substituting (2.1) in (2.2), then, we achieve a

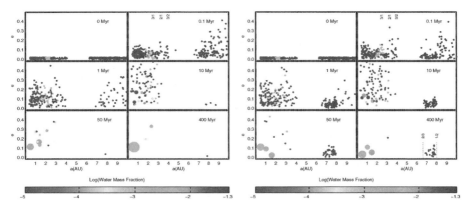

Figure 1. *Left panel*: (a) Snapshot of simulation 2a with $M_{Saturn} = 5M_\oplus$. The total mass of embryos is $2.4M_\oplus$, the masses of planetesimals inside Jupiter are $0.0317M_\oplus$, and those outside Jupiter are $0.0375M_\oplus$. Planetesimals among Jupiter and Saturn were nonself-gravitational. Note the size of each object is relative, and the value bar is log of water mass fraction. *Right panel*: (b) snapshot for similuation 2b.

relation law between the mass of embryos and the parameter Λ as

$$M_{embryo} \propto r\Sigma(r)R_{H,m} \propto r^{3/4}\Lambda^{3/2}\Sigma^{3/2} \tag{2.4}$$

Here, we equally set the masses of planetesimals inside and outside Jupiter, respectively. Consequently, the number distribution of the planetesimals is simply required to meet $N \propto r^{-1/2}$. Additionally, we remain the total number of planetesimals and embryos inside Jupiter, and the number of planetesimals outside Jupiter both equal to 200. The mass inside and outside Jupiter is equal to be $7.5M_\oplus$. The eccentricities and inclinations vary in $(0-0.02)$ and $(0-0.05°)$, respectively. The mass of Saturn in simulations 1a/1b, 2a/2b and 3a/3b are $0.5M_\oplus$, $5M_\oplus$, $50M_\oplus$ respectively. Each simulations marked by label a (or b) is run to consider (not consider) self-gravitation of planetesimals among giants.

We use the hybrid symplectic integrator (Chambers (1999)) in MERCURY package to integrate all the simulations. In addition, we adopt 6 days as the length of time step, which is a twentieth period of the innermost body at 0.5 AU. All runs are carried out over 400 Myr time scale. At the end of the intergration, the changes of energy and angular momenta are 10^{-3} and 10^{-11} respectively. Six simulations are performed on a workstation composed of 12 CPUs with 1.2 GHz, and each costs roughly 45 days.

3. Results

Fig. 1(a) is a snapshot of simulation 2a. At 0.1 Myr, it is clear that the planetesimals are excited at the $3:2$ (3.97 AU),$2:1$ (3.28 AU) and $3:1$ (2.5 AU) resonance positions with Jupiter, and this is quite similar to the Kirkwood gaps of the main asteroidal belt in solar system. For about 1 Myr, planetesimals and embryos are deeply intermixed, where most of the bodies have stirred to be large eccentricities. Collisions and accretions frequently emerge among planetesimals and embryos. This process continues until about 50 Myr, and the planetary embryos are mostly generated. The formation timescale of embryos is in accordance with that of Ida & Lin (2004). Finally, inside Jupiter, 3 terrestrial planets are formed with masses of $0.15-3.6M_\oplus$. However, at the outer region, the planetesimals are continuously scattered out of the system at 0.1 Myr. For about 10 Myr, there are no survivals except at some resonances with the giant planet. As shown in the Figure, there is a small body at the $1:2$ resonance with Jupiter. Due to scattering amongst

Table 1. Properties of terrestrial planets from different systems

System	accretion rate	n	$\bar{m}(m_\oplus)$	concentration	$\bar{e}$	$\bar{i}(°)$
1a	73.2518%	3	1.8313	0.4606	0.1381	7.6963
1b	80.3853%	3	2.0096	0.4262	0.0937	1.7790
2a	59.8322%	3	1.4958	0.8116	0.2108	16.9117
2b	72.9779%	4	1.3683	0.4299	0.0999	5.1415
3a	65.1098%	3	1.6277	0.5337	0.2063	5.9153
3b	66.9694%	3	1.6742	0.5040	0.1839	5.2447
1a-3b	69.7544%	3.2	1.6678	0.5276	0.1554	7.1148
solar	-	4	0.4943	0.5058	0.0764	3.0624

planetesimals, Jupiter (Saturn) migrates inward (outward) 0.13 AU (1.19 AU) toward the sun respectively. Such kind of migration agrees with the work of Fernandez & Ip (1984). Hence, the 2 : 5 mean motion resonance is destroyed, then the ratios of periods between Jupiter and Saturn degenerate to 1 : 3. Therefore, the ratio of periods for Jupiter, small body and Saturn is approximate to 1 : 2 : 3. In the 0.5 − 4 AU, when the eccentricities of planetesimals are excited, planetesimals are able to accrete material from wide radial direction. The plenty of water material of the terrestrial planet in the Habitable Zone may be transferred from the farther places by this mechanism.

Fig. 1(b) is illustrated for simulation 2b. In comparison with Fig. 1(a), it is apparent that planetesimals are excited more quickly at the 3 : 2 (3.97 AU), 2 : 1 (3.28 AU) and 3 : 1 (2.5 AU) resonance location with Jupiter. The several characteristic timescales are the same as simulation 2a for the bodies within Jupiter. 4 planets are formed in simulation 2b, the changes of position of Jupiter and Saturn behaves like simulation 2a. We point out that simulations 2a and 2b share the initial conditions, but the only difference in them is whether we consider the self-gravitation among the outer planetesimals. There is a little gathered planetesimals survival over 400 Myr among 7 − 8 AU, located in the area of 2 : 3 (6.63 AU) and 1 : 2 (8.03 AU) resonances with Jupiter. The detailed results for whole simulations that the reader may refer to Zhang & Ji (2009).

The production efficiency of the terrestrial planet in our model is high, and the accretion rate inside Jupiter is 60% − 80% in the simulations. 3 − 4 terrestrial planets formed in 50 Myr. 5 of 6 simulations have a terrestrial planet in the Habitable Zone (0.8 − 1.5 AU). The planetary systems are formed to have nearly circular orbit and coplanarity, similar to the solar system (see Table 1). We suppose that the above characteristics are correlated with the initial small eccentricities and inclinations. The concentration in Table 1 means the ratio of maximum terrestrial planet formed in the simulation and the total terrestrial planets mass. It represents different capability on accretion. The average value of this parameter is similar to the solar system. Considering the self-gravitation of planetesimals among Jupiter and Saturn, the system has a better viscosity, so that the planetesimals will be excited slower. The consideration of self-gravitation may not change the formation time scale of terrestrial planets, but will affect the initial accretion speed and the eventual accretion rate.

4. Summary and Discussion

We simulate terrestrial planet formation by using two-planet model. In the simulations, the variations of the mass of outer planet, the initial eccentricities and inclinations of embryos and planetesimals are also considered. The results show that, during the terrestrial planet formation, planets can accrete material from various areas inside Jupiter.

Among $0.5 - 4.2$ AU, the accretion rate of terrestrial planet is $60\% - 80\%$, i.e., about $20\% - 40\%$ initial mass is removed during the progress. The planetesimals will improve efficiency of accretion rate for certain initial eccentricities and inclinations, and this also makes the newly-born terrestrial planets have lower orbital eccentricities. It also indicates that in the planet formation that water-rich terrestrial planet may be formed in the Habitable Zone. Most of the planetesimals among Jupiter and Saturn are scattered out of the system, and such migration induced by scattering (Fernandez & Ip (1984)) or long-term orbital evolution can make smaller bodies capture at some mean motion resonance location. Accretion could also happen a few times between two planets if the outer planet owns a moderate mass, and a small terrestrial planet could survive at some resonances over time scale of 10^8 yr. The outcomes further reveal that the outer planet has little effect on dynamical architecture inside Jupiter.

Acknowledgements

This work is financially supported by the National Natural Science Foundations of China (Grants 10973044, 10833001, 10573040, 10673006), the joint project by the Academy of Finland and NSFC, and the Foundation of Minor Planets of Purple Mountain Observatory.

References

Briceño, C. *et al.*, 2001, *Science*, 291, 93
Chambers, J. E. 1999, *MNRAS*, 304, 793
Chambers, J. E. 2001, *Icarus*, 152, 205
Fernandez, J. A. & Ip, W. H. 1984, *Icarus*, 58, 109
Ida, S. & Lin, D. N. C. 2004, *ApJ*, 604, 388
Ji, J. H., *et al.* 2003, *ApJ*, 585, L139
Kokubo, E. & Ida, S. 2002, *ApJ*, 581, 666
Lee, M. H. & Peale, S. J. 2002, *ApJ*, 567, 596
Lissauer, J. J. 1993, *ARAA*, 31, 129
Nagasawa, M. & Ida, S. 2000, *AJ*, 120, 3311
Raymond, S. N., Quinn, T., & Lunine, J. I. 2004, *Icarus*, 168, 1
Raymond, S. N., Quinn, T., & Lunine, J. I. 2006, *Icarus*, 183, 265
Raymond, S. N., Mandell, A. M., & Sigurdsson, S. 2006, *Science*, 313, 1413
Safronov, V. S. 1969, *Evolution of the Protoplanetary Cloud and Formation of the Earth and the Planets*, (Moscow:Nauka)
Mayor, M. & Queloz, D. 1995, *Nature*, 378, 355
Wetherill, G. W. 1990, *Ann. Rev. Earth Planet Sci.*, 18, 205
Zhang, N. & Ji, J. 2009, *Science in China Series G*, 52(5), 794

Icy Bodies of the Solar System
Proceedings IAU Symposium No. S263, 2009
J.A. Fernández, D. Lazzaro, D. Prialnik & R. Schulz, eds.

© International Astronomical Union 2010
doi:10.1017/S1743921310001493

Interaction between gas and ice phase in the three periods of the solar nebula

Carmen Tornow[1], Ekkehard Kührt[1], Stefan Kupper[1], and Uwe Motschmann[2]

[1] Inst. of Planetary Research,
Rutherfordstr. 2, D-12489 Berlin, Germany
email: carmen.tornow@dlr.de,ekkehard.kuehrt@dlr.de, and stefan.kupper@dlr.de

[2] Inst. of Theoretical Physics, Technical University Braunschweig,
Mendelssohnstr. 3, D-38106 Braunschweig
email: u.motschmann@tu-braunschweig.de

Abstract. We simulate the chemical processes in the three evolution periods of the solar nebula, which are (i) the quasi-stationary prestellar cloud core, (ii) the gravitationally collapsing protostellar core, and (iii) the evolving gas-dust disk. Our purpose is to identify chemical parameters which reflect special aspects of the interactions between the gas and ice phase in the different periods, e.g. isotopic or molecular ratios. In this study we derive the D/H and ^{15}N/^{14}N ratio of selected compounds as well as the CO_2/H_2O ratio to measure the fraction of non-polar to polar ice in the grain mantles. The chosen ratios depend on the depletion-enrichment relation between the ice and gas phases driven by the thermal evolution in each period, especially during the collapse. Hence, we have made great efforts in order to derive realistic and compact hydrodynamic models to describe the evolutionary periods of the solar nebula.

Keywords. Astrochemistry, star formation, comets, hydrodynamics, molecular processes

1. Introduction

Stars form from gas clumps with a small dust amount ($\approx 1\%$ in mass) which are density fluctuations on an extended size range of a large turbulent hydrogen cloud. We assume that this was true for our sun as well. In the following sections we will describe each evolution period of the solar nebula in more detail and calculate the time dependency of its chemical composition in the gas and ice phase. The complete time interval of the three different periods of the solar nebula extends over about 40 million years. The chemical evolution is studied by means of ratios defined for selected species. At the beginning of the simulation process the gas phase contains 321 different species denoted by $x(i)$ and the ice phase is formed from 50 compounds symbolised by $x^*(i)$. The isotope ratios are derived from water and HCN in the case of the D/H ratio and from N_2H^+ for the ^{15}N/^{14}N ratio. Note, that the latter is a gas phase ion, since the ^{15}N chemistry is not contained in our ice phase model so far. The polar to non-polar ice fraction is measured by $x^*(CO_2)/x^*(H_2O)$ where H_2O and CO_2 are major species of the interstellar and cometary ice.

Essentially, one could propose more complex ratios in order to measure the corresponding material fractions, but due to feasibility restrictions in modelling methods and the amount of available observational results of ice phase abundancies for a future verification of our results we have considered simple and commonly occurring molecules, especially in comets. Since they consists of relatively pristine material the comets can be seen as important representatives for the ice phase of the solar nebula.

2. Simulation of the quasi-stationary prestellar cloud core

The quasi-stationary evolution of a prestellar core is modelled with a linear time dependency of the temperature and density. Systematic flow processes are not considered. The negligence of flows and unsteady evolution events such as shock waves or cloud collisions is justified since the temperature and density of the cloud core change over the large time interval of nearly 30 million years. The initial and final temperature and density at the beginning and end of the evolution interval are given by

$$T_{init} = 70\,\text{K} \quad and \quad \rho_{init} = 3.1 \times 10^{-21} g/cm^3 \quad \text{for} \quad t = 0$$

$$T_{final} = 10\,K \quad and \quad \rho_{final} = 1.8 \times 10^{-18} g/cm^3 \quad \text{for} \quad t = 3 \cdot 10^7 years$$

The relative abundances $x(i)$ and $x^*(i)$ are calculated from a set of kinetic equations. The rates for the corresponding chemical reactions are calculated from data published in Woodall *et al.* (2007) and Aikawa *et al.* (1997). Table 1 contains the initial abundances. We have restricted our set of species to compounds having no more than seven atoms. In the ice phase only hydrogenation reactions are considered.

Table 1. Initial abundances of the elements in the gas phase, all other initial data are zero.

H	H_2	D	He	O	C^+	N	^{15}N	Si, Mg, Fe
0.9	0.1	1.5×10^{-5}	0.14	1.8×10^{-4}	7.3×10^{-5}	2.1×10^{-5}	1.1×10^{-7}	6.0×10^{-11}

From the calculated abundance evolution we obtain the time dependence of the ratios shown in Fig. 1. One recognises an increasing amount of non-polar ice and bounded heavy isotopes in the course of the prestellar core evolution. A large H_2 to H ratio seems to be advantageous for the formation of CO_2 relative to H_2O.

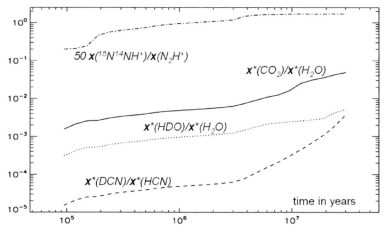

Figure 1. Time dependency of the nitrogen isotope ratio in the gas phase (note the factor of 50 to present all curves in the same figure), the D/H ratios in the ice phase and the CO_2-H_2O ratio for the polar to non-polar ice fraction calculated for the slowly evolving quasi-stationary prestellar core.

3. Simulation of the gravitationally collapsing protostellar cloud core

The gravitational collapse of a cloud core causes the central density to increase over more than 15-16 orders of magnitudes. At the end of this process a stellar core, the T

Tauri star, and a young disk have formed in the centre of the solar nebula. Therefore, a numeric simulation of this type of collapse is a complex task. We have derived an analytical solution to solve the continuity, momentum and Poisson equation for a collapsing cloud core that is valid for spherical symmetry. According to Saigo *et al.* (2008) this restriction has no serious drawbacks as long as the rotation rate is low ($\approx 10^{-15}$ s^{-1}). The mathematics of this solution will be described in a different publication.

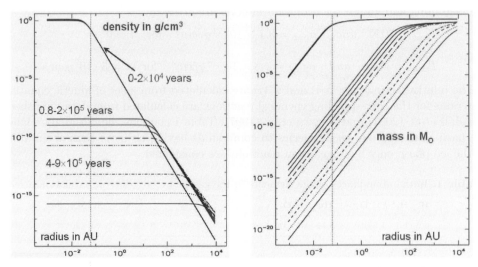

Figure 2. Density and mass profiles derived from the analytical model of the collapsing cloud core. The protostar is the so-called second core (Saigo *et al.* (2008)). The collapse takes place at the time t = 0.

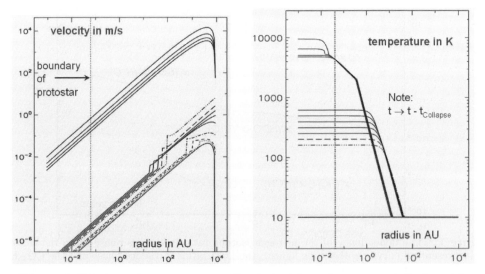

Figure 3. Velocity and temperature profiles derived from the same analytical model.

In Fig. 2 and Fig. 3 we present the calculated radial density, mass, velocity, and temperature profiles at different times. In order to include the influence of the formed protostellar disk we have coupled our collapse solution to the disk model derived by Stahler *et al.* (1994). The values of the four radial profiles in Fig. 2 and Fig. 3 are given for an Eulerian grid. However, the computation of the chemical abundance evolution of the gas and ice phase following from the continuity equation of each species can be simplified if one uses a transformation to a Lagrangian grid defined by the initial positions of the gas-ice parcels at the beginning of the collapse. The resulting total time dependencies of the density and temperature are calculated for an inner gas parcel moving from 2.5 to 1.3 AU. In this case the temperatures are high enough to guarantee the loss of the ice phase due to the evaporation of the icy grain mantles. In order to study the temporal progress of depletion of the ice phase species we have computed the ratio of the current to the initial abundance for selected compounds. The obtained values are presented in Fig. 4.

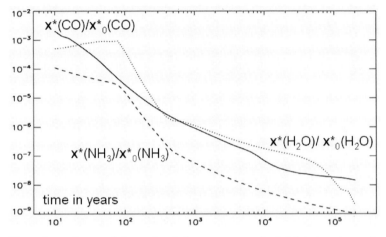

Figure 4. Time dependency of the ratio of the current to the initial abundance for CO, H_2O, and NH_3 calculated for the period of the collapsing protostellar core.

4. Simulation of the gravitationally collapsing protostellar cloud core

The disk model of Stahler *et al.*, 1994, is valid for a young disk only. In order to study the chemical evolution of the gas and ice species in a mature disk we have used the non-stationary model of Davis *et al.* (2003).

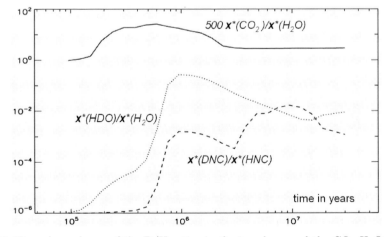

Figure 5. Time dependency of the D/H ratios in the ice phase and the CO_2-H_2O molecular ratio for the polar to non-polar ice fraction calculated for the evolving disk.

This model describes the disk cooling and depletion in the course of its evolution. Due to the gas flow we have to switch to the Lagrangian grid again in order to compute the abundance values. The necessary initial data follow from the final abundance results calculated for the collapse period. In contrast to our collapse model the Davis model is based on axial symmetry. In order to keep a simple radial dependency without angular variations, the relative abundances are derived with respect to the column density. For time intervals much larger than 10^7 the corresponding number density would be less then 0.01 cm^{-3}, i.e. a gas disk is not existent anymore. Therefore, at most 10 million years are of physical interest only. Fig. 5 shows the time behaviour of the same ice ratios as seen in Fig. 1. However, one recognizes clear differences although in both cases the ice phase abundancies are growing with respect of their initial values. For the evolving disk, there is a superposition of the time dynamics of the disk parameter itself and the time dynamics of the chemical processes. Therefore the shapes of the disk related abundance ratios versus time are less monotonic than the same curves of the prestellar core. In addition the disk density of the considered gas parcel decreases whereas the core density increases slowly.

5. Conclusion and outlook

The collapse of a rotating quasi-stationary cloud core into a young dense nebula and finally into a cooling disk is modelled in order to study the chemical evolution of the volatile gas and ice phase species. We have developed an analytical solution for the collapse period that gives the chance to simulate this process very efficiently. In addition we have studied the feasibility of merging the evolution periods of the solar nebula using the transition from an Eulerian to a Lagrangian grid. However, the transition from the spherically collapsing cloud core to the disk is complicated and further research needs to be done for the transition between the different temperature models. From our chemical calculations a distinct difference between the disk and the prestellar core chemistry becomes conspicuously. It is related to the higher dynamics in the disk on the one hand and to its complex initial chemical state on the other. The effects of both phenomenons are entangled and further research needs to be done to investigate their influence independently.

References

Aikawa, Y., Umebayashi, T., Nakano, T., & Miyama, S. M. 1997, *ApJ Letters*, 486, L51
Davis, S. S. 2003, *ApJ*, 592, 1193
Saigo, K., Tomisaka, K., & Matsumoto, T. 2008, *ApJ*, 674, 997
Stahler, Steven W., Korycansky, D. G., Brothers, M. J., & Touma, J. 1994, *ApJ*, 431, 341
Woodall, J., Agndez, M., Markwick-Kemper, A. J., & Millar, T. J. 2005, *A&A*, 466, 1197

Part III

Dynamical Aspects of Icy Bodies. The Oort Cloud

Icy Bodies of the Solar System
Proceedings IAU Symposium No. 263, 2009
J.A. Fernández, D. Lazzaro, D. Prialnik, & R. Schulz, eds.
© International Astronomical Union 2010
doi:10.1017/S174392131000150X

Galactic environment and cometary flux from the Oort cloud

Marc Fouchard

LAL-IMCCE/Université de Lille 1
1 Impasse de l'Observatoire
59 000 Lille, France
email:fouchard@imcce.fr

Abstract. The Oort cloud, which corresponds to the furthest boundary of our Solar System, is considered as the main reservoir of long period comets. This cloud is likely a residual of the Solar System formation due to the gravitational effects of the young planets on the remaining planetesimals. Given that the cloud extends to large distances from the Sun (several times 10 000 AU), the bodies in this region have their trajectories affected by the Galactic environment of the Solar System. This environment is responsible for the re-injection of the Oort cloud comets into the planetary region of the Solar System. Such comets, also called "new comets", are the best candidates to become Halley type or "old" long period comets under the influence of the planetary gravitational attractions. Consequently, the flux of new comets represents the first stage of the long trip from the Oort cloud to the observable populations of comets. This is why so many studies are still devoted to this flux.

The different perturbers related to the Galactic environment of the Solar System, which have to be taken into account to explain the flux are reviewed. Special attention will be paid to the gravitational effects of stars passing close to the Sun and to the Galactic tides resulting from the difference of the gravitational attraction of the Galaxy on the Sun and on a comet. The synergy which takes place between these two perturbers is also described.

Keywords. celestial mechanics, comets: general, Oort Cloud, solar system: general

1. Introduction

The first person who proposed the existence of a cloud of objects surrounding the Solar system was Öpik (1932). Studying the gravitational influence of the passing stars on very elongated orbit around the Sun as those of comets and meteors, he showed that their mean effect was to increase the perihelion distance. Because these orbits are prevent from any planetary perturbations once their perihelion has been raised enough, a cloud of objects at more than 10 000 AU might exist. From this preliminary study, Öpik concluded, however, that such cloud is likely not observable because of the preference of stellar perturbations to rise the perihelion rather than to decrease it.

Later on, looking at the original semi-major axis of 19 well observed long period comets, Oort (1950) showed that the orbital energy of these comets picked toward zero, with semi-major axes between 50 000 and 100 000 AU: the *Oort peak*.

Oort argued that these comets were entering the planetary region of the Solar System for the first time, and should form a reservoir surrounding the Sun between 10^4 and 10^5 AU : the *Oort cloud*. As Öpik, Oort took into consideration the effects of passing stars on elongated orbits. However, he noticed that these perturbations were able to move the perihelion of an Oort cloud comet close enough to the Sun for the comet to be observable.

More generally, at such distance from the Sun the Galactical environment of the Sun is able to modify the heliocentric trajectories of the Oort cloud comets. The present review is dedicated to the effects of these *external perturbers* on the flux of Oort cloud comets.

The estimation on population, shape and size of the Oort cloud will be first given in Sec. 2. Then, in Sec. 3 the three main external perturbers will be described: the passing stars (Sec. 3.2), the giant molecular clouds (Sec. 3.3) and the Galactic tides (Sec. 3.4). The efficiency of each perturber may be evaluated using a tool called the *loss cone* defined in Sec. 3.1. Section 4 is devoted to the synergy between the passing stars and the Galactic tides. The conclusions are given in Sec. 5.

2. Size, population and shape of the Oort cloud

The main goal of any study on the cometary flux from the Oort cloud is: how does the Oort cloud look like? Indeed, being not directly observable from Earth, one has to deduce the informations from the few comets which come close enough to the Sun to be observed.

One has to make a distinction between an inner Oort cloud, and the observable Oort cloud. No strict definition of these regions exists in the literature, however the following definitions may be acceptable. The inner border of the *inner Oort cloud* is the threshold from which the time scale for the Galactic tide to change the perihelion distance of a comet becomes comparable to the time scale for the planets to change its orbital energy (Duncan *et al.* 1987); and the inner border of the observable region of the Oort cloud - also called *outer Oort cloud* - is the threshold from which the Galactic tides are able to move the perihelion of a comet from outside the planetary region of the Solar system to inside the orbit of Jupiter in less than one orbital period. Both frontiers depend on the parameters used, however from numerical experiments (*e.g.* Duncan *et al.* 1987; Dones *et al.* 2004), the inner border of the Oort cloud should be around $a \sim 3\,000$ AU, and the inner border of the observable Oort cloud around $a \sim 20\,000$ AU. From now on, the region with $3\,000 < a < 20\,000$ AU will be called the inner Oort cloud, and the region with $a > 20\,000$ AU the outer Oort cloud. An innermost cloud, with $a \lesssim 3\,000$ AU, may also be defined (Brasser 2008).

The size. The parameter which is the easiest to obtain, *a priori*, is the size of the Oort cloud. Indeed, it is enough to reconstruct the original semi-major axis of the observed comets, *i.e.* the semi-major axis before the comet enters the planetary region of the Solar System. This method allows mainly to deduce the size of the outer Oort cloud. This was made by Oort (1950), and he concluded that the Oort cloud should be at heliocentric distance between 50 000 and 150 000 AU. It appeared, however, that the non gravitational forces induced by the out-gazing of a comet when it is close to the Sun, change the determination of the original semi-major axis. Using a set of long period comets with large perihelion distance for which non-gravitational forces are weak, Marsden & Sekanina (1973) showed that the Oort peak is rather around 25 000 AU. A more recent study made by Królikowska (2006) gives an even smaller value with a peak around 17 000 AU but with a more spread distribution.

The population. Among the parameters which defined the Oort cloud, its population is probably the most important. Indeed, this parameter constrains the scenarios of the formation of the solar system. The population of the cloud may be estimated only by numerical experiments and making comparison with the observations. Consequently, one is mainly able to evaluate the population of the outer Oort cloud. The more recent estimates of these population range between 5×10^{11} and 50×10^{11} (Weissman 1996; Emel'yanenko *et al.* 2007). As regards the inner Oort cloud, it is supposed to be from 1 (Dones *et al.* 2004) to 4 times (Duncan *et al.* 1987) the population of the outer Oort cloud. These estimations are however uncertain because of the difficulty to evaluate the absolute magnitude of a comet. Using an averaged comet mass of 4×10^{16} g, Weissman (1996) estimated the mass of the outer Oort cloud from 6 to 7 Earth masses, however this quantity is poorly constrained.

The shape. The orientation of the perihelion of the observed long period comets does not show a clear preference (even if some gathering might be observed, but this point will be discussed later). Oort (1950) and Yabushita *et al.* (1982) showed that the distribution of the heliocentric velocity of Oort cloud comets with semi-major axis greater than 20 000 AU should be isotropic, with an eccentricity distribution function proportional to the square of the eccentricity (Hills 1981). Numerical experiments showed that the threshold from which the Oort cloud becomes isotropic is rather at $a \sim 10\,000$ AU. For smaller heliocentric distances, the orbit planes should concentrate around the ecliptic (*e.g.* Levison *et al.* 2001; Emel'yanenko *et al.* 2007), keeping a memory of their origin. The density profile of the heliocentric distance for the outer Oort cloud, may be obtained only from numerical experiments of the Oort cloud formation. Such a problem is out of the scope of the present review. However, all the studies seem to agree for a density profile $n(r) \propto r^{-s}$ with $s \sim 3.5$ (Duncan *et al.* 1987; Dones *et al.* 2004; Brasser *et al.* 2006; Emel'yanenko *et al.* 2007).

3. The external perturbers

3.1. *The loss cone*

When a quasi parabolic comet enters the planetary region of the Solar System, its trajectory will be affected by the planets. The Fig 1 (From Fernández 1981) shows the typical energy change of a comet passing through the planetary region of the Solar System versus the perihelion distance for 6 different ranges of initial inclination. The figure hilights a property already used by Oort (1950) and Weissman (1980) among others: a comet with a perihelion distance smaller than $10-15$ AU will be likely removed from the Oort cloud by planetary perturbations, being sent on a more tidily orbit to the Sun or ejected from the Solar System. This consideration allows Oort to postulate that the observed long period comets in the Oort peak are new. Because the perturbations are mainly due by Jupiter and Saturn, the threshold around 15 AU has been called the *Jupiter-Saturn barrier.*

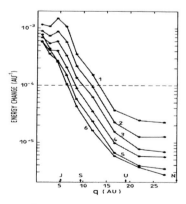

Figure 1. Typical energy change of comet passing through the planetary region as a function of its perihelion distance q. Each curve corresponds to a certain range of inclinations such that: curve 1 for $0° < i < 30°$, ..., curve 6 for $150° < i < 180°$. Credit: Fernández, A&A, 96, 26, 1981, reproduced with the permission of © ESO.

Now, for quasi-parabolic ($q \ll r$) comets, one has:

$$v_t = \frac{\sqrt{2\,\mu\,M_\odot\,q}}{r},$$ (3.1)

where v_t is the tangential velocity of the comets, r the heliocentric distance, q the perihelion distance, μ the universal gravitational constant and $M_\odot$ the mass of the Sun.

The Jupiter-Saturn barrier may be modelled by a tangential velocity $v_t = \sqrt{2\mu M_\odot 15}/r$, which defines a cone with decreasing width, given by v_t, for increasing heliocentric distance. Inside the loss cone, an observable cone may be defined in a similar way, by a perihelion distance equal to 5 AU for instance (Oort used 1.5 AU).

Just outside the planetary region going away from the Sun, the loss cone is empty. To observe a flux of comets from the Oort cloud, the loss cone on the way inward to the Sun has to be filled such that the observable cone contains also some comets. For such event to occur, the external perturbers, *i.e.* passing stars, Galactic tides, giant molecular clouds, must fill completely the loss cone.

An external perturber affects mainly the angular momentum of a comet, *i.e.* its tangential velocity v_t. The typical perturbation of v_t increases with the heliocentric distance, which defines also a cone: the *smear cone* (Hills 1981). The efficiency of an external perturber is at its maximum when the size of its smear cone becomes greater than the size of the loss cone. Indeed, for such, or greater, heliocentric distances the loss cone, and consequently the observable cone, are filled completely. Thus the flux of Oort cloud comets may be directly estimated by the size of the observable cone.

Preliminary experiments have shown that the stars are able to filled completely the loss cone when $r > 50\,000$ AU (Oort 1950; Rickman 1976). We will now discuss in more detailed the efficiency and characteristics of each external perturber.

3.2. *The passing stars*

Once in the Oort cloud, since the comets are far from the Sun, Oort considered that only perturbations from random passing stars, can change significantly the angular momenta of comets and send some of them into the loss cone. From this time, and for more than three decades, stellar perturbations were almost the only mechanism considered to produce observable comets. Many studies were devoted to this transport (e.g. , Rickman 1976; Weismann 1979; Fernández 1980; Hills 1981; Remy & Mignard 1985).

It appeared from all the experiments that this flux may be divided into two components: *(i)* a quasi constant background flux from the outer part of the cloud (Rickman 1976; Hills 1981; Heisler *et al.* 1987), *(ii)* a sporadic flux characterised by eventually strong comets showers extending toward the inner Oort cloud (Hills 1981). The former is due by frequent stellar encounters at large heliocentric distances filling the loss cone at any time; whereas the latter is due by rare but close stellar encounters able to fill temporally the loss cone at small heliocentric distance, depending on the impact distance of the star with the Sun.

This phenomenon is illustrated in Fig. 2 (from Heisler *et al.* 1987), where the flux of comets at heliocentric distance smaller than 10 AU per period of 10^6 yr for three different values of initial semi-major axis versus time is given. One observes: *(i)* for $a_0 = 10\,000$ AU, few but very strong comets showers, *(ii)* for $a_0 = 30\,000$ AU, a background flux with large fluctuations due to comets showers and *(iii)* for $a_0 = 40\,000$ AU, the loss cone is always filled yielding an almost constant background flux.

Some studies were devoted to the problem of determining whether or not we are experiencing a comets shower caused by a recent close stellar encounter with the Sun. Dybczyński (2002) has shown that such encounter should induce a strong asymmetry

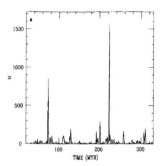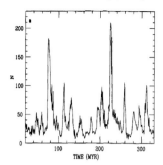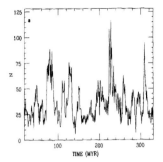

Figure 2. Number of comets per million years than pass through a perihelion less than 10 AU and have initial semi-major axis $a_0 = 10\,000$, $30\,000$ and $40\,000$ AU, from left to right. From Heisler *et al.* (1987).

in the distribution of the aphelion direction of the comets directly injected by the passing star, the asymmetry being characterised by an accumulation of perihelion direction toward the anti-perihelion direction of the stellar path.

Because the distribution of perihelion directions of the observed long period comets does not show any accumulation consistent with a recent stellar encounter it has been concluded that we are not experiencing a comets shower.

3.3. *The giant molecular clouds*

When the Solar System travels around the Galactic centre, it may also encounter some molecular clouds. These objects may be huge ($\gtrsim 20$ pc) and massive ($\sim 5 \times 10^5 \, M_\odot$). However, these quantities as well as the structure of a giant molecular cloud are poorly defined. The preliminary studies have shown that an encounter with a giant molecular cloud would have devastating effects on the Oort cloud (Biermann 1978) and that only the inner part of the Oort cloud could survived (Napier & Staniucha 1982; Clube & Napier 1982; Bailey 1983). Hut & Tremaine (1985) were less dramatic and showed that 2/3 of the comets should survived at $25\,000$ AU, with a destructive effect comparable to that of passing stars. A recent work by Jakubík & Neslušan (2008) shows that one should have a maximal erosion of the outer part of the cloud with 22% of the comets ejected.

The main conclusion one may retain from these studies is that the outer part of the Oort cloud is likely not primordial. However, because these encounters are difficult to model and should consist in rare event during the life of the Solar system (Bailey 1983), it has been of common use not to take them into account, even on long time span simulation.

3.4. *The Galactic tides*

The first time that the Galaxy was taken into account in the frame of Oort cloud comets dynamics, it has been modelled as a point mass (Chebotarev 1964, and Byl 1983). Later, in 1984-1986, the tides induced by the Galactic disk was also introduced in a series of papers (*e.g.*, Smoluchowski & Torbett 1984; Morris & Muller 1986; Byl 1986). It turned out that this *normal component* to the Galactic disk was very efficient to change the perihelion distance of an Oort cloud comet in one orbital period.

Nowadays, the Galactic tides model commonly used is the one defined by Heisler & Tremaine (1986) where the tides are supposed to be axisymmetric and the Sun moving on a circular orbit around the Galactic centre in the Galactic plane.

The particularity of the dynamics generated by the Galactic tides is that it is completely integrable when two approximations are made: *(i)* neglecting the *radial component*, *i.e.* the component of the tides which lies in the Galactic disk, with respect to the normal one (the common used values shows that there is almost one order of magnitude between the two components, see Levison *et al.* 2001); and, *(ii)* averaging the equations of motion with respect to the mean anomaly of the comet (Heisler & Tremaine 1986; Matese & Whitman 1989, 1992; Breiter *et al.* 1996; Breiter & Ratajczak 2005). In this case, the equations of motion write:

$$\left\langle \frac{dG}{dt} \right\rangle = -\mathcal{G}_3 \frac{5L^2}{4\mu^2} \left(L^2 - G^2\right) \left(1 - \frac{H^2}{G^2}\right) \sin 2g, \qquad (3.2)$$

$$\left\langle \frac{dg}{dt} \right\rangle = \mathcal{G}_3 \frac{L^2 G}{2\mu^2} \left[1 - 5\sin^2 g \left(1 - \frac{L^2 H^2}{G^4}\right)\right], \qquad (3.3)$$

where $L = \sqrt{\mu a}$, $G = \sqrt{\mu a(1 - e^2)}$, $H = \sqrt{\mu a(1 - e^2)} \cos i$, e is the comet eccentricity, i and g are the comet inclination and argument of perihelion with respect to the galactic plane respectively, and $\mathcal{G}_3 = 4\pi\mu\rho_0$, where $\rho_0 = 0.1\, M_\odot \mathrm{pc}^{-3}$ is the density of the Galactic disk in the solar neighbourhood (Levison *et al.* 2001).

Figure 3 shows the dynamics in the $(g - G/L)$ space generated by Eqs. 3.2 and 3.3, with $|H|/L = 0.585$.

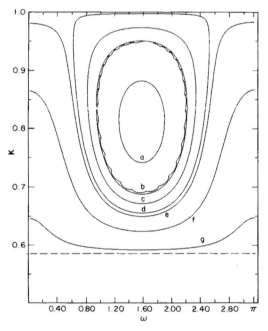

Figure 3. A family of trajectories in the $(g - K)$ space with $K = G/L$ for $|H|/L = 0.585$. From Heisler & Tremaine (1986).

The main characteristics of this model are: *(i)* the semi-major axis of the comet and the third component of the angular momentum are constant, *(ii)* the motion of the perihelion is strictly periodic with a period minored by $(4\sqrt{5\mu a^3 \rho_0})^{-1}$; *(iii)* from Eq. 3.2 the efficiency of the Galactic tides to reduce the angular momentum over one orbital period is at its maximum for $g = \pi/4 \bmod(\pi)$ and increases as $a^{7/2}$.

Some studies were also devoted to the quasi-integrable system where either the radial component is included or the averaging of the equations of motion is not performed (Brasser 2001; Breiter *et al.* 2008). It was also pointed out that the radial component of the tide should be included precisely because it breaks the integrability of the system (Matese & Whitmire 1996; Fouchard 2004).

It turned out that the tide is twice as efficient as the passing stars in injecting comets into the loss-cone (Torbett 1986; Heisler & Tremaine 1986; Bailey 1986). In addition, Duncan *et al.* (1987) have shown that the characteristic timescale for changing the perihelion distance, whatever is the semi major axis, is shorter for the Galactic tides than for the stellar perturbations.

Ultimately, an observational confirmation of the action of the vertical Galactic tides was pointed out by Delsemme (1987), who studied the distribution of the galactic latitudes of perihelia of 152 known original orbits of comets and found that these new Oort Cloud comets present a double-peaked distribution that is a characteristic of the disk tide (see also Wiegert & Tremaine 1999 for a more recent discussion one the distributions of orbital parameters of long period comets). Consequently from that time, stellar perturbations have been neglected when cometary injection is concerned.

4. The synergy

It appears that a synergy is at work when both the Galactic tides and the passing stars are at work as shown in Rickman *et al.* (2008). Figure 4 (from Rickman *et al.* 2008) shows a histogram plot of the number of comets injected into the observable region (defined by heliocentric distance smaller than 5 AU) as a function of time from the beginning till the end of the simulation. Three histograms are shown together: the one in black corresponds to a model with only Galactic tides, and the grey one to a model including only stellar perturbations. Finally, the top, white histogram is for the combined model that includes both tides and stars.

At first glance, it is evident that, at least after the first Gyr, the flux induced by the combined model is more than the some of the flux obtained with only the Galactic tides and the flux obtained with only the stellar perturbations. At the very beginning an anti-synergy is observed: the sum of the separate fluxes is larger than the combined flux. This phenomenon was found by Matese & Lissauer (2002), whose calculations were limited to only 5 Myr, and as they explained, it is typical of a situation where both tides and stars individually are able to fill the loss cone to a high degree.

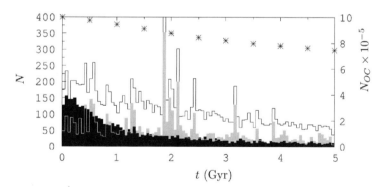

Figure 4. The upper diagram shows the number of comets entering the observable zone per 50 Myr versus time. The white histogram corresponds to the combined model, the black histogram to the Galactic tide alone, and the grey histogram to the passing stars alone. The asterisks indicate the number of comets remaining in our simulation for the combined model at every 500 Myr with scale bars to the right. From Rickman *et al.* (2008)

In order to quantify the synergy, Rickman *et al.* (2008) computed the filling factor f_{lc} of the loss cone for two different quasi quiescent periods, *i.e.* with no strong comets showers: one at the beginning and one at the end of the simulations. The filling factor

$\Delta(1/a)$	f_{lc} (Beginning)			f_{lc} (End)		
$(10^{-5}\ \mathrm{AU}^{-1})$	Tidal	Stellar	Combined	Tidal	Stellar	Combined
$(2-3)$	36%	10%	62%	0.6%	10%	52%
$(3-4)$	6.5%	1.1%	18%	–	2.3%	13%
$(4-5)$	0.09%	0.3%	2.2%	–	0.5%	2.0%
$(5-10)$	–	0.1%	0.15%	–	0.06%	0.11%
>10	–	0.0006%	0.0009%	–	0.0008%	0.0016%

Table 1. Filling factors for the observable part of the loss cone, computed for different ranges of semi-major axis and separately for the three dynamical models (tides-only, stars-only, and combined). From Rickman *et al.* (2008)

was computed for different ranges of orbital energy. Tab. 1 reproduces a selection of the results of Rickman *et al.* (2008).

The values of f_{lc} in the combined model are much larger than the sum of the two other entries, especially when $a > 25\,000$ AU even if a synergy is also at work in the inner Oort cloud. The loss cone was completely filled for smaller values of semi-major axis in Heilser (1990). This is likely due to lower value for the Galactic mid-plane density and somewhat higher stellar velocities used in Rickman *et al.* (2008).

The most important synergy mechanism of the Galactic tide and stellar perturbations is that the latter are able to repopulate the critical phase space trajectories that in the quasi-regular dynamics imposed by the tide lead into the loss cone (Dybczyński 2002; Fernández 2005). But note in Fig. 4 that the initial flux of the model with tides only is not matched by the white areas in the later part of the simulation. Thus, even though there is an ongoing replenishment of the tidal infeed trajectories due to the randomising effect of stellar encounters, this replenishment is not complete. Further experiments, not published yet, show that the massive stars have a key role in this replenishment.

The extension of the synergy to energy range where the tides are not able to inject comets into the observable region is rather explained by a 'constructive interference' mechanism for which the stellar perturbations is added to the tidal one.

Considering the distributions of orbital energy and direction of perihelions of the injected comets during the last quiescent period for the three models (see Fig. 5) Rickman *et al.* (2008) showed that: *(i)* the tide at the end of the integration is able to inject only few comets from the outermost part of the Oort cloud due to the non-integrable part of the tide, and the distribution of perihelion is typical of the tides imprint; *(ii)* the passing stars alone are poorly efficient to inject comets but the range of orbital energy extends to smaller semi-major axis, and the distribution of perihelion is nearly isotropic, *(iii)* the combined model yields a distribution of orbital energy as wide as for the stellar perturbations alone, with an increase on the number of comets per energy range consistent with the values of f_{lc} given in Tab. 1, and a perihelion direction distributions which carries, to some extend, the imprint of the tides.

Rickman *et al.* (2008) showed that this imprint might be even observed during a moderate comets shower. Consequently, one cannot rely on the perihelion distribution of the observable comets to determine whether or not we are in a comets shower.

5. Conclusion

The Galactical environment of the Sun affects the injection flux of comets from the Oort cloud under the influence of three main perturbers: *(i)* the passing stars, *(ii)* the giant molecular clouds, and *(iii)* the Galactic tides. The perturbations due to passing stars and tide are now well understood with a strong synergy between the two effects.

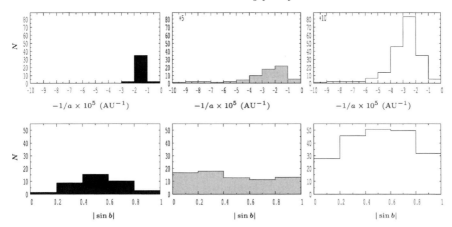

Figure 5. Distributions of $-1/a$, where a is the semi-major axis (top panels) and $|\sin b|$, where b is the Galactic latitude of perihelion (bottom panels), for the comets entering the observable region during 170 Myr near the end of the simulation. When present, numbers in the top-left corners of $-1/a$ distribution panels correspond to comets with $-1/a < -1 \times 10^{-4}$ AU^{-1}. The left column corresponds to the model with Galactic tide alone, the middle column to passing stars alone, and the right column to the model with both effects. From Rickman *et al.* (2008)

The modelling of the giant molecular clouds is still a gap in the long term study of Oort cloud comets dynamics, most of all because their effects are supposed to be destructive.

As regards the stellar and tidal effects, new improvements will be made after spatial mission as GAIA, which will give us a much better picture of the Galactical environment of the Solar System, and of the Solar motion within the Galaxy. Indeed, such motion should induce a variation of both the tidal strength and of the stellar population in the neighbourhood of the Sun.

Any realistic simulations of the cometary flux from the Oort cloud should include the planetary perturbations in a more realistic way than the loss cone technic. These perturbations were not considered in this review. The main difference will be that a dynamical path through the planetary region outside or near the Jupiter-Saturn barrier is allowed, and might even be quite common according to Kaib & Quinn (2009).

The problem of the cometary flux from the Oort cloud is related to the formation of the Solar System. This was out of the scope of the present review, however, it is still an open problem to built an Oort cloud consistent with both the actual flux of long period comets and the existence of objects as Sedna on one hand, and with the actual mass of the Kuiper belt (Charnoz & Morbidelli 2007). Many studies are devoted to this problem of the Oort cloud formation, the reader is referred to, *e.g.*, Brasser *et al.* (2006, 2007), Emel'yanenko *et al.* (2007), Kaib & Quinn (2008), Brasser (2008) among others, to have an idea on the state of art of this topic.

Acknowledgements

I thank the SF2A for financial support to participate at the Symposium. I am also indebted to Hans Rickman, Giovanni Valsecchi and Christiane Froeschlé for their constant scientific and friendly support throughout these years.

References

Bailey, M. E. 1986, *MNRAS*, 218, 1
Bailey, M. E. 1983, *MNRAS*, 204, 603

Biermann, L. 1978, in: A. Reiz & T. Andersen (eds.), *Astronomical Papers Dedicated to Bengt Stromgren*, p. 327

Brasser, R. 2001, *MNRAS*, 324, 1109

Brasser, R. 2008, *A&A*, 492, 251

Brasser, R., Duncan, M. J., & Levison, H. F. 2006, *Icarus*, 184, 59

Brasser, R., Duncan, M. J., & Levison, H. F. 2007, *Icarus*, 191, 413

Breiter, S., Dybczyński, P., & Elipe, A. 1996, *A&A*, 315, 618

Breiter, S. & Ratajczak, R. 2005, *MNRAS*, 364, 1222

Breiter, S., Fouchard, M., & Ratajczak, R. 2008, *MNRAS*, 383, 200

Byl, J. 1983, *Earth Moon and Planets*, 29, 121

Byl, J. 1986, *Earth Moon and Planets*, 36, 263

Chebotarev, G. A. 1964, *Soviet Astronomy - AJ*, 10,341

Clube, S. V. M. & Napier, W. M. 1982, *QJRAS*, 23, 45

Delsemme, A. H. 1987, *A&A*, 187, 913

Dones, L., Weissman, P. R., Levison, H. F., & Duncan, M. J. 2004, *Comets II*, p. 153

Duncan, M., Quinn, T., & Tremaine, S. 1987, *AJ*, 94, 1330

Dybczyński, P. A. 2002, *A&A*, 396, 283

Emel'yanenko, V. V., Asher, D. J., & Bailey, M. E. 2007, *MNRAS*, 381, 779

Fernández, J. A. 1980, *Icarus*, 42, 406

Fernández, J. A. 1981, *A&A*, 96, 26

Ferández, J. A. 2005, in: Fernández, J. A. (ed.), *Comets - Nature, Dynamics, Origin and their Cosmological Relevance*, volume 328 of *Astrophysics and Space Science Library*

Fouchard, M. 2004, *MNRAS*, 349, 347

Heisler, J. 1990, *Icarus*, 88, 104

Heisler, J. & Tremaine, S. 1986, *Icarus*, 65, 13

Heisler, J., Tremaine, S., & Alcock, C. 1987, *Icarus*, 70, 269

Hills, J. G. 1981, *AJ*, 86, 1730

Hut, P. & Tremaine, S. 1985, *AJ*, 90, 1548

Jakubík, M. & Neslušan, L. 2008, *Contrib. Astron. Obs. Skalnaté Pleso*, 38, 33

Kaib, N. A. & Quinn, T. 2008, *Icarus*, 197, 221

Kaib, N. A. & Quinn, T. 2009, *Science*, to be published

Królikowska, M. 2006, *AcA*, 56, 385

Levison, H. F., Dones, L., & Duncan, M. J. 2001, *AJ*, 121, 2253

Marsden, B. G. & Sekanina, Z. 1973, *AJ*, 78, 1118

Matese, J. J. & Whitman, P. G. 1992, *Celestial Mechanics and Dynamical Astronomy*, 54, 13

Matese, J. J. & Whitman, P. G. 1989, *Icarus*, 82, 389

Matese, J. J. & Whitmire, D. 1996, *ApJ Letters*, 472, L41

Matese, J. J. & Lissauer, J. J. 2002, *Icarus*, 157, 228

Morris, D. E. & Muller, R. A. 1986, *Icarus*, 65, 1

Napier, W. M. & Staniucha, M. 1982, *MNRAS*, 198, 723

Oort, J. H. 1950, *Bull. Astron. Inst. Neth.*, 11, 91

Öpik, E. 1932, *Proceedings of the American Academy of Arts and Science*

Remy, F. & Mignard, F. 1985 *Icarus*, 63, 1

Rickman, H. 1976, *Bulletin of the Astronomical Institutes of Czechoslovakia*, 27, 92

Rickman, H., Fouchard, M., Froeschlé, Ch., & Valsecchi, G. B. 2008, *Celestial Mechanics and Dynamical Astronomy*, 102, 111

Smoluchowski, R. & Torbett, M. 1984 *Nature*, 311, 38

Torbett, M. 1986, *MNRAS*, 223, 885

Weissman, P. R. 1979, in: R. L. Duncombe (ed.), *Dynamics of the Solar System*, Proc. IAU Symposium No. 81, p. 277

Weissman, P. R. 1980, *Nature*, 288, 242

Weissman, P. R. 1996, in: T. Rettig & J. M. Hahn (eds.), *Completing the Inventory of the Solar System*, Astronomical Society of the Pacific Conference Series, Vol. 107, p. 265

Wiegert, P. & Tremaine, S. 1999, *Icarus*, 137, 84

Yabushita, S., Hasegawa, I., & Kobayashi, K. 1982, *MNRAS*, 200, 661

Icy Bodies of the Solar System
Proceedings IAU Symposium No. 263, 2009
J.A. Fernández, D. Lazzaro, D. Prialnik & R. Schulz, eds.
© International Astronomical Union 2010
doi:10.1017/S1743921310001511

Sedna, 2004 VN112 and 2000 CR105: the tip of an iceberg

Rodney S. Gomes and Jean S. Soares

Observatório Nacional
Rua General José Cristino 77, CEP 20921-400, Rio de Janeiro, RJ, Brazil
email: `rodney@on.br`

Abstract. We review two main scenarios that may have implanted Sedna, 2004 VN112 and 2000 CR105 on their current peculiar orbits. These scenarios are based on perihelion lifting mechanisms that acted upon primordial scattered icy bodies. Supposing that the Sun was formed in a dense star cluster and that the gas giants were also forming while the cluster was still dense, an inner Oort cloud that includes Sedna at its inner edge could have been formed by the circularization of icy leftovers orbits scattered by the gas giants. A putative planetary mass solar companion can also produce a similar population of icy bodies through a perihelion lifting mechanism induced by secular resonances from the companion. A third scenario also dependent on a primordial dense cluster may contribute to adding a significant number of extrasolar icy bodies to the main solar component of the population created by the cluster model. These extrasolar objects are transferred to Sun orbits from the scattered disk of passing stars that were numerous in the dense primordial environment. We compare the scenarios as to the orbital distribution of the induced populations as well as their total mass. We conclude that both the cluster model and the solar companion model can produce icy body populations consistent with Sedna's orbit. It is also quite possible that this inner Oort cloud may be composed of roughly one tenth of extrasolar objects.

Keywords. Sedna, Oort cloud, star cluster, solar companion

1. Introduction

Gladman *et al.* (2008) presents a comprehensive nomenclature for trans-Neptunian orbits by which Sedna, 2004 VN112 and 2000 CR105 are classified as detached objects. These icy bodies are those nonscattering transneptunian objects with large eccentricities ($e > 0.24$) and not so far away that external influences are important to their current dynamics ($a < 2000\,AU$). On the same nomenclature a scattering object is that which is currently scattering actively off Neptune. Although it is usually accepted that a detached object was once a scattering object that for some dynamical process had its perihelion lifted (see Gomes *et al.* 2005a, Gladman *et al.* 2002), it is also possible that some detached objects may have another origin. Nevertheless Gomes *et al.* (2005a) show that resonant perturbations from the outermost planets are very effective in lifting scattered objects perihelia, for those with semimajor axes not larger than roughly $200AU$. With this in mind we might conjecture with some confidence that detached objects are formerly scattered objects whose perihelia were lifted by the coupling of mean motion and Kozai resonances with an outer planet, usually Neptune. If we thus define a detached object, we are forced to leave undefined three trans-Neptunian objects that cannot (or can hardly, in the case of 2000 CR105) experience an increase of their perihelion by the sole perturbations of the outer planets. A scenario for their origin must therefore be sought elsewhere beyond the perturbations of the known Solar System. The main scenarios invoke either conditions that prevailed in a primordial Solar System or conditions that may still be present in the Solar System. The latter case stands for the solar companion scenario (Gomes et al. 2006) and the former case refers to the scenario of a primordial dense star cluster in which the Sun would be embedded. In this case torques from this dense environment would increase the perihelia of icy bodies being scattered outwards by the giant planets (Brasser et al. 2006, 2007, 2008) . This latter scenario is also consistent with the transfer of several extrasolar icy bodies from other stars to circumsolar orbits. For this we just have to assume that the other stars were experiencing a similar process of planetesimals scattering by their icy giants, which is a quite reasonable hypothesis. In this sense, this capture scenario will be considered as a third one and described in Section 2, in which we also review the first two scenarios. With respect to the cluster model, we have produced our own data following Brasser et al. (2006). In Section 3, we compare scenarios so as to assess their relative plausibilities. We estimate the mass of the populations produced by each scenario in Seccion 4 and conclusions are drawn in Section 5.

2. Scenarios for the formation of Sedna population

Galactic tides induce strong long period variations to the eccentricities of objects in distant orbits around the Sun. This process can thus create a cloud of comets in the Oort cloud as well as send an icy body back to the inner Solar System as a long period comet. The closest distance where galactic tide can be effective in increasing an icy body perihelion to that of Sedna is roughly 4000 AU. A much stronger tide would be necessary to raise Sedna's perihelion, with a semimajor axis at 510 AU. This condition could be satisfied by the torque from close passing stars and the gas in a primordial star cluster dense environment where the Sun would be embedded (Brasser et al. 2006). The inner edge of the thus formed 'inner Oort cloud' will be as closer to the Sun as larger is the density of the cluster. Figure 1 shows the distribution of semimajor axes and perihelion distances of planetesimals scattered by Jupiter and Saturn with raised perihelia by the tidal effect of a star cluster where the Solar System would be embedded. In a primordial star cluster, experiencing gravitational effects from passing stars and the gas, many planetesimals will experience perihelion increases. To build this figure, we followed the procedures described in Brasser et al. 2006, which we recommend for details on the method. In principle, from an observational basis, one would expect that the right population created by the primordial cluster should have Sedna and 2004 VN112 (eventually 2000 CR105) at the population inner edge. From Fig. 1, we estimate that the best central density that could create such a population would be between 10^4 and 10^5 M_{Sun}/pc^3. One should note however that those $a \times q$ distributions also depend on the orbit of the Sun inside the cluster. A Sun that inhabited the densest parts of the cluster would produce in the end a Sedna population closer to the Sun than a Sun with an orbit in a less dense part of the cluster. The cluster model yields a fairly likely scenario since it is probable that the Sun like most stars also formed in such a dense environment. Another reasonable supposition implied by this scenario is that at this primordial time the Solar System formed its gas giants and planetesimals leftovers would experience deep encounters with the planets, thus increasing their semimajor axes and placing them at the right distance to have their perihelia increased by the tides from the cluster.

Another reasonable conjecture related to the scenario of the Sun in a primordial dense star cluster is that, like the Sun, the other cluster stars must experience a similar planetary formation and scattering of icy leftovers. A cloud of scattered (lifted perihelia)

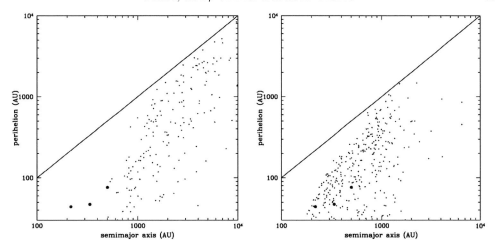

Figure 1. Distribution of semimajor axes and perihelion distances for a population of icy bodies created by the cluster model. Left panel: cluster central density 10^4 M_{Sun}/pc^3, right panel: cluster density 10^5 M_{Sun}/pc^3. Sedna, 2004 VN112 and 2000 CR105 are represented by large circles.

planetesimals around a close passing star must transfer some of its objects to a Sun orbit. Fig. 2 shows the distribution of semimajor axes and perihelia of planetesimals (gray circles) transferred from a passing star orbit to a Sun orbit. The star has one solar mass and comes as close as 1000 AU to the Sun with $v_\infty = 0.6\,km\,s^{-1}$. The dots in this figure comes from the same simulation as those in Fig. 1 (less dense case). We have however taken particles from more than one time in the original simulation to increase the number of transferred objects and get a nicer statistics for the $a \times q$ distribution for the extrasolar bodies. So Fig. 2 allows for the comparison of the relative amount of planetesimals mass transferred to the Oort cloud from the inner Solar System with the mass from an extrasolar origin. We usually get roughly 10% of extrasolar objects with respect to solar planetesimals, for a star like the one considered in this example. On the other hand, the extrasolar icy bodies population usually have an inner edge closer to the Sun than that from solar origin. It must be noted that the star considered here is consistent with a cluster star orbit that induced the distribution of icy bodies orbits from the inner Solar System, so that the dots and gray circles in Fig. 2 are comparable. This example is mostly a proof of concept since we must do a more complete simulation with a series of close passages of stars with their scattered disks. Each star may leave some extra material around the Sun but also eject some of the already captured planetesimals from other stars.

A third scenario is based on the perturbation of a putative planetary mass solar companion (Gomes *et al.* 2006). At remote distances from the Sun, the angular displacement of objects around the Sun is very slow. This induces secular resonances on other distant objects. Figure 3 shows the distribution of semimajor axes and perihelion distances of planetesimals scattered by the major planets and gravitationally perturbed by a solar companion with 10^{-4} Earth mass, and semimajor axis, eccentricity and inclination respectively $1500\,AU$, 0.4 and 40°. This example is taken from a simulation where particles have mass and induce a migration on the giant planets. Different solar companions yield similar effects provided $\rho = M/b^3$ is about the same, where b is the companion semiminor axis and M its mass. In other words, for a distant massive solar companion, there is a not so distant and less massive companion that yield similar induced $a \times q$ distributions. The

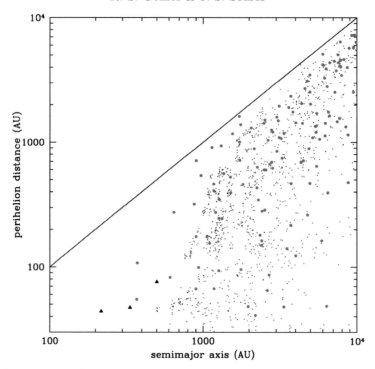

Figure 2. Distribution of semimajor axes and perihelion distances for a population of icy bodies created by the cluster model (small dots) added with a population of extrasolar objects transferred to a Sun orbit from the scattered disk of a passing star (gray circles). Sedna, 2004 VN112 and 2000 CR105 are represented by triangles.

solar companion model, differently from the cluster model, requires that the companion perturbations be effective for a much longer time than a star cluster's lifetime. This is not a problem, since it is supposed that the solar companion is there for the solar system age. On the other hand, existing for the solar system age is not necessarily required for the solar companion. Some hundreds of million years are sufficient to increase the perihelia of scattered objects. Thus a putative scattered planet (of roughly an Earth mass) could increase scattered objects perihelia to produce a Sedna population. This scattered planet could be eventually ejected from the Solar System. This hypothesis was also invoked by Lykawka and Mukai (2008) to explain other features of transneptunian objects orbits. However the probability of a specific scattered object staying for some hundred million years as a scattered/detached object is quite low (roughly $< 1\%$). Another possible origin of a solar companion comes from the star cluster model. In fact one of the objects scattered by the gas giants in the primordial cluster embedded solar system could be a planet. If there was originally one such planet the probability that it is still orbiting the Sun is given by the relative amount of perihelion increased bodies, which is near 10%. Also the extrasolar objects eventually captured by the Sun as described by scenario 2 above might also include an extrasolar planetary size body, with the same probability as the total amount of extrasolar bodies transferred to the Sun from a star to the amount of scattered objects by the same star. And finally, a floating planet could be captured by the Sun in the dense cluster environment in a three-body encounter (Sun + star + planet). This floating planet might have just been ejected by a star system, what is a likely event in a primordial planetary system.

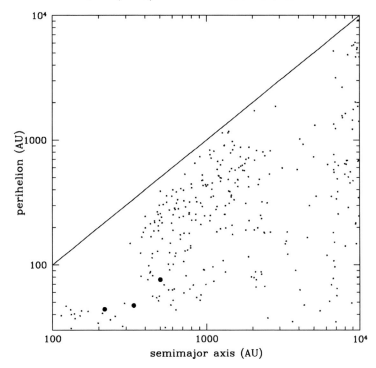

Figure 3. Distribution of semimajor axes and perihelion distances for a population of icy bodies created by the solar companion model. The companion mass is 10^{-4} solar mass and its semimajor axis, eccentricity and inclination are $1500\,AU$, 0.4 and $40°$. Sedna, 2004 VN112 and 2000 CR105 are represented by large circles.

It must be also noted that differently from the cluster scenario, the objects scattered by the solar system planets and deposited in the inner Oort cloud belong to a not so primordial time. These icy bodies would have been preferentially scattered by the icy giants during planetary migration in a planetesimal disk, possibly after the Late Heavy Bombardment (Gomes *et al.* 2005b). This is an important difference since in this case, there not being a gas disk around the Sun, small (visible comet size) bodies could also fill the inner Oort cloud, what would be prevented in a more primordial scenario where gas drag would not allow smaller icy bodies ($< 2km$ radius) to reach distances as far as Sedna's distance from the Sun (Brasser *et al.* 2007). This is an interesting feature of the companion model, since Kaib & Quinn (2009) claims that the inner Oort cloud could be an important reservoir of LPC's. The mechanism by which these comets follow a path that will make them eventually visible could in principle also be followed by an object from the much inner Oort cloud represented by the putative Sedna population. In the solar companion model Sedna's population is not fossilized but their objects are still moving in perihelion, what turns them eventually Neptune crossers again.

3. Comparison between scenarios

Comparing Figs. 1 and 3, it is noticeable that both distributions are similar. Although the inner edges of the populations in Fig. 1 and Fig. 3 are different, a suitable choice of cluster densities can place the inner edge from the cluster model at a more consistent position with the observation of Sedna. So in principle both models are undifferentiated if

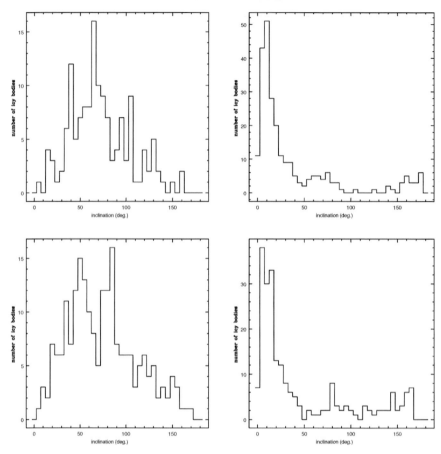

Figure 4. Histograms showing the distribution of inclinations of the icy bodies: upper left: from the cluster model with central density 10^4 M_{Sun}/pc^3, upper right: companion model with companion's inclination $i_c = 0°$, lower left: companion model for $i_c = 45°$, lower right: companion model for $i_c = 90°$.

we just compare their distributions in semimajor axis and perihelion distances. Another important orbital parameter for the icy bodies orbital distribution is the inclination. Fig. 4 shows a histogram of inclinations for the cluster model and for the companion model, in which case we consider three different inclinations for the companion, 0°, 45° and 90°. We see that both for the cluster model and the companion model with the companion inclination at 45° the distributions of inclinations do not essentially differ peaking for inclinations just below 90°. However if the companion has an orbital inclination near 0° or 90° the distribution of inclination for Sedna's population peaks for $i < 20°$. This is an interesting feature to be taken into account when new Sedna-like objects are discovered. It is noteworthy that Sedna's inclination is 12° and was discovered in a all-sky survey (Brown *et al.* 2004), what would suggest a low inclination Sedna population. 2004 VN112 and 2000 CR105 have inclinations respectively at 25° and 23°, which are also fairly low. We must however await for new discoveries for a better judgment of this point.

We have developed a basic observational simulator in order to search for the most consistent model and the best parameters for that model. The simulator works as follows:

   - We choose a population created by one of the above scenarios.
   - We randomly choose an orbit within the population.

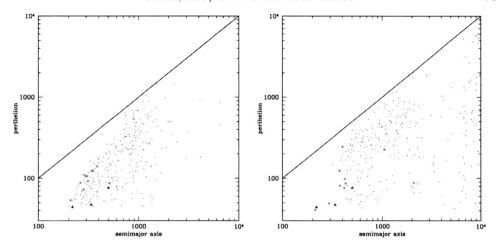

Figure 5. First ten 'observed objects' by the application of the observational simulator (gray circles). Left: for the cluster model with central density 10^5 M_{Sun}/pc^3, right: for the companion model with parameters as in Fig. 3. Sedna, 2004 VN112 and 2000 CR105 are represented by large triangles.

- We randomly choose a mean longitude for that orbit.
- We randomly choose a size for the body from a size distribution.
- Given an albedo, and the already computed size and distances of the body from the Sun and the Earth, we have its visual magnitude.
- If this magnitude is smaller that a given upper limit, the object is considered as observed.
- We can also limit the range of latitudes from which the observation was made.

Fig. 5 shows the first 10 observed objects from the cluster and companion models. For compatibility, we considered all objects with $a > 200\,AU$ and $q > 45\,AU$. We note that the simulator observations better accounts for the companion model. This should not be taken too rigorously since for the moment we have more examples with the companion model than for the cluster model, and a more fine-tuned cluster density may yield a better $a \times q$ distribution for the cluster model.

Another interesting application of the simulator can be made for the extrasolar objects model. In this case we put together the population created by the cluster model with an extrasolar population transferred from a star that passed near the Sun with orbital parameters compatible with the cluster parameters that generated the solar objects, so that both populations can be consistently placed together for an observational simulation. Applying the simulator several times we notice that although the extrasolar component of the population has several members closer to the Sun than the closest member from the solar component, an extrasolar object is first observed in about one time out of ten, showing that Sedna has most probably a solar origin.

4. The size of Sedna population

When Sedna was discovered (Brown *et al.* 2004) a mass of about 5 Earth masses was estimated for the population to which Sedna belongs. A more recent estimate predicts a total mass of about ten times the Kuiper belt mass (Schwamb *et al.* 2009). This yields roughly 0.1 to 1 Earth mass for Sedna population but possibly closer to the lower limit. Since these estimates are deduced from observational arguments, this inferred mass refers in fact to objects in orbits similar to Sedna's (with respect to semimajor axis and perihelion distance), so that would only represent the inner portion of the population coming from one of the theoretical models above presented. On the other hand, those models produce about 1-2 Earth masses in Sedna's population. We can thus conclude that the most recent observational based estimate of Sedna's population is more consistent with the populations created by any of the available models. In fact if we consider just the objects with $400\,AU < a < 800\,AU$ and $50\,AU < q < 200\,AU$, in the population shown in Fig. 3, that will stand for 3% of the total population, thus about 0.03 Earth masses, roughly one order of magnitude below the observational mass estimate. In any case, observational considerations associated with formation models suggest that with respect to Sedna, 2004 VN112 and 2000 CR105, we presently see just the tip of an iceberg. Since models consistently create a population distributed in a wide range of semimajor axes and perihelion distances and we now observe just three objects presumably at the inner edge of the population, it is thus likely that we are now just seeing a very small portion of the total population.

5. Conclusions

The orbits of Sedna, 2004 VN112 and to a lesser degree 2000 CR105 demand an explanation that goes beyond the perturbations from the known Solar System. These orbits might be considered close to the inner edge of an inward extended Oort cloud. Differently from the classical Oort cloud whose inner edge is formed at about 4000 AU due to galactic tides, Sedna's population would start at roughly 300 - 500 AU. Galactic tides cannot account for such a close perihelion lifting, but an early dense environment where the Sun was supposedly formed can produce a tidal effect of much larger magnitude that could explain Sedna's orbit. The high plausibility of the Sun forming in such a high density environment associated with an induced orbital distribution of perihelion-lifted icy bodies consistent with Sedna's orbit gives the dense star cluster model a high degree of plausibility. A competing scenario requires that a planetary-mass solar companion would orbit the Sun in a distant orbit. Although also yielding a consistent orbital distribution for the icy bodies, it is not a strong supposition that a companion should exist or have existed orbiting the Sun. It must be noted on the other hand, that present observational methods or other indirect detection methods cannot presently discard a solar companion with the mass/distance compatible with the production of Sedna's population. Also the dense primordial star cluster scenario could be responsible for the placement of a planet at a distant orbit around the Sun, both by a solar origin or through an extrasolar origin. Although we do not need to invoke a solar companion to explain Sedna's orbit, its putative existence, although not very likely, is however far from negligibly probable, say some 10% chance of existing just for the known processes that can place a companion in a Sun orbit. So the companion scenario must be considered seriously as new Sedna-like bodies are discovered. The dense cluster scenario can also produce a subpopulation of extrasolar icy bodies. We just have to suppose that the passing stars, like the Sun, also carried a scattered/detached population of icy bodies. A simulation of a Sun mass star coming as close as 1000 AU to the Sun and with $v_\infty = 0.6\,km\,s^{-1}$ showed that roughly 10% of its scattering/detached population (simulated like the Sun scattered population at that time) is transferred into orbits around the Sun. This subpopulation has its closest members to the Sun consistently inside the inner edge of the solar population. More complete simulations are needed to better determine the real fraction of extrasolar bodies with respect to solar ones and also their relative orbital distributions. Although this extrasolar component is not particularly large, it is however far from negligible and has a likeliness comparable to that of the solar component for the same scenario, so that one would expect to find one extrasolar icy body in Sedna's putative population after 10 solar icy bodies discoveries. Undoubtedly, new discoveries of transneptunian objects with large semimajor axes and perihelion distances will be extremely helpful is disclosing the invisible part of the 'iceberg' so that we can better understand the origin of Sedna and its correlates.

References

Brasser, R., Duncan, M. J., & Levison, H. F. 2006, *Icarus*, 184, 59

Brasser, R., Duncan, M. J., & Levison, H. F. 2007, *Icarus*, 191, 413

Brasser, R., Duncan, M. J., & Levison, H. F. 2008, *Icarus*, 196, 274

Brown, M. E., Trujillo, C., & Rabinowitz, D. 2004, *ApJ*, 617, 645

Gladman, B., Holman, M., Grav, T., Kavelaars, J., Nicholson, P., Aksnes, K., & Petit, J-M. 2002, *Icarus*, 157, 269

Gladman, B., Marsden, B. G., & VanLaerhoven, C. 2008, in: M. A. Barucci, H. Boehnhardt, D. P. Cruikshank & A. Morbidelli (eds.), *The Solar System Beyond Neptune* (Tucson: The University of Arizona Press), p. 43

Gomes, R. S., Gallardo, T., Fernàndez, J. A., & Brunini, A. 2005, *CeMDA*, 91, 109

Gomes, R., Levison, H. F., Tsiganis, K., & Morbidelli, A. 2005, *Nature*, 7041, 466

Gomes, R. S., Matese, J., & Lissauer, J. 2006, *Icarus*, 184, 589

Kaib, A. K. & Quinn, T. 2009, *Science*, 325, 1234

Likawka, P. K. & Mukai, T. 2008, *AJ*, 135, 1161

Schwamb, M. E., Brown, M. E., & Rabinowitz, D. 2009, *ApJ*, 694, L45

Icy Bodies of the Solar System
Proceedings IAU Symposium No. 263, 2009
J.A. Fernández, D. Lazzaro, D. Prialnik & R. Schulz, eds.
© International Astronomical Union 2010
doi:10.1017/S1743921310001523

The discovery rate of new comets in the age of large surveys. Trends, statistics, and an updated evaluation of the comet flux

Julio A. Fernández

Departamento de Astronomía, Facultad de Ciencias,
Iguá 4225, 11400 Montevideo, Uruguay
email: julio@fisica.edu.uy

Abstract. We analyze a sample of 58 Oort cloud comets (OCCs) (original orbital energies x in the range $0 < x < 100$, in units of 10^{-6} AU^{-1}), plus 45 long-period comets with negative orbital energies or poorly determined or undetermined x, discovered during the period 1999-2007. To analyze the degree of completeness of the sample, we use Everhart's (1967 Astr. J 72, 716) concept of "excess magnitude" (in magnitudes × days), defined as the integrated magnitude excess that a given comet presents over the time above a threshold magnitude for detection. This quantity is a measure of the likelihood that the comet will be finally detected. We define two sub-samples of OCCs : 1) *new comets* (orbital energies $0 < x < 30$) as those whose perihelia can shift from outside to the inner planetary region in a single revolution; and 2) *inner cloud comets* (orbital energies $30 \leqslant x < 100$), that come from the inner region of the Oort cloud, and for which external perturbers (essentially galactic tidal forces and passing stars) are not strong enough to allow them to overshoot the Jupiter-Saturn barrier. From the observed comet flux and making allowance for missed discoveries, we find a flux of OCCs brighter than absolute total magnitude 9 of $\simeq 0.65 \pm 0.18$ per year within Earth's orbit. From this flux, about two-thirds corresponds to new comets and the rest to inner cloud comets. We find striking differences in the q-distribution of these two samples: while new comets appear to follow an uniform q-distribution, inner cloud comets show an increase in the rate of perihelion passages with q.

Keywords. comets: general, Oort cloud, celestial mechanics

1. Introduction

Oort cloud comets (OCCs) are injected into the inner planetary region by the combined action of galactic tidal forces and passing stars (e.g. Rickman *et al.* 2008). External perturbers produce a smooth drift of the perihelia of near-parabolic comets in such a way that they can either be removed from or injected into the planetary region. In the latter case they can evolve dynamically over short time scales under the perturbing action of the Jovian planets until they are ejected to interstellar space. Comets with orbital periods $P > 200$ yr are generically referred to as long-period comets (LPCs).

The relative change in the comet's perihelion distance per orbital revolution as due to the tidal force of the galactic disk, $\Delta q/q$, goes as $a^{7/2}$ (e.g. Fernández 2005). For semimajor axes $a \sim 3.0 - 3.5 \times 10^4$ AU (orbital energy $x \sim 30$), the relative change in q of comets in very eccentric orbits ($q << a$) becomes $\Delta q/q \sim 1$, i.e. the comet can decrease its perihelion distance to near-zero values.

We can then distinguish two subclasses of OCCs: *new comets* with energies in the range $0 \lesssim x \lesssim 30$, which are those that can overshoot the Jupiter-Saturn barrier in a revolution, and the *inner cloud comets* with energies in the range $30 \lesssim x \lesssim 100$ which come from the inner Oort cloud, and that can reach the inner planetary region only after a smooth diffusion of their perihelion distances. Therefore, inner cloud comets are not new in a strict sense, since they probably visited the trans-jovian region before, though they may be new in the inner planetary region.

The estimation of the comet flux has been a very difficult task basically due to the incompleteness of comet discoveries, mainly for greater perihelion distances. Kresák and Pittich (1978) attempted to estimate the comet flux based on those comets that approached the Earth within 0.2 AU which they estimated to be a highly complete sample. They found a best fit for a comet flux $\nu(q)$ increasing with q, according to the law $\nu(q) \propto q^{1/2}$, at least valid within 4 AU of the Sun. Therefore, a very interesting point is to define how the rate of comet passages varies with q.

2. The discovery conditions

Our sample consists of 22 new comets and 36 inner Oort cloud comets discovered during 1999-2007, taken from Marsden and Williams's (2008) catalogue. For these comets we computed the difference ΔT between the discovery time and the time of perihelion passage, and estimated the absolute total magnitude H_o.

We can see in Fig. 1 the apparent total magnitudes and elongations of the Oort cloud comets discovered during the period 1999-2007. Most apparent total magnitudes at the time of discovery are in the range 17-20, whereas most elongations are greater than $90°$, in contrast with discoveries by amateurs where elongations are usually smaller than $90°$ and brighter (total magnitudes $\sim 8 - 12$).

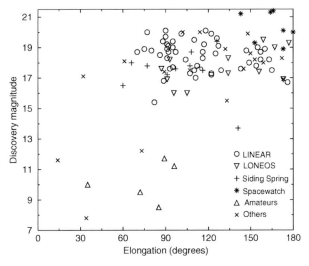

Figure 1. The discovery conditions of our sample of 103 comets. The absolute magnitudes were derived from the photometric data provided by the IAUCs and MPECs.

As seen in Fig. 2, there are many discoveries of faint comets near the Sun (total magnitudes $H_o \sim 10 - 19$), but only the brightest ones ($H_o \lesssim 10$) are discovered among the distant comets ($q \gtrsim 5$ AU). Even though the large surveys have led to the discovery of some faint comets ($H_o \gtrsim 10$), the number is nevertheless quite low suggesting that the population of faint, small LPCs is quite scarce, in agreement with previous results (Kresák and Pittich 1978, Sekanina and Yeomans 1984). We also plot those comets with undetermined or poorly known original orbital energies x, which mostly have low q-values, as the difficulty for determining accurate values of x arises from the strong perturbing action of nongravitational forces. We also find that the number of inner cloud comets increases within the range $3 \lesssim q \lesssim 7$ AU, whereas that for new comets keeps more or less constant within the considered q-range.

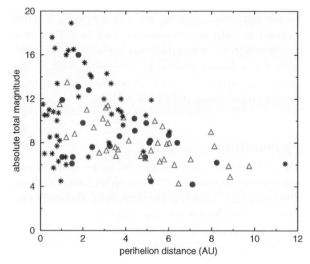

Figure 2. Absolute total magnitudes versus perihelion distances of: new comets (triangles), inner cloud comets (black dots), and comets with undetermined/innacurate original energies, though potential Oort cloud comets (stars).

3. The discovery probability

Any attempt to make a bias correction for completeness in the comet discovery record requires to know how likely is that a given comet be detected during its perihelion passage as a function of its absolute total magnitude, its perihelion distance, and the geometric configuration Sun-Earth-comet. One of the best attempts so far to compute the discovery probability has been performed by Everhart (1967). He defined the "excess magnitude" (S) as a quantity proportional to the discovery probability. The excess magnitude is expressed in units of magnitudes × days that the comet stays above a certain threshold magnitude H_{th}, and is expressed as

$$S = \int (H_{th} - H) dt \qquad (3.1)$$

where H is the apparent total magnitude defined as $H = H_o + 5 \log \Delta + 2.5n \log r$, where Δ is the geocentric distance, and r the heliocentric distance of the comet. We have adopted for the index $n = 3$ which is suitable for very active OCCs which usually brighten rapidly at great heliocentric distances followed by a much slower increase near perihelion (Whipple 1978). We adopted $H_{th} = 20$, which corresponds to the average bottom line of faintest comets that are discovered (see Fig. 1).

We find that the larger S, the earlier the comet is discovered on average (Fig. 3), which is an obvious consequence of its large brightness. We can say that for an excess $S \gtrsim 8000$ mag × days the great majority of comets are discovered before perihelion, which suggests that such comets will hardly pass undetected. Therefore, we can associate empirically excess magnitudes $S \gtrsim 8000$ mag × days with a discovery probablity $P \sim 1$. Furthermore, for $S \sim 2000 - 8000$ most comets are still discovered before perihelion which points to a high discovery probability P.

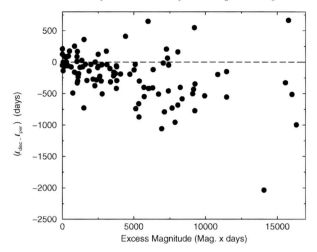

Figure 3. The discovery time (with respect to the time of perihelion passage) of the sample of 103 comets versus the excess magnitude.

4. The computed flux of Oort cloud comets. The distribution of perihelion distances

4.1. *New comets*

We have 11 comets with $q < 7$ AU and $H_o < 9$ discovered during 1999-2007. This sample shows excess magnitudes $\gtrsim 2000$ for most of the q-range, so we can expect for it a high degree of completeness. We may presume that some of the missing new comets (mainly in the range $0 < q < 1$ AU) may correspond to some of those whose original orbits could no be determined accurately due to the uncertain effect of nongravitational forces. This is the group that we have labelled as LPCs with unknown original x. Even considering missing detections and comets whose original x could not be determined, it does not seem to be more than a factor of two in the corrected population with respect to the observed one with computed original x. An educated guess would place the corrected population at about 25 ± 5 comets. The corrected distribution may still be compatible with an uniform one. Considering that 25 ± 5 new comets have passed perihelion in the range $0 < q < 7$ AU, distributed more or less uniformly in q, during a nine-year time span (1999-2007), the rate of passages of new comets can be estimated to be

$$\dot{n}_{new} = \frac{25 \pm 5}{9 \times 7} \simeq 0.4 \pm 0.08 \text{ yr}^{-1} \text{ AU}^{-1} \tag{4.1}$$

4.2. *Inner cloud comets*

The q-distribution shows in this case an increase with q, for the sample with $H_o < 9$, as well as for the most restricted sample of brighter comets with $H_o < 7$. The incoming flux in the innermost zone (say $q \lesssim 3$ AU) is well below that of new comets, say about half, with a large error bar given the smallness of the sample. We can then conclude that the ratio inner cloud/new comets should be about 0.5-0.7 within the zone close to the Sun ($q \lesssim 3$ AU), i.e.

$$\dot{n}_{inner} \simeq 0.25 \pm 0.10 \text{ yr}^{-1} \text{ AU}^{-1} \tag{4.2}$$

The discovery rate increases by a factor 3-4 in Jupiter's region.

The fact that the q-distribution of new comets is compatible with an uniform one is consistent with the injection of comets from a thermalized random population (the Oort cloud). On the other hand, the increase of inner cloud comets with q is a signature of the action of the Jupiter-Saturn barrier which hinders most comets coming from the inner Oort cloud to reach the Sun's vicinity. On the contrary, new comets coming from the outer Oort cloud, and thus subject to large changes in q by external perturbers, can easily overshoot the Jupiter-Saturn barrier thus explaining their rather uniform q-distribution.

References

Everhart, E. 1967, *Astron. J.*, 72, 716.

Fernández, J. A. 2005, *Comets. Nature, Dynamics, Origin, and their Cosmogonical Relevance*, (Dordrecht: Springer)

Kresák, L. & Pittich, E. M. 1978, *Bull. Astron. Inst. Czech.*, 29, 299.

Marsden, B. G. & Williams, G. W. 2008, *Catalogue of Cometary Orbits*, 17th Edition, IAU, MPC/CBAT, Cambridge

Rickman, H., Fouchard, M., Froeschlé, Ch., & Valsecchi, G. B. 2008, *Cel. Mech. Dynam. Astron.*, 102, 111.

Sekanina, Z. & Yeomans, D. 1984, *Astron. J.*, 89, 154.

Whipple, F. L. 1978, *Moon Planets* 18, 343.

Icy Bodies of the Solar System
Proceedings IAU Symposium No. 263, 2009
J.A. Fernández, D. Lazzaro, D. Prialnik & R. Schulz, eds.

© International Astronomical Union 2010
doi:10.1017/S1743921310001535

On hyperbolic comets

A. S. Guliyev and A. S. Dadashov

Nassriddin Tussi Astrophysical Observatory of National Academy of Sciences of Azerbaidjan
Republic, ShAO.
Mammadaliyev settlement, Pirgulu, Shamaky, AZ5613, Azerbaijan Republic,
email: ayyub54@yahoo.com

Abstract. Preliminary results of an investigation on the problem of hyperbolic comets are presented and discussed.

Keywords. comets: general, hyperbolic orbit, inter-orbital distance, perihelion concentration

1. Introduction

According to the existing practice, comets are considered to be hyperbolic ones if their "original" $1/a$ (hereinafter $1/a$) are negative. Number of such comets, according to the catalogue by Marsden and Williams (2008), is equal to 34. Negative $1/a$ for three other long-period comets were calculated later. So, the present work contains results of the analysis of 37 long-period comets.

As is known, there are various works on hyperbolic comets (hereinafter HCs) in the scientific literature. Existing widely accepted and less popular versions regarding to the origin of HCs could be classified as follows:

1. Hyperbolic excess of eccentricity is formed as a result of physical processes occurring on comet nucleus;

2. Hyperbolic excess of eccentricities could be results of errors in determination of orbits, or "original" ones;

3. HCs might have an interstellar origin;

4. Hyperbolic excesses of eccentricities are result of unaccounted factors.

Authors' methodological approach in study of HCs is specific. We consider HCs not only as set of individual long-period comets, but as a whole system. This system can have specific characteristics, which are not inherent for all cometary system.

During the preliminary analysis we have found following specific characteristics of HC:

1. Values of $1/a$ of HCs vary from -1 up to -772 (1 unit $= 10^{-6}$ a.u.$^{-1}$); they are distributed not chaotically and exponentially (fig. 1). Thus distributions of $1/a$ of HC for class 1 and 2 (Marsden and Williams, 2008) are slightly different.

2. The ratio of direct and retrograde orbits is $16 : 21; 18$ from $37HC$'s perihelion are concentrated in two narrow and opposite intervals of longitudes, namely in $49^0 - 102^0$ and $248^0 - 292^0$;

3. Values of absolute brightness for 20 HCs are determined: they are distributed in the interval from $5^m.2$ to $12^m.4$. Hence, comets are not unusual in sense of brightness. There is an inverse correlation between parameters H_{10} and $1/a$ for 20 considered comets (correlation coefficient is -0.45).

4. Noticeable concentration of HC's perihelion close to the solar apex, solar equator's and galaxy plane does not exist.

5. The fraction of HCs gradually grows in the total list of comets. It excludes the version that they are results of mistakes of calculations;

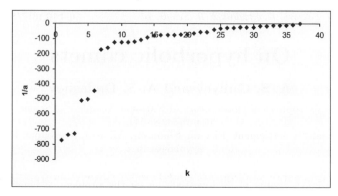

Figure 1. Increasing order of values of $1/a$ of hyperbolic comets.

6. HCs practically do not differ from other long-periodic comets on parameters L, q and i.

It is necessary note that absence of distinction of HC from general ensemble on parameters q and H_{10} (probably absence of connection with solar equator also) creates difficulties for the version about only non-gravitational nature of hyperbolic excess of HCs velocity. At the same time some features noted in items $1, 4$ and 5 creates difficulties for the assumption of an interstellar origin of such comets.

Now we will state some results of our analysis and calculations on HCs:

1. Some previous researches (Guliyev, 1999,2007) showed that perihelion of long-period comets, except for some groups with small q, are concentrated near the plane with parameters $I = 86^0$ and $\Omega = 273^0$. In addition it was found, that remote nodes of long-period comets orbits concerning this plane have overpopulation in the interval $250 - 400$ a.u. Some excess of aphelion number on this interval exists also (Guliyev 2008). One of authors in his relevant works has advanced hypothesis that in this plane and on the distances $250 - 400$ a.u. there might be very big kuiper body. This body could be a source of generation of comet nucleus from transneptunian zones in sphere of visibility. Developing this hypothesis, we can assume that this hypothetical planet body might be reason of acceleration of heliocentric velocity of some HCs . Our analogical calculations show that perihelion of HCs have very noticeable concentration near the plane

$$I = 87^0; \qquad \Omega = 274^0. \tag{1.1}$$

Perihelion of $15HC$ are located in 10^0 latitude of this plane (fig.2). This plane is very close to the previous one. However, degree of concentration in the case of HCs is sharper.

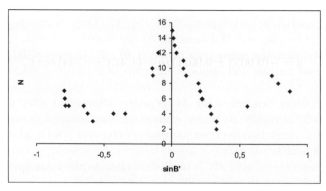

Figure 2. Distribution of hyperbolic comets' perihelion on intervals $\sin B'$ (lengths of interlacing intervals are equal to 0.33) concerning the plane (1.1).

2. The ratio of direct and retrograde movements concerning the plane (1.1) is 23 : 14 and it differs from the elliptical one.

3. Authors have analyzed the data $(1/a, \cos I'$ and $q)$ of 37 comets, entering the Tisseran's criterion for checking an above-mentioned idea. Our calculations showed that there was some multiple correlation dependence between them. The following empirical formula has been found

$$1/a = (-155 \pm 66)\cos I' - (40 \pm 15)q$$

(I' is inclination of comets orbit concerning accordingly plane hereinafter). Level of determinacy of this expression and its confidentiality probability are 0.26 and 0.99 accordingly. For the class of comets $1A$ and $1B$ (Marsden and Williams, 2008) empirical expressions become more certain. It means that 15 comets, or some part of them, can have the certain community connected with the plane (1.1).

4. Authors tried to check up the hypothesis that perihelion of HC are concentrated near several planes. Indeed, if we will exclude 15 comets from list of HCs, other 22 ones will tend to have concentration of perihelion near the plane with parameters

$$I = 18.6^0; \qquad \Omega = 287.2^0. \tag{1.2}$$

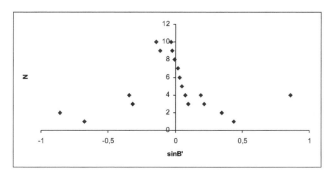

Figure 3. Distribution of hyperbolic comets' perihelion on intervals $\sin B'$ (lengths of interlacing intervals are equally to 0.25) regarding the plane (2.2).

Perihelion of $10HC$ are located in 10^0 latitude of this plane (fig.3). Multiple correlation dependence between parameters $1/a$, q and $\cos I$ in this system exists also. The following empirical formula for $22HC$ has been obtained

$$q = -(0.0053 \pm 0.0021)/a - (1.473 \pm 0.744) \cos I'.$$

The ratio of direct and retrograde movements concerning this plane is 7 : 15. We tried to find distinctive features of this plane. After relevant calculations we have found that remote nodes of long-period comets regarding this plane have some excess in the interval $122 - 137$ A.U. We have used the method of testing used in (Guliyev, 2007) for a basing of this prospective excess. Particularly keeping interval $122 - 137$ A.U., parameters $\Omega(0^0, 30^0, 60^0 \ldots 330^0)$ and $i(0^0, 9^0.49, 19^0.47, 30^0, 41^0.81, 56^0.44$ and $90^0)$ were varied in such way, that poles of corresponding 67 planes were equally removed from each other. Number of comet nodes in fixed interval was found for each plane in the next step of calculations. Consequently we had conducted comparisons using methods of statistics and obtained next values:

$$N = 21; \quad n = 7.73; \quad \sigma = 2.57; \quad t = 5.2; \quad \alpha = 0.999,$$

where N, n, σ, t and α are number of nodes regarding (2.2), midrange value on other planes, normalized difference $(t = (N - n)/\sigma)$, rms deviation and confidential probability of t accordingly. These figures give some bases to assume, that in a plane (2.2) on distances $122 - -137$ A.U. might be a latent source of comets. The same source could cause increase in heliocentric velocity of some long-period comets.

5. We have calculated minimal inter-orbital distances of HCs from 4 dwarf planets also. 13 of them have distances less than 2 a.u. (Pluto-5, Makemake-1, Eris-6 and Haumea-1). We have found that 8 other comets have such inter-orbital distances from four bright TNO (Orcus-4, Varuna-1, Ixion -1, Quaoar-1). It means that possible influence of these and other TNOs in discussion of the question on hyperbolic comets should be taken into account.

2. Conclusions

1. Rate of growth HCs exceeds rate of growth of other long-period comets. It means at least that system of such comets really exists.

2. Absence of concentration of HCs perihelion near the Galaxy's plane, solar equator plane and the solar apex, as well as patterns in distribution of $1/a$ excludes the version of interstellar origin of HCs.

3. Absence of distinction of HCs from general ensemble on parameters q and H_{10} creates some difficulties for the version about **only** non-gravitational nature of hyperbolic excess of HCs velocity.

4. Perihelion of 25 from 37 HCs are concentrated near two planes: $(\Omega = 273^0; I = 86^0)$ and; $(\Omega = 287^0; I = 19^0)$. Some HCs could obtain an excess of velocity from hypothetical planet bodies moving in these planes.

5. Hyperbolic comets, by their removed units and inter-orbital distances, could have approaching to dwarf planets and some big kuiper bodies in the past.

References

Marsden, B. G. & Williams, G. V. 2005, *Catalogue of Cometary Orbits, 16-th edition. (SAO, Cambridge.)*, 207P

Guliyev, A. S. 2007, *Astronomical letters*, 33, 8, pp. 562–570

Guliyev, A. S. 1999, *Kinematics and physics of celestial bodies*, 15, 1, pp. 85–92

Guliyev, A. S. 2008, *Azerbaijani Astronomical Journal*, 2, 1–2, pp. 5–9

Icy Bodies of the Solar System
Proceedings IAU Symposium No. 263, 2009
J.A. Fernández, D. Lazzaro, D.Prialnik & R. Schulz, eds.

© International Astronomical Union 2010
doi:10.1017/S1743921310001547

Non-gravitational forces and masses of some long-period comets. The cases of Hale-Bopp and Hyakutake

Andrea Sosa and Julio A. Fernández

Departamento de Astronomía, Facultad de Ciencias,
Igua 4225, 11400, Montevideo, Uruguay
email: asosa@fisica.edu.uy, julio@fisica.edu.uy

Abstract. By means of a simple non-gravitational force model of the cometary nucleus, which relies on the observed light curves assumed to be a good representation of the water sublimation rate, we estimate the masses of a sample of long-period comets (LPCs).

A critical issue of our method is the assumption of a correlation between visual heliocentric magnitudes and water production rates. This is a necessary assumption because of the sparse observational data of gas production rates (with the exception of very few comets like Hale-Bopp or Hyakutake). In this regard we present here a new correlation for LPCs. We also present the preliminary results for the masses of comets Hale-Bopp and Hyakutake.

Keywords. Comets: general, methods: analytical, data analysis.

1. Introduction

The main non-gravitational effect that can be detected in a periodic comet is a change in its orbital period, with respect to that derived from purely gravitational theory. Whipple (1950) proposed an icy conglomerate model for the cometary nuclei, and showed that the momentum transferred to the nucleus by the outgassing could cause the observed non-gravitational effect.

A method for computing cometary masses based on the non-gravitational effect was introduced by Rickman (1986, 1989), Rickman *et al.* (1987), and Sagdeev *et al.* (1987). Similar approaches were presented later by Rickman *et al.* (1991), and more recently by Szutowicz *et al.* (2002a, 2002b), Farnham & Cochran (2002), Szutowicz & Rickman (2006), and Sosa & Fernández (2009).

Most of the studies of comet masses are for Jupiter family comets (JFCs). Unlike the short-period comets, the evaluation of non-gravitational forces is more difficult for LPCs, since these have not been observed in a second apparition to check for a change in the orbital period attributable to non-gravitational forces. Nevertheless, non-gravitational terms have been fitted to the equations of motion of several LPCs leading to more satisfactory orbital solutions.

A total of 30 comets, observed between 1973 and 2007, have been used in our study, selected from the known population of LPCs with orbital periods P > 1000 yr and short perihelion distances (q < 2 AU), by demanding that each comet has measured water production rates and a data set of photometric observations good enough to determine a light curve. We use the visual magnitudes from the International Comet Quaterly database, while the gas production data are collected from the literature and other public sources, like the Nançay database of OH production rates, and the International Astronomical Union Circulars. The orbital and non-gravitational parameters are extracted from the Catalogue of Cometary Orbits (Marsden & Williams 2008).

2. The method

According to Whipple (1950), we can relate the non-gravitational acceleration $\vec{J}$ and the comet's mass M by means of the conservation of momentum i.e.

$$M\vec{J} = -Qm\vec{u}, \qquad (2.1)$$

where Q is the gas production rate, $\vec{u}$ is the effective outflow velocity, and m is the average molecular mass.

The non-gravitational acceleration (which is in the opposite direction to that of the net outgassing) can be described in terms of its radial (i.e. in the antisolar direction), transverse (i.e. perpendicular to the radial component in the direction of the motion), and normal (i.e. perpendicular to the orbital plane) components $\vec{J}_r$, $\vec{J}_t$ and $\vec{J}_n$, respectively. According to the standard symmetric model developed by Marsden et al. (1973), we have $J_r = A_1 g$, $J_t = A_2 g$, where the parameters A_1, A_2 represent the radial and transverse components of the non-gravitational acceleration at 1 AU from the Sun, respectively, and $g = g(r)$ is an empirical function which describes the variation of the water snow sublimation rate with respect to the heliocentric distance. The NG parameters (A_1, A_2) arise from the best-fitting orbital solution.

Finally, by taking absolutes values for the vectors in eq. (2.1), and solving for the mass, we derive the following expression:

$$M = \frac{Qm <u>}{J}, \qquad (2.2)$$

where $J = \sqrt{A_1^2 + A_2^2} \times g$, and $<u>$ represents an average value for all escaping molecules during the passage of the comet in the inner solar system ($r < 3$ AU). As an educated guess, we chose $<u> = (0.27 \pm 0.1)$ km s^{-1} (see Sosa & Fernández 2009 for further details). At heliocentric distances $r < 3$ AU the cometary activity is governed by the sublimation of water ice, so we take m as the water molecular mass. As we can see from eq. (2.2), the computation of the cometary mass requires to know from observations the shape of the curve Q. Due to the scarce measurements of gas production rates we assume an empirical law introduced by Festou (1986), to approximately convert total visual heliocentric magnitudes m_h (i.e. the apparent magnitudes m corrected for the geocentric distance Δ: $m_h = m - 5\log \Delta$) into water production rates,

$$\log Q = a_1 \times m_h + a_0 \qquad (2.3)$$

The light curve $m_h(t)$ is defined by a polynomial fit to the upper envelope of the ensemble of photometric observations (details of the procedure can be found in Sosa & Fernández 2009).

3. The results and concluding remarks

Fig. 1 shows the light curves obtained as plots of the total heliocentric visual magnitude as a function of the time relative to the perihelion passage, for comets Hale-Bopp and Hyakutake. We performed a linear regression between the heliocentric magnitudes (as inferred from the light curves), and the measured water production rates, by an iterative procedure that removes the observations outside 3-σ of the residuals. We obtained a sample of 585 data points for 30 LP comets, for which the heliocentric distances varied between 0.20 and 2.98 AU, while the geocentric distances spanned from to 0.10 to 3.04

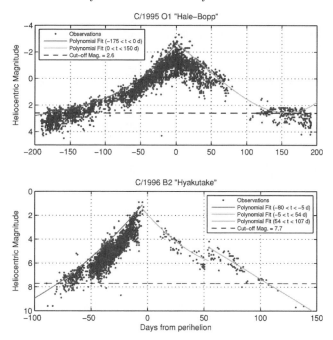

Figure 1. Heliocentric total visual magnitudes as a function of time, for comets Hale-Bopp (up) and Hyakutake (bottom). The polynomial fit $m_h(t)$ to the upper envelope of the broad distribution of photometric measurements is shown. The horizontal line indicate the *cut-off* magnitude m_{hC}. A total of 4576 and 3101 observations were used for HB and Hyakutake, respectively.

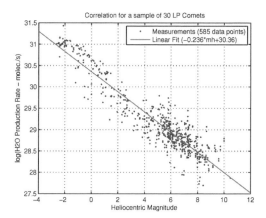

Figure 2. The points indicate the data (logarithm of the measured water production rate vs. the estimated visual heliocentric magnitude from the light curves), while the line shows the correlation law.

AU. For this sample, the best fitted coefficients were $a_1 = -0.236$, $a_0 = 30.26$, with a regression coefficient of -0.94 (Fig. 2).

With our work still in progress, we present here only the preliminary results for comets Hale-Bopp and Hyakutake. We note that the shape of $Q(r)$ differs from $g(r)$, as it is evidenced by the light curves of Fig. 1, which show certain degree of asymmetry with respect to the perihelion. Hence, we will have different values of M, as r varies. Since we are interested in an average value for M (neglecting the loss of mass due to the outgassing), we consider the variations in M due to such difference as a source of uncertainty inherent to the method; the more the ratio Q/g approaches unity, the more constrained will be the range of possible values for M. Our preliminary results lead to a mass estimate within the range $\sim [1.6 - 2.7] \times 10^{14}$ kg for comet Hale-Bopp , while the mass estimate for comet Hyakutake would be within the range $\sim [1.3 - 3.3] \times 10^{12}$ kg.

We find a strong correlation between visual magnitudes and water production rates, like previous works. Our results are similar to that of Jorda *et al.* (2008), practically with the same slope (they derived $a_1 = -0.2453$), but with a slightly smaller value for a_0 (they derived $a_0 = 30.675$), possibly due to the method chosen to derive the visual magnitudes (e.g. we privilege the brightest observations over the faintest, while Jorda *et al.* average the observations). Other main differences with respect to the work of Jorda *et al.* (besides the method for determining the visual magnitudes) are that they consider only the Nançay database (with a smaller sample size), and that they did not distinguish between dynamical classes, while we consider all water measurements (obtained by the different techniques), and restrict our study to LPCs. As future work, it remains a more complete statistical analysis of the correlation, by including the error estimates.

The mass estimates for comets Hale-Bopp and Hyakutake are in good agreement with previous works of Szutowicz *et al.* (2002a) and Szutowicz *et al.* (2002b), respectively. In comparison with the masses derived by us (Sosa & Fernández 2009) for a sample of JFCs, comet H-B would be a "massive" comet, belonging to the same mass rank as comets Halley and Tempel 2, while comet Hyakutake would be a "lighter" comet, in the same mass rank as comets d'Arrest and Borrelly.

References

Farnham, T. L. & Cochran, A. L. 2002, *Icarus*, 160, 398
Festou, M. 1986, in: Proc. Asteroids, Comets, Meteors 1985
Jorda, L., Crovisier, J., & Green, D. W. E. 2008, in: Proc. Asteroids, Comets, Meteors 2008
Marsden, B. G., Sekanina, Z., & Yeomans, D. K. 1973, *AJ*, 78, 211
Marsden, B. G. & Williams, G. V. 2008, *Catalogue of Cometary Orbits, 17th ed.*, Smithsonian Astrophysical Observatory
Rickman, H. 1986, in: ESA SP 249
Rickman, H. 1989, *Adv. Sp. Res.*, 9, 59
Rickman, H., Kamél, L., Festou M., & Froeschlé, C. 1987, in: ESA SP 278
Rickman, H., Festou, M. C., Tancredi, G., & Kamél, L. 1991, in: Proc. Asteroids, Comets, Meteors 1991, p. 509
Sagdeev, R. Z., Elyasberg, P. E., & Moroz, V. L. 1987, *Sov. Astron. Lett.*, 13, 259
Sosa, A. & Fernández, J. A. 2009, *MNRAS*, 393, 192
Szutowicz, S. & Rickman, H. 2006, *Icarus*, 185, 223
Szutowicz, S., Królikowska, M., & Sitarski, G. 2002a, *Earth, Moon and Planets*, 90, 119
Szutowicz, S., Królikowska, M., & Sitarski, G. 2002b, in: Proc. Asteroids, Comets, Meteors 2002
Whipple, F. L. 1950, *AJ*, 111, 375

Icy Bodies of the Solar System
Proceedings IAU Symposium No. 263, 2010
J.A. Fernández, D. Lazzaro, D. Prialnik & R. Schulz, eds.

© International Astronomical Union 2010
doi:10.1017/S1743921310001559

The contribution of plutinos to the Centaur population

Romina P. Di Sisto, Adrián Brunini, and Gonzalo C. de Elía

Facultad de Ciencias Astronómicas y Geofísicas - UNLP,
IALP - CONICET, Paseo del Bosque *S/N*, La Plata. Argentina
email: `romina@fcaglp.unlp.edu.ar`, `gdeelia@fcaglp.unlp.edu.ar`,
`abrunini@fcaglp.unlp.edu.ar`

Abstract. We present a study of the dynamical evolution of plutinos recently escaped from the resonance through numerical simulations. It was shown in previous works the existence of weakly chaotic orbits in the plutino population that diffuse very slowly finally diving into a strong chaotic region. These orbits correspond to long-term plutino escapers and then represent the plutinos that are escaping from the resonance at present. Then, we divided the numerical simulation in two parts. First, we develop a numerical simulation of 20,000 test particles in the resonance in order to detect the long-term escapers. We set the initial orbital elements such that cover the present observational range of orbital elements of plutinos. Second, we perform a numerical simulation of the selected escaped plutinos in order to study their dynamical post escaped behavior. We describe and characterize the routes of escape of plutinos and their evolution in the Centaur zone. Also, we obtain that Centaurs coming from plutinos would represent a fraction of less than 6% of the total Centaur population.

Keywords. Kuiper Belt, methods: numerical.

1. Introduction

Plutinos are the resonant transneptunian population most densely populated. They are trapped into the 2:3 mean motion resonance with Neptune, being Pluto its most representative member. They are in a stable configuration, with the critical angle $\sigma = 3\lambda - 2\lambda_N - \varpi$ librating around $180°$, where λ and ϖ are the mean longitude and the longitude of perihelion of the particle, and λ_N is the mean longitude of Neptune. Duncan *et al.* (1995), showed that the boundaries of the 2:3 long-lived mean motion resonance have a time scale for instability of the order of the age of the Solar System and they may be related to the origin of the observed JFCs. Morbidelli (1997) studied the dynamical structure of the 2:3 resonance in order to analyze possible diffusive phenomena and their relation to the existence of long-term escape trajectories. From his work it could be deduced a present rate of escape of 1 plutino with $R > 1$ km every 20 years. Yu & Tremaine (1999) and Nesvorný *et al.* (2000) analyzed the effect of Pluto on the 2:3 resonant orbits. The latest found similar values for the escape rate as the ones obtained by Morbidelly (1997) without Pluto. Melita & Brunini (2000) suggested that the existence of plutinos in very unstable regions can be explained by physical collisions or gravitational encounters with other plutinos. Nesvorný & Roig (2001) explored the dynamics of the 2:3 resonance and showed that the regions of orbital stability do not seem to be well sampled by observed plutinos, which may be a dynamically primordial feature or a consequence of collisions and mutual gravitational scattering. De Elía *et al.* (2008) performed a collisional evolution of plutinos and obtained a plutino removal by their collisional evolution of 2 plutinos with $R > 1$ km every 10000 years. Since, from that previous works, plutino removal by "dynamic" is much greater than plutino removal by collisions we perform "dynamical" numerical simulations. We refer to "dynamical" numerical simulations, those that take into account the gravitational forces to follow up the evolution of a particle ocasionally causing the remotion. Our goal is to describe and characterize the routes of escape of plutinos and their contribution to the Centaur population.

2. The numerical runs

We need to identify the plutinos that have recently escaped from the resonance. Morbidelli (1997) showed the existence of weakly chaotic orbits that diffuse very slowly finally diving into a strong chaotic region. These orbits corresponds then to long-term escapers i.e. plutinos recently escaped from the resonance. Then we divided the numerical simulation in two parts. First: the pre-runs where we developed numerical simulations of plutinos in the resonance in order to detect those that escape at present. Second: the post-runs where we performed a numerical simulation of the selected escaped plutinos in order to study their dynamical post escaped behavior.

2.1. Pre-runs. The integration in the resonance

We performed a numerical integration following the study of Morbidelli (1997). We integrated here 20,000 test particles under the gravitational influence of the Sun and the four giant planets over 4.5 Gy with an integration step of 0.5 years using EVORB (Fernández et al. (2002)). We set the initial semimajor axis of the particles equal to the exact value of the resonance $a_i = 39.5$ AU. The initial argument of perihelion ω_i, longitude of node Ω_i and the mean anomaly M_i have been chosen at random in the range of $[0°, 360°]$ so the critical angle σ remains between $180°$ and $330°$. Since σ librates around $180°$, and given the relation between a and σ in the resonance, the election of the previous orbital elements covers the 2:3 mean motion resonance (see Morbidelli (1997) for a complete explanation). The initial eccentricity and inclination of the particles have been randomly chosen in the intervals [0,0.35] and $[0°, 45°]$, respectively, covering the present observational ranges for the orbital elements of Plutinos. The test particles where integrated up to the first encounter within the Hill sphere of a giant planet, collision onto a planet or ejection. Such final states represent the escape conditions of plutinos. So we can calculate the rate of escape of particles from the resonance.

In Fig. 1 we plot the cummulative number of escape particles from the resonance (N_e) with respect to the number of the remaining particles N_s, where $N_s = 20000 - N_e$, as a function of t. It can be seen that the number of escaped particles raise quickly at the begining up to $t \sim 1.5$ Gy. In this point the slope of the curve changes and behaves roughly as a linear relation. This change of slope was already noticed by Morbildelli (1997) and is related with the time when the strongly chaotic region is completely depleted and

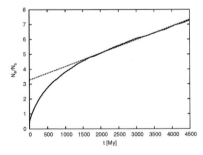

Figure 1. Cummulative number of escape particles from the resonance (N_e) with respect to the number of the remaining particles N_s versus time.

the weakly chaotic region starts to be the dominant source of Neptune-encountering bodies. We fit to the plot for $t > 1.5$ Gy, a linear relation $N_e/N_s = at + b$ where: $a = 9.00713 \times 10^{-10} \pm 4.735 \times 10^{-13} yr^{-1}$ and $b = 3.28079 \pm 0.001347$. The slope of the linear relation, a, represents the present rate of escape of the particles from the 2:3 mean motion resonance.

From our initial particles, we have 1183 that escape from the resonance after $t = 1.5$ Gy. Since those particles would represent the present escaped plutinos, we identify their original orbital elements for the second integration.

2.2. *Post-runs. The Post escape integration.*

We numerically integrated again from $t = 0$, the 1183 particles that escape at $t > 1.5$ Gy, in order now, to obtain their post escape evolution. We integrated those particles with the same computing conditions than the previous integration but now we followed the integration for 10 Gy. The particles were removed from the simulation when they either collide with a planet or the Sun, or they reached a semimajor axis greater than 1000 AU or they enter the region inside Jupiter's orbit ($r < 5.2$ AU) where the perturbations of the terrestrial planets are not negligible. We recorded the orbital elements of the particles and the planets every 10^4 years.

3. Results

In Fig. 2 we plot the time-weigthed distribution of escaped plutinos in the orbital element space. These plots assume time-invariability, so they don't represent the real case where plutinos are continously entering and leavig the zone, but they help to identify the densest and empties regions. The great majority of plutinos have encounters with Neptune, so this planet is the main responsible for their post escape evolution. This behavior can be seen in the plots as the densest zone near Neptune's perihelion. The densest zone in the other orbital elements of escaped plutinos corresponds to the ranges $30 < a < 100$ AU and $5° < i < 40°$. When escaped plutinos are transferred to the SD (this is the zone of perihelion distances $q < 30$ AU) they are quickly locked into a mean motion resonance with Neptune (similar to the behavior of SDOs analized by Fernández *et al.* (2004) and Gallardo (2006)). In the Centaur zone (this is the zone of $q < 30$ AU) the distribution of escaped plutinos is similar to that of SDOs in the Centaur zone obtained by Di Sisto and Brunini (2007).

The escaped plutinos have a mean lifetime in the Centaur zone of $l_C = 108$ My, greater than that of Centaurs from SD of $l_C = 72$ My . Escaped-plutinos live more time than SDOs in the greater-perihelion Centaur zone, causing a slower diffusion to the inner Solar System of escaped-plutino orbits than of SDOs orbits.

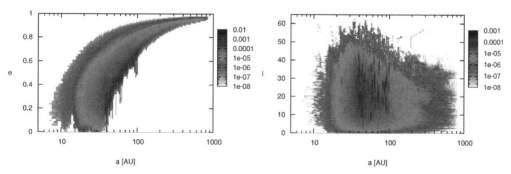

Figure 2. Time-weighted distribution of escaped plutinos in the orbital element space.

In order to calculate the number of escaped plutinos located at present in the Centaur population, we calculate the present rate of injection of escaped plutinos into the Centaur zone. As we have mentioned we consider that the long-term escapers from the plutino population, i.e., those that escape after $t = 1.5$ Gy, represent the present plutino espapers. Then the plot of the cummulative number of escaped plutinos injected into the Centaur zone (N_c) with respect to the number of the remaining plutinos N_p, as a function of time is well fitted by the linear relation: $\frac{N_c}{N_p} = ct + d$, where $c = 1.62076 \times 10^{-10} \pm 8.502 \times 10^{-14} yr^{-1}$ and $b = -0.228442 \pm 0.000242$. The slope of the linear relation, c, represents the present rate of injection of plutinos into the Centaur zone. Assuming that the present number of plutinos, N_p, is constant with time, we can calculate the present number of escaped plutinos in the Centaur population as $N_c = c\,N_p\,l_C$ where l_C is expressed in years. The number of small objects in the plutino population is not well determined, since the present observational surveys can't cover all the small-sized objects. The size distribution of plutinos is calculated from different surveys up to a given size, typically up to $R \sim 30$ km. For objects with radius less than this size, the population could have a break (Kenyon et $al.$ (2007), Bernstein et $al.$ (2004), Elliot et $al.$ (2005)). Considering the possible existence of a break and its magnitude, the number of plutinos $N_p(R > 1km)$ could be between 10^8 and 10^9. Then the present number of plutino-Centaurs with $R > 1$ km would be between $1.8 \times 10^6 - 1.8 \times 10^7$. Di Sisto and Brunini (2007) estimated $N_C(R > 1km) \sim 2.8 \times 10^8$ coming from the SD, then Centaurs coming from plutinos would represent a fraction of less than 6% of the total population being then a secondary source of Centaurs.

References

Bernstein, G. M., Trilling, D. E., Allen, R. L., et $al.$ 2004, AJ, 128, 1364

de Elía, G. C. & Brunini, A. 2008, $A\&A$, 490, 835

Di Sisto, R. P. & Brunini, A. 2007, $Icarus$, 190, 224

Duncan, M. J., Levison, H. F., & Budd, S. M. 1995, AJ, 110, 3073

Elliot, J. L., Kern, S. D., Clancy, K. B., et $al.$ 2005, AJ, 129, 1117

Fernández, J. A., Gallardo, T., & Brunini, A. 2002, $Icarus$, 159, 358

Fernández, J. A., Gallardo, T., & Brunini, A. 2004, $Icarus$, 172, 372

Gallardo, T. 2006, $Icarus$, 184, 29.

Kenyon, S. J., Bromley, B. C., O'Brien, D. P., & Davis, D. R. 2008, in: M. A. Barucci, H. Boehnhardt, D. Cruikshank, & A. Morbidelli (eds.), The $Solar$ $System$ $Beyond$ $Neptune$, (Tucson: University of Arizona Press), p. 293

Melita, M. D. & Brunini, A. 2000, $Icarus$, 147, 205

Morbidelli, A. 1997, $Icarus$, 127, 1

Nesvorný, D., Roig, F., & Ferraz-Mello, S. 2000, AJ, 119, 953

Nesvorný, D. & Roig, F. 2000, $Icarus$, 148, 282

Yu, Q. & Tremaine, S. 1999, AJ, 118, 1873

Icy Bodies of the Solar System
Proceedings IAU Symposium No. 263, 2009
J.A. Fernández, D. Lazzaro, D. Prialnik & R. Schulz, eds.
© International Astronomical Union 2010
doi:10.1017/S1743921310001560

Influence of trans-neptunian objects on motion of major planets and limitation on the total TNO mass from planet and spacecraft ranging

E. V. Pitjeva

Institute of Applied astronomy RAS,
Kutuzov quay 10, 191187 St. Petersburg, Russia
email: evp@ipa.nw.ru

Abstract. Perturbations from asteroids and Trans-Neptunian Objects affect significantly on the orbits of planets and should be taken into account when high-accuracy planetary ephemerides are constructed. On the other hand, from an analysis of motion of the major planets by processing of precise measurements of spacecraft a limitation on the total TNO mass may be obtained. To estimate influence of TNO on motion of planets the largest 21 TNO have been included into the process of simultaneous numerical integration, and positions of planets obtained with taking for TNO have been compared with positions of planets of numerical EPM ephemeris of IAA RAS constructed without these objects. The perturbations of other TNO have been modeled by the perturbation from a circular ring having a radius 43 AU and different masses. It has been shown that all the test masses of the TNO ring except the minimum mass ($5.26 \cdot 10^{-8} M_{\odot}$) are too large and make the data residuals worse. Thus, the upper limit of the total mass of all TNO including Pluto, the 21 largest TNO and the TNO ring (with the 43 AU radius) should not exceed $8.04 \cdot 10^{-8} M_{\odot}$.

Keywords. Celestial mechanics, ephemerides, radar astronomy, solar system

1. Introduction: a precision of observations and dynamical models

At present, the accuracy of radar observations of spacecraft orbiting near planets reaches a meter precision in distances between the Earth and planets, that is the twelfth figure in distances. The construction of high-precision ephemerides of major planets, which corresponds to meter accuracy of ranging, requires the creation of the adequate mathematical and dynamical model of the motion of planets on the base of General Relativity and that takes into account all perturbating factors.

Perturbations from asteroids and Trans-Neptunian Objects (TNO) affect significantly the orbits of planets and should be taken into account when high-accuracy planetary ephemerides are constructed. On the other hand, from analysis of these perturbations it appears possible to derive values of some physical parameters of the asteroids and TNO, including a limitation on the total TNO mass by processing of precise measurements of spacecraft.

The EPM ephemerides (Ephemerides of planets and the Moon) of IAA RAS originated in the seventies of the last century to support space flights at about the same time as DE ephemerides and have been developed since that time. The EPM2008 ephemerides have been used for data analysis.

2. EPM2008 ephemerides

For the construction of the EPM2008 ephemerides a numerical integration of the equations of motion of the major planets, the Sun, the Moon and the lunar physical libration, asteroids and TNO, taking into consideration perturbations from solar oblateness has been performed in the Parameterized Post-Newtonian metric for the harmonic coordinates $\alpha = 0$ and General Relativity values $\beta = \gamma = 1$.

A serious problem in the construction of planetary ephemerides arises due to the necessity to take into account the perturbations caused by minor planets. The experiment showed that the fitting of ephemerides accounted for the perturbations from only several biggest asteroids (DE200 and EPM87) to the Viking lander data was poor. The perturbations from 300 and more asteroids have been taken into account in the ephemerides starting with DE403, and EPM98. However, masses of many of these asteroids are quite poorly known, and the accuracy of the planetary ephemerides deteriorates due to this factor.

Masses of most massive asteroids which more strongly affect Mars and the Earth can be estimated from observations of martian landers and spacecraft orbiting Mars. The five of the 300 large asteroids proved to be double and their masses are known now. The masses of Ida(243), Gaspra(951), Eros(433) and Mathilda(253) have been derived by perturbations of the spacecraft during the NEAR flyby. Unfortunately, the classical method of determining masses of asteroids for which close encounters occur can give a accurate determination of asteroid masses only for separate cases when very close encounters are provided with useful data before and after encounters. The masses of the rest of the 301 large asteroids have been estimated by the astrophysical method from analysis of data concerning their diameters and spectral classes estimating the mean densities of the three asteroid taxonomical classes (C, S, M) from ranging observations. However, many of these objects are too small to be observed from the Earth, but their total mass is large enough to affect the orbits of the major planets. The total contribution of all remaining small asteroids is modeled as the acceleration caused by a solid ring with the constant mass distribution in the ecliptic plane (Krasinsky et al., 2002).

At present hundreds of large TNO were revealed including Eris which surpasses Pluto in the mass. The updated model of EPM2008 includes the 21 largest TNO (Eris, Haumea, Makemake, Sedna, Quaoar, Orcus, Varana, Ixion and others) into the process of the simultaneous numerical integration. In order to investigate the influence of trans-neptunian objects on motion of planets, positions of planets obtained with using the two versions of EPM ephemeris with and without the 21 TNO have been compared. The maximum differences obtained in right ascension $(d\alpha)$, declination $(d\delta)$ and heliocentric distance (dR) on the time interval 1913–2020 are shown in Table 1. These differences are small, they are less on the order of the magnitude than the formal uncertainties of planet positions. After adjusting ephemerides to the present set of observational data it wasn't found the difference in residuals for these versions of ephemerides (with and without TNO). However, it turns out that the total shift of the barycenter of the solar system due to the 21 largest trans-Neptunian objects is 6140 m within the lifetime of GAIA (2011-2020).

Some tests have been made for estimating the effect of other TNO on the motion of planets. Their perturbations have been modeled by the perturbation from a circular ring having a radius of 43 AU and different masses. The minimum mass (EPM_{1-TNO}) of this ring is equal to the mass of 100000 bodies with 100 km in diameter and density is equal to $2g/cm^3$, it amounts to 110 masses of Ceres. The maximum mass (EPM_{5-TNO}) of the ring is expected to be 100 times the minimum mass. Masses of EPM_{2-TNO}, EPM_{3-TNO}, EPM_{4-TNO} amount 25%, 50%, and 75% of the maximum mass respectively. The effect

Table 1. The maximum differences in $d\alpha$, $d\delta$ and dR, 1913–2020, for ephemerides with and without the 21 largest TNO.

Planet	$d\alpha[mas]$	$d\delta[mas]$	$dR[m]$
Mercury	0.006	0.003	2.4
Venus	0.006	0.010	6.1
Mars	0.068	0.025	15.5
Jupiter	0.260	0.110	199
Saturn	0.298	0.103	311
Uranus	1.688	0.744	5932
Neptune	2.321	0.279	50332
Pluto	2.578	1.064	60821

Table 2. The rms residuals in m and the weight unit errors σ_0 for EPM ephemerides with different masses of the TNO ring.

Observations Interval Numbers n.p.	Martian landers 1976-1997 1348	Martian spacecraft 1998-2008 13903	Venus Express 2006-2007 547	Spacecraft at Jupiter 1973-2001 7	Spacecraft at Saturn 1979-2006 34	σ_0 1913-2008 97101
EPM2008	11.82	2.04	2.59	13.09	3.04	0.876
EPM_{1-TNO}	12.00	1.87	2.59	13.08	3.02	0.876
EPM_{2-TNO}	13.16	1.90	2.63	13.20	64.35	1.000
EPM_{3-TNO}	13.35	2.00	2.68	13.29	129.7	1.305
EPM_{4-TNO}	13.83	2.05	2.74	17.64	195.0	1.696
EPM_{5-TNO}	14.26	2.06	2.80	27.43	200.3	2.126

of the ring is only noticeable for more accurate observations — the spacecraft data. The rms residuals and the weight unit errors for these data after fitting the standard and test EPM ephemerides are given in Table 2.

It is seen from the above that all the masses of the TNO ring except the minimum mass ($EPM_{1-TNO}=5.26\cdot10^{-8}M_\odot$) are too large and make the data residuals worse. These results give a possibility to estimate the upper limit of the total mass of all TNO and to include the mass value of the TNO ring into the set of the adjusted parameters.

Thus, the dynamical model of EPM2008 ephemerides takes into account the following:
- mutual perturbations from major planets, the Sun, the Moon and 5 more massive asteroids;
- perturbations from the other 296 asteroids chosen due to their strong perturbations on Mars and the Earth;
- perturbation from the massive asteroid ring with constant mass distribution in the ecliptic plane;
- perturbations from the 21 largest TNO;
- perturbation from a massive ring of TNO with the radius of 43 AU;
- perturbations due to the solar oblateness $J_2 = 2 \cdot 10^{-7}$.

The modern EPM2008 ephemerides have resulted from a least squares adjustment to observational data totaling about 550000 position observations of different types including different American and Russian radiometric observations of planets and spacecraft (VEX, MGS, Odyssey, MRO, Cassini, etc.) 1961-2008, CCD astrometric observations of outer planets and their satellites, meridian transits and photographic observations of the

Table 3. Masses of Ceres, Pallas, Juno, Vesta, Iris, Bamberga in $10^{-10} M_\odot$.

	Ceres	Pallas	Juno	Vesta	Iris	Bamberga
	4.71	1.06	0.129	1.32	0.040	0.046
σ_{formal}	±0.007	±0.003	±0.003	±0.001	±0.001	±0.001
σ_{real}	±0.03	±0.03	±0.008	±0.03	±0.008	±0.008

XX-th century, as well as the VLBI spacecraft data. Data used for the production of ephemerides were taken from databases of the JPL website (http:/ssd.jpl.nasa.gov/iau-comm4/) created by Standish and continuing by Folkner, and extended to include Russian radar observations of planets (//www.ipa.nw.ru/ PAGE/DEPFUND/LEA/ENG/englea.htm). The significance of the high precise radiometric observations of planets beginning in 1961 (and afterward spacecraft) continuing with the increasing accuracy it should be stressed. It has been these observations that have made it possible to determine and improve a broad set of astronomical constants. The detailed discription of the EPM2008 ephemerides are given in the papers by Pitjeva (2009, 2010).

3. Values of the adjusted parameters

More than 260 parameters have been determined while improving the planetary part of EPM2008. In addition to the orbital elements of all the planets and the main satellites of the outer planets, different physical constants have were estimated including parameters of a surface topography of planets and the rotation of Mars, the ratio masses of the Earth and the Moon, masses of ten asteroids that perturb Mars most strongly, mean densities for three taxonomic classes of asteroids (C, S, M), the mass and the radius of the asteroid ring, the mass of the TNO ring.

In Table 3 and further the adjusted value of several of these parameters are presented. The obtained values of parameters are shown with their real uncertainties estimated by comparing the values obtained in dozens of different test LS solutions that differed by the sets of observations, their weights, and the sets of parameters included in the solution, as well as by comparing parameter values produced by independent groups. The discussion of real uncertainties is given in the paper by Pitjeva & Standish, 2009.

Two parameters that characterize the ring modeling the effect from the rest of small asteroids (its mass M_{ring} and radius R_{ring}) have been determined:

$$M_{ring} = (0.87 \pm 0.35) \cdot 10^{-10} M_\odot, \; R_{ring} = (3.13 \pm 0.05) AU.$$

Thus, the estimation of the total mass of the main belt asteroids represented by the sum masses of 301 asteroids and the asteroid ring is:

$$M_{belt} = (13 \pm 2) \cdot 10^{-10} M_\odot \; (about \; 3M_{Ceres}).$$

The mass value of the ring of TNO has been obtained:

$$M_{TNOring} = (498 \pm 14) \cdot 10^{-10} M_\odot (5\sigma).$$

Thus, the total mass of all TNO including Pluto, the 21 largest TNO and the TNO ring of other TNO objects with the 43 AU radius is:

$$M_{TNO} = 775 \cdot 10^{-10} M_{\odot} \; (about \; 164 M_{Ceres} \; or \; 2 M_{Moon}).$$

References

Krasinsky, G. A., Pitjeva, E. V., Vasilyev, M. V., & Yagudina, E. I. 2002, *Icarus*, 158, 98

Pitjeva, E. V. 2009, in: M. Soffel & N. Capitane (eds.), *Astrometry, Geodynamics and Astronomical Reference Systems*, Proc. JOURNEES-2008 (Dresden), p. 57

Pitjeva, E. V. & Standish, E. M. 2009, *Celest. Mech. Dyn. Astr.*, 103, 365

Pitjeva, E. V. 2010, in: S. Klioner, P. K. Seidelmann & M. Soffel (eds.), *Relativity in fundamental astronomy*, IAUS 261, (Cambridge: Cambridge University Press), in print

Icy Bodies of the Solar System
Proceedings IAU Symposium No. 263, 2009
J.A. Fernández, D. Lazzaro, D. Prialnik & R. Schulz, eds.
© International Astronomical Union 2010
doi:10.1017/S1743921310001572

Impactor Flux on the Pluto-Charon System

Gonzalo C. de Elía, Romina P. Di Sisto and Adrián Brunini

Facultad de Ciencias Astronómicas y Geofísicas - UNLP
IALP - CCT La Plata - CONICET
Paseo del Bosque S/N (1900), La Plata, Buenos Aires, Argentina
email: gdeelia@fcaglp.unlp.edu.ar, romina@fcaglp.unlp.edu.ar
abrunini@fcaglp.unlp.edu.ar

Abstract. In this work, we study the impactor flux on Pluto and Charon due to the collisional evolution of Plutinos. To do this, we develop a statistical code that includes catastrophic collisions and cratering events, and takes into account the stability and instability zones of the 3:2 mean motion resonance with Neptune. Our results suggest that if 1 Pluto-sized object is in this resonance, the flux of $D = 2$ km Plutinos on Pluto is ~4–24 percent of the flux of $D = 2$ km Kuiper Belt projectiles on Pluto. However, with 5 Pluto-sized objects in the resonance, the contribution of the Plutino population to the impactor flux on Pluto may be comparable to that of the Kuiper Belt. As for Charon, if 1 Pluto-sized object is in the 3:2 resonance, the flux of $D = 2$ km Plutinos is ~10–63 percent of the flux of $D = 2$ km impactors coming from the Kuiper Belt. However, with 5 Pluto-sized objects, the Plutino population may be a primary source of the impactor flux on Charon. We conclude that it is necessary to specify the Plutino size distribution and the number of Pluto-sized objects in the 3:2 Neptune resonance in order to determine if the Plutino population is a primary source of impactors on the Pluto-Charon system.

Keywords. Kuiper Belt, Solar System: General, Methods: Numerical

1. Introduction

The 3:2 mean motion resonance with Neptune, located at ~39.5 AU, is the most densely populated one in the Kuiper Belt. The residents of this resonant region are usually called Plutinos because of the analogy of their orbits with that of Pluto, which is its most representative member. Aside from Pluto and its largest moon Charon, the Minor Planet Center (MPC) database contains ~200 Plutino candidates.

Weissman & Stern (1994) estimated current impact rates of comets on Pluto and Charon. They showed that cratering is dominated by Kuiper Belt and inner Oort cloud comets. Then, Durda & Stern (2000) calculated collision rates in the Kuiper Belt and Centaur region. They estimated that the flux of Kuiper Belt projectiles onto Pluto and Charon is ~3–5 times that of Weissman & Stern (1994). Later, Zahnle *et al.* (2003) studied the cratering rates for the moons of the jovian planets and Pluto produced mainly by ecliptic comets, obtaining results consistent with previous estimates.

In this work, we study the impactor flux on Pluto and Charon due to the collisional evolution of Plutinos. Our main goal is to determine if the Plutino population can be considered a primary source of impactors on the Pluto-Charon system.

2. The Full Model

In order to simulate the collisional and dynamical evolution of the Plutino population, we use the statistical code developed by de Elía *et al.* (2008). This algorithm considers catastrophic collisions and cratering events, and takes into account the main dynamical characteristic associated to the 3:2 mean motion resonance with Neptune.

2.1. *Initial Populations*

From Kenyon *et al.* (2008), the cumulative size distribution of the resonant population of the trans-Neptunian region shows a break at a diameter D near 40–80 km. Moreover, for larger resonant objects, the population seems to have a shallow size distribution with a cumulative power-law index of $\sim$3. On the other hand, Kenyon *et al.* (2008) suggested that the resonant population has $\sim$0.01–0.05 $M_{\oplus}$ in objects with $D \gtrsim$ 20–40 km. From these estimates, it is possible to infer the existence of 5 Pluto-sized objects in the whole resonant population. If these 5 Pluto-sized objects were all in the 3:2 Neptune resonance we would have an upper limit for the large objects in this resonance.

From this, the general form of the cumulative initial population used in our model to study the collisional and dynamical evolution of the Plutinos can be written as follows

$$N(> D) = C_1 \left(\frac{1\mathrm{km}}{D} \right)^p \text{ for } D \leqslant 60\mathrm{km},$$

$$N(> D) = C_2 \left(\frac{1\mathrm{km}}{D} \right)^3 \text{ for } D > 60\mathrm{km}, \tag{2.1}$$

where C_2 adopts values of 7.9×10^9 and 3.9×10^{10} for 1 and 5 Pluto-sized objects in the 3:2 Neptune resonance, respectively. Given the uncertainty in the parameters of the Plutino size distribution at small sizes, we decide to use in our model three different initial populations, which are defined as follows

- Initial Population 1, with a cumulative power-law index p of 3.0 for $D \leqslant 60$ km,
- Initial Population 2, with a cumulative power-law index p of 2.7 for $D \leqslant 60$ km,
- Initial Population 3, with a cumulative power-law index p of 2.4 for $D \leqslant 60$ km.

2.2. *Collisional Parameters*

Here, we adopt constant values of the intrinsic collision probability $\langle Pi_c \rangle$ and the mean impact velocity $\langle V \rangle$ for Plutinos derived by Dell'Oro *et al.* (2001). Based on a sample of 46 Plutinos, these authors computed values of $\langle Pi_c \rangle$ and $\langle V \rangle$ of $4.44 \pm 0.04 \times 10^{-22}$ km^{-2} yr^{-1} and 1.44 ± 0.71 km s^{-1}, respectively.

As for the impact strength, we use a combination of the shattering impact specific energy Q_S and the inelasticity parameter f_{ke} that yield the impact energy required for dispersal Q_D derived by Benz & Asphaug (1999) for icy bodies at 3 km s^{-1}. The Q_S law used in our simulations can be represented by an expression of the form

$$Q_S = C_1 D^{-\lambda_1} (1 + (C_2 D)^{\lambda_2}), \tag{2.2}$$

where C_1, C_2, λ_1, and λ_2 are constant coefficients whose values are 24, 1.2, 0.39 and 1.75, respectively. Once the Q_S law is specified, we adjust the ineslaticity parameter f_{ke} to get the Benz & Asphaug (1999) Q_D law. According to O'Brien & Greenberg (2005), we express the parameter f_{ke} as

$$f_{ke} = f_{ke_0} \left(\frac{D}{1\,000\,km} \right)^{\gamma}, \tag{2.3}$$

where f_{ke_0} is the value of f_{ke} at $1\,000$ km and γ is a given exponent. Our simulations indicate that the Q_D law from Benz & Asphaug (1999) for icy bodies at 3 km s^{-1} is obtained with good accuracy from the combination of the selected Q_S law and f_{ke}, with $f_{ke_0} = 0.27$ and $\gamma = 0.7$. Such values are consistent with those from Davis *et al.* (1989).

2.3. *Dynamical Considerations*

To study the orbital space occupied by the Plutino population, we develop a numerical integration of 197 Plutino candidates extracted from the Minor Planet Center database with semimajor axes between 39 and 40 AU. These objects are assumed to be mass-less particles subject to the gravitational field of the Sun (including the masses of the terrestrial planets) and the perturbations of the four giant planets. The simulation is performed with the simplectic code EVORB from Fernández *et al.* (2002). The evolution of the test particles is followed for 10^7 years which is a timescale greater than any secular period found in this resonance (Morbidelli 1997). From this, we build maps of the distribution of Plutinos in the orbital element planes (a,e) and (a,i), which allows us to determine the main stability regions of the 3:2 Neptune resonance. Such maps are used to assign a characteristic orbit for every colliding Plutino and to specify the final fates of the different fragments generated in the collisional evolution (see de Elía *et al.* 2008).

3. Results

Figures 1 a) and b) show the number of Plutino impacts onto Pluto over 4.5 Gyr as a function of impactor diameter, when 1 and 5 Pluto-sized objects are present in the 3:2 Neptune resonance, respectively. From the Initial Populations 3, 2 and 1, the total number of $D = 2$ km Plutinos impacting Pluto over 4.5 Gyr is $\sim$4.6 $\times$ 10^2, 1.1 $\times$ 10^3, and 2.7 $\times$ 10^3, respectively, assuming 1 Pluto-sized object in the resonance. With 5 Pluto-sized objects, the number of such impacts ranges from $\sim$1.8 $\times$ 10^3 to 9.2 $\times$ 10^3. On the other hand, our simulations indicate that the largest Plutino that impacted Pluto over the age of the Solar System had a diameter of $\sim$26–32 km, if 1 Pluto-sized object is assumed to be present in the resonance. With 5 Pluto-sized objects, the diameter of the largest Plutino that impacted Pluto over 4.5 Gyr is $\sim$44–51 km. Figures 1 a) and b) also show the number of Kuiper Belt object (KBO) impacts onto Pluto over 4.5 Gyr derived by Durda & Stern (2000). According to this work, the number of $D = 2$ km KBOs striking Pluto over 4.5 Gyr is $\sim 10^4$, while the largest KBO expected to have impacted Pluto over the Solar System history had a diameter of $\sim$90 km. Thus, if 1 Pluto-sized object is present in the resonance, the flux of $D = 2$ km Plutinos on Pluto is found to be $\sim$ 4–24 percent of the flux of $D = 2$ km Kuiper Belt projectiles on Pluto. However, if 5 Pluto-sized objects are assumed to be present in the resonance, the contribution of the Plutino population to the impactor flux on Pluto may be comparable to that of the Kuiper Belt.

Figures 2 a) and b) show the number of Plutino impacts onto Charon over 4.5 Gyr as a function of impactor diameter, when 1 and 5 Pluto-sized objects are present in the 3:2

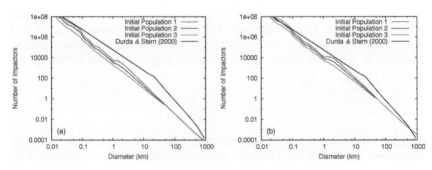

Figure 1. Number of Plutino impacts on Pluto over 4.5 Gyr as a function of impactor diameter, assuming 1 a) and 5 b) Pluto-sized objects in the 3:2 Neptune resonance.

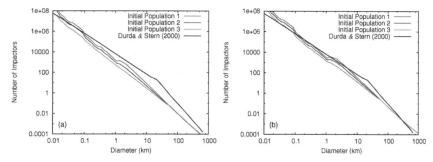

Figure 2. Number of Plutino impacts on Charon over 4.5 Gyr as a function of impactor diameter, assuming 1 a) and 5 b) Pluto-sized objects in the 3:2 Neptune resonance.

Neptune resonance, respectively. With 1 Pluto-sized object, the total number of $D = 2$ km Plutinos impacting Charon over 4.5 Gyr is $\sim 1.3 \times 10^2$, 3.1×10^2 and 7.6×10^2, from the Initial Populations 3, 2 and 1, respectively. With 5 Pluto-sized objects, the number of such impacts ranges from $\sim 5.1 \times 10^2$ to 2.6×10^3. Moreover, our simulations indicate that if 1 Pluto-sized object is in the resonance, the diameter of the largest Plutino that impacted Charon over 4.5 Gyr is $\sim$ 15–20 km. With 5 Pluto-sized objects, the largest Plutino that impacted Charon over the Solar System history had a diameter of $\sim$30–35 km. Figures 2 a) and b) also show the number of Kuiper Belt object (KBO) impacts onto Charon over 4.5 Gyr from Durda & Stern (2000). From this, the number of $D = 2$ km KBOs striking Charon over 4.5 Gyr is $\sim 1.2 \times 10^3$, while the largest KBO expected to have impacted Charon over the Solar System history had a diameter of $\sim$ 50 km. Thus, if 1 Pluto-sized object is present in the resonance, the flux of $D = 2$ km Plutinos on Charon is found to be $\sim$10–63 percent of the flux of $D = 2$ km impactors on Charon coming from the Kuiper Belt. However, if 5 Pluto-sized objects are assumed in the resonance, the Plutino population may be a primary source of the impactor flux on Charon.

Our results depend strongly on the initial size distribution and primarily on the number of Pluto-sized objects present in the 3:2 Neptune resonance. Thus, we conclude that it is very necessary to know the Plutino size distribution and the number of Pluto-sized objects in this resonant region in order to determine if the Plutino population can be considered a primary source of impactors on the Pluto-Charon system.

References

Benz, W. & Asphaug, E. 1999, *Icarus*, 142, 5

Davis, D. R., Weidenschilling, S. J., Farinella, P., Paolicchi, P., & Binzel, R. P. 1989, in: R. P. Binzel, T. Gehrels & M. S. Matthews (eds.), *Asteroids II*, (Tucson: University of Arizona Press), p. 805

de Elía, G. C., Brunini, A., & Di Sisto, R. P. 2008, *A&A*, 490, 835

Dell'Oro, A., Marzari, F., Paolicchi, P., & Vanzani, V. 2001, *A&A*, 366, 1053

Durda, D. D. & Stern, S. A. 2000, *Icarus*, 145, 220

Fernández, J. A., Gallardo, T., & Brunini, A. 2002, *Icarus*, 159, 358

Kenyon, S. J., Bromley, B. C., O'Brien, D. P., & Davis, D. R. 2008, in: M. A. Barucci, H. Boehnhardt, D. P. Cruikshank, & A. Morbidelli (eds.), *The Solar System Beyond Neptune*, (Tucson: University of Arizona Press), p. 293

Morbidelli, A. 1997, *Icarus*, 127, 1

O'Brien, D. P. & Greenberg, R. 2005, *Icarus*, 178, 179

Zahnle, K., Schenk, P., Levison, H., & Dones, L. 2003, *Icarus*, 163, 263

Icy Bodies of the Solar System
Proceedings IAU Symposium No. 263, 2009
J.A Fernández, D. Lazzaro, D. Prialnik & R. Schulz, eds.

© International Astronomical Union 2010
doi:10.1017/S1743921310001584

Numerical simulations of Jupiter Family Comets; physical and dynamical effects

Romina P. Di Sisto[1], Julio A. Fernández[2] and Adrián Brunini[1]

[1]Facultad de Ciencias Astronómicas y Geofísicas - UNLP,
IALP - CONICET, Paseo del Bosque S/N, La Plata. Argentina
email: romina@fcaglp.unlp.edu.ar, abrunini@fcaglp.unlp.edu.ar

[2]Departamento de Astronomía, Facultad de Ciencias, Iguá 4225, 11400, Montevideo. Uruguay.
email: julio@fisica.edu.uy

Abstract. We present results from numerical simulations of a Jupiter family comet (JFC) population (orbital periods $P < 20$ yr and Tisserand parameters in the range $2 < T < 3.1$) originated in the Scattered Disk and transferred to the Jupiter's zone through gravitational interactions with the Jovian planets. We shall call 'non-JFCs' those comets coming from the same source, but that do not fulfill the previous criteria (mainly because they have periods $P > 20$ yr). We have carried out series of numerical simulations of fictitious comets with a purely dynamical model and also with a more complete dynamical - physical model that includes besides nongravitational forces, sublimation and splitting mechanisms. We obtained better fits with models including physical effects, and in particular our best fits are for four splitting models with a relative weak dependence on q, and a mass loss in every splitting event that is smaller/greater for higher/lower frequencies respectively. The mean lifetime of JFCs with radii $R > 1$ km and $q < 1.5$ AU is found to be of about 150-200 revolutions ($\sim 10^3$ yr). We find a total population of JFCs with radii $R > 1$ km within Jupiter's zone of 450 ± 50 and a mean lifetime of about 150-200 revolutions ($\sim 10^3$ yr) for those getting $q < 1.5$ AU. The population of JFCs + non-JFCs with radii $R > 1$ km in Jupiter-crossing orbits may be about $\sim 2,250 \pm 250$. Most of non-JFCs have perihelia close to Jupiter's orbit.

We also present maps of the densest zones of JFCs in the orbital element space.

Keywords. comets: general, methods: numerical.

1. The observed population

Tancredi et al. (2006) showed that the cumulative luminosity function (CLF) of JFCs can be well fitted by a straight line with a slope 0.54 ± 0.05 down to $H_N = 16.7$, and then starts to flatten for fainter comets. Then the sample of observed JFCs seems to be complete down to $H_N = 16.7$. The JFCs with $q < 1.5$ AU and $H_N < 16.7$ are 9 and there are 17 for $H_N < 17.6$. This last population is incomplete. Fernández & Morbidelli (2006) found that the CLF for the total magnitudes of JFCs follows a bimodal distribution, where the slope drops for fainter comets. They found a break in the distribution at a nuclear magnitude $H_N \sim 16.5 - 17.5$. Then we calculated, following those previous studies, the total number of JFCs with $q < 1.5$ AU and $H_N < 17.6$ at $N(17.6) = 25 \pm 5$. From these 25 JFCs we have 17 observed JFCs and 8 whose orbital elements would be allocated in some way. Then we follow appropriate criteria for the incompleteness to assign the orbital elements of the remaining 8 JFCs. We will use the whole sample of 25 JFCs as our canonical data to compare with the numerical results obtained from our simulations.

2. The model

Di Sisto & Brunini (2007) performed numerical integrations of 1000 Scattered Disk (SD) objects (95 real + 905 fictitious) with the aim of study their dynamical evolution as Centaurs. They follow the simulation under the gravitational action of the Sun and the four giant planets and stored in a file the orbital elements of the particles when they crossed Jupiter's orbit for the first time. Also they stored the orbital elements of the giant planets at every crossing. Here we considered as our initial particles the 218 objects that crossed Jupiter's orbit and performed series of numerical simulations but now adding the perturbations of the terrestrial planets, taking their mean anomalies at random. We made two groups of numerical runs:

First we numerically integrated the 218 initial objects and three more clones for each particle under the gravitational action of the Sun and the planets with EVORB (Fernández *et al.* 2002) with an integration step of 0.01 yr for 10^8 yr or until collision with the Sun or a planet, or ejection.

Second we added in other three simulations the action of nongravitational forces, sublimation of ices and splitting effects. We used the 218 initial objects plus 218 clones, assigning an initial radius of 1 km, 5 km and 10 km to each particle in each integration. We followed the simulations until the comet reached a minimum radius of 100 m, or collided with the Sun or a planet, or reached $q > 6.3$ AU (transferred to the Centaur zone), or was ejected. We recorded the orbital elements of the test comets at perihelion.

The nongravitational forces were included as in Fernández *et al.* (2002). For the sublimation we fit a polynomial approximation to theoretical computations by Tancredi *et al.* (2006) of the mass loss per orbital revolution as a function of q, for $q < 2.5$ AU. We considered a standard value of the fraction of the active zone of the comet nucleus surface, $\nu = 0.15$. Observations suggest that non-tidal splittings follow a certain trend to decrease for larger q. Then we try laws for the frequency of comet splittings of the kind $f = f_0 (q/q_0)^{-\beta}$, where $\beta = 0, 0.5, 1, 1.5$ and 2, $q_0 = 0.5$ AU and let f_0 as a free parameter. We propose a mass loss of the form $\Delta M = s\, M$ where $s = s(R) = \frac{s_0}{R/R_0}$. s_0 is the mass fraction of a comet with radius R_0 that is lost in a splitting event. The whole splitting model has then three free parameters: β, s_0 and f_0. We fix $R_0 = 10$ km and $q_0 = 0.5$ AU. We compared the orbital element distributions of our computed comet sample for each splitting model with the observed orbital element distribution of JFCs in order to define the free parameters.

3. Results and conclusions

In Fig. 1, we compare the orbital element distributions of JFCs obtained from the purely dynamical model, with observations. We can see that the test JFCs reach short perihelion distances and an inclination distribution broader than the observed one. Those features are artificial and a byproduct of the long lifetime of comets. For those JFCs we obtained mean lifetimes of 42,300 yr for $q < 5.2$ AU, 16,300 yr for $q < 2.5$ AU and 7,600 in the zone $q < 1.5$ AU.

Then we included a complete model that takes into account nongravitational forces, sublimation and splitting. For the real size distribution of JFCs we adopted a bimodal cumulative size distribution for JFCs with $q < 2.5$ AU given by

$$N_1(>R) = C_1\ R^{-1.3}, \qquad 0.1 \text{ km} < R \leqslant 1 \text{ km} \qquad (3.1)$$
$$N_2(>R) = C_2\ R^{-2.7}, \qquad 1 \text{ km} \leqslant R \leqslant 10 \text{ km} \qquad (3.2)$$

We joined next the three dynamical - physical simulations picking out the JFCs with $q < 2.5$ AU and assigning to each one of the outputs a weight in such a way that the whole sample follows the size distribution of JFCs. The weights w_i so allocated represent the contribution of each instant comet, i.e. each comet revolution, to the whole sample of comet sizes. We shall call this the *assembled sample*. Comets of the assembled sample have radii between 0.1 km and 10 km and follow the observed size distribution of JFCs.

We run 52 splitting models with different sets of frequencies and mass losses, and compared the orbital element distribution to the observed sample. We have four best fits models where $\beta = 1$ in two of them and 0.5 in the other two, f_0 goes from 1/6 to 1 and s_0 is between 0.01 and 0.001. We plot one of the best fit models as an example in Fig. 2. These models reflect a relative weak to moderate dependence on q, and a mass loss in every splitting event that is smaller/greater for higher/lower frequencies respectively. These frequencies of comet splittings are relatively high with respect to the observed major splittings. However since smaller splitting events may have passed undetected, they could increase the real frequency to make observations consistent with our results. Then, our results strongly suggest that splittings must be a frequent phenomenon in comets.

With regard to the end states, we find that comets with 1-km initial radius mainly reach the minimum radius. This is due to the quick erosion that small comets suffer in our models reaching the minimum radius before they could evolve into other dynamical states. The most common final state of comets with 5-km and 10-km initial radius is the return to the Centaur region. We found that all the collisions occur with Jupiter.

In Fig. 3 we show maps that present the densest region, in the plane of orbital elements, where we would expect a greater number of JFCs. Those regions are the zones of $3 < a < 4$ AU, $0.25 < e < 0.7$, $0° < i < 25°$, and $2.7 < T < 3.05$. The observed JFCs tend to overlap the denser zones of our model.

With the complete dynamical - physical model we obtained mean lifetime of JFCs far smaller than the ones obtained through the dynamical simulation as shown above. For JFCs with radii $R > 1$ km we have a mean lifetime of $\sim$ 150-200 revolutions ($\sim 10^3$ yr) for $q < 1.5$ AU ($\sim$4.6 times smaller), while for $q < 2.5$ AU we have $\sim$ 300-450 revolutions ($\sim 3 \times 10^3$ yr) ($\sim$5.7 times smaller). This is a product of the dominant role played by physical effects in the lifetimes of low-q comets.

In Fig. 4 we represent the cumulative number of JFCs and of JFCs plus non-JFCs depending on q. We have found that the number of JFCs with radii $R > 1$ km inside

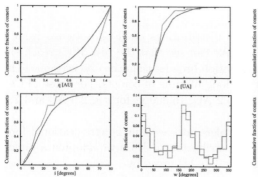
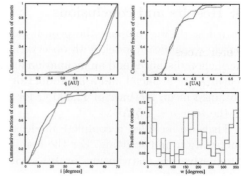

Figure 1. Orbital element distributions of JFCs from the dynamical model (solid line) and of JFCs with $q < 1.5$ AU and $H_N < 17.6$ (dashed line) (Di Sisto *et al.* (2009), Fig.5).

Figure 2. Orbital element distributions of JFCs from one of the best fit models (solid line) and of JFCs with $q < 1.5$AU, $H_N < 17.6$ (dashed line).

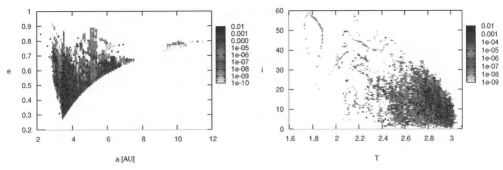

Figure 3. Time-weighted distribution of comets with $q < 2.5$ AU obtained from from one of the best fit models in the orbital element space. The black circles are the observed JFCs.

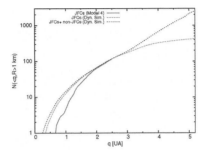

Figure 4. Cumulative number of JFCs and of JFCs+non-JFCs.(Di Sisto *et al.* (2009), Fig.8).

Jupiter's orbit is 450 ± 50. This is smaller by about one order of magnitude than some previous estimates (e.g. Fernández et al. 1999). For non-JFCs ($P > 20$ yr) in Jupiter-crossing orbits we obtained a number ~ 4 times greater that the number of JFCs (1800 ± 200). The perihelia of most of the non-JFCs is close to Jupiter's orbit.

The possibility that the observed JFCs with Tisserand parameters $T < 2$ could come straight from the SD cannot be ruled out since we have some few JFCs that could reach a $T < 2$ state through a peculiar long dynamical evolution. However this constitutes an unlikely event that requires long survival times as active comets.

In our simulations most comets that reach $q < 2.5$ AU have orbital periods $P < 20$ yr. Then we can conclude that Halley-type comets should have a different dynamical history, presumably requiring the passage by the Oort cloud prior to their plunging into the inner planetary region.

References

Di Sisto, R. P., Fernández, J. A., & Brunini, A. 2009, *Icarus*, 203, 140

Di Sisto, R. P. & Brunini, A. 2007, *Icarus*, 190, 224

Fernández, J. A., Gallardo, T., & Brunini, A. 2002, *Icarus*, 159, 358

Fernández, J. A. & Morbidelli, A. 2006, *Icarus*, 185, 211

Fernández, J. A., Tancredi, G., Rickman, H., & Licandro, J. 1999, *A&A*, 352, 327

Tancredi, G., Fernández, J. A., Rickman, H., & Licandro, J. 2006, *Icarus*, 182, 527

Icy Bodies of the Solar System
Proceedings IAU Symposium No. 263, 2009
J.A. Fernández, D. Lazzaro, D. Prialnik & R. Schulz, eds.

© International Astronomical Union 2010
doi:10.1017/S1743921310001596

How to take into account the relativistic effects in dynamical studies of comets

Julia Venturini[1] and Tabaré Gallardo[2]

[1]Departamento de Astronomía, Facultad de Ciencias, Universidad de la República,
Iguá 4225, 11400 Montevideo, Uruguay
email: jventurini@fisica.edu.uy

[2]Departamento de Astronomía, Facultad de Ciencias, Universidad de la República,
Iguá 4225, 11400 Montevideo, Uruguay
email: gallardo@fisica.edu.uy

Abstract. Comet-like orbits with low perihelion distances tend to be affected by relativistic effects. In this work we discuss the origin of the relativistic corrections, how they affect the orbital evolution and how to implement these corrections in a numerical integrator. We also propose a model that mimics the principal relativistic effects and, contrary to the original "exact" formula, has low computational cost. Our model is appropriated for numerical simulations but not for precise ephemeris computations.

Keywords. relativity, celestial mechanics, methods: n-body simulations, comets: general

1. Introduction

The problem of computing relativistic effects in the Solar System is of increasing importance for study of low perihelia and low semimajor axis populations. It is a usual practice in textbooks to show the relativistic effect in a planet's perihelion starting from a simplified problem that gives rise to a unique radial perturbation to the Newtonian potential conserving the angular momentum. But, a more rigorous analysis (Brumberg 1991, Shahid-Saless & Yeomans 1994) gives rise also to a transverse component so the angular moment is not more conserved and the expression for the relativistic perturbation becomes:

$$\Delta \ddot{\mathbf{r}} = \frac{\mu}{r^3 c^2} \left[\left(\frac{4\mu}{r} - \mathbf{v}^2 \right) \mathbf{r} + 4(\mathbf{v}.\mathbf{r})\mathbf{v} \right] \tag{1.1}$$

where $\mu = k^2 m_\odot$ (Anderson *et al.* 1975, Quinn *et al.* 1991). This expression, that has no normal component, is valid when considering the relativistic effects generated only by a spherically symmetric and non rotating Sun. When the rotation of the Sun is considered, a new effect called gravitomagnetic Lense-Thirring effect appears (Soffel 1989, Iorio 2005a). Furthermore, if all bodies have relativistic contributions a more complex expression should be used as explained in, for example, Benitez & Gallardo (2008). The time in Eq. (1.1) should be considered as the *coordinate time*; in order to compare with observables it is necessary to transform, for example, to Barycentric Coordinate Time (TCB) or Terrestrial Time (TT) (Shahid-Saless & Yeomans 1994, Soffel *et al.* 2003).

In order to avoid either the computational cost or the non symplectic form of Eq. (1.1), some simpler models that only depend on r have been proposed which mimic or reproduce the correct relativistic shift in the argument of the perihelia of the bodies under the gravitational effect of a central mass (Nobili & Roxburgh 1986, Saha & Tremaine 1992). But, as pointed out by Saha & Tremaine (1994), these models correctly recover the perihelion motion but fail in reproducing the evolution of the mean anomaly, M, point that we will discuss in this work.

These r-dependent secular models have a disadvantage when used to compute high eccentricity low perihelion orbits in constant time-steps algorithms of the type "advance-impulse-advance", as is the usual case in sympletic algorithms. In this case, the relativistic perturbation when computed as an impulse, generates spurious results near the perihelion where the relativistic perturbation is stronger and poorly sampled. This is because the expression for the acceleration has a term that goes with $1/r^3$ and for low r values this impulse can be excessively high. For example, in the Solar System the relativistic perturbation for an object at $r < 0.1$ AU is greater than the gravitational perturbation by Jupiter. Saha & Tremaine (1994), in a sympletic method for planetary integrations, incorporate part of the relativistic correction (1.1) in the "advance" part of the algorithm but an acceleration proportional to $1/r^3$ remains as part of the "impulse".

In this work we review the relativistic effects on orbital elements with special attention to mean anomaly (Rubincam 1977, Calura *et al.* 1997, Iorio 2007) and we propose an r-independent model that correctly reproduces the secular evolution of the argument of the perihelion even for high eccentricity orbits which is very useful for constant timestep algorithms.

2. Secular Relativistic Effects due to the Sun

By means of the Gauss planetary equations, averaged on an orbital period, one can compute the secular variations of orbital elements produced by Eq. (1.1). The only non zero variations obtained for this acceleration are those corresponding to the argument of perihelion and the mean anomaly:

$$< \dot{\omega} > = \frac{3}{c^2 (1 - e^2)} \sqrt{\frac{\mu^3}{a^5}} \tag{2.1}$$

$$< \dot{M} > = \frac{3}{c^2} \sqrt{\frac{\mu^3}{a^5}} \left(2 - \frac{5}{\sqrt{1 - e^2}} \right) \tag{2.2}$$

where units are radians per day. Eq. (2.1) is the well known relativistic effect on the argument of the perihelion and, for small eccentricities, Eq. (2.2) coincides, for example, with the approximate expressions for the secular drift in M given by Iorio (2005b) and with the secular drift in M generated by the relativistic model for low eccentricities used by Vitagliano (1997). The exact formula, valid for all eccentricities, is Eq. (2.2) and can be checked numerically integrating a particle using the classic model and then the relativistic one. The effect in M is several times greater than in ω.

3. Models that Mimic the Secular Relativistic Effects

3.1. *Proposed Models*

If we are interested in computing the relativistic effects due to the Sun on a small body, we could just introduce Eq. (1.1) into an integrator. The problem is that this acceleration depends on both vectors position and velocity of the particle, so the speed of the integrator may be slowed down in order to calculate accurately the vectorial products at small heliocentric distances. To overcome this difficulty, some alternative simpler models have been created in the last two decades. The relativistic precession of the argument of perihelion is correctly reproduced defining the radial perturbation:

$$R = -\frac{6\mu^2}{c^2 r^3} \tag{3.1}$$

(Nobili & Roxburgh 1986). Inserting this R in the time averaged Gauss planetary equations, the exact secular drifts generated by Eq. (1.1) are recovered except for mean anomaly. Saha & Tremaine (1992) added one more term into (3.1) in order to account for both ω and M drifts:

$$R = -\frac{6\mu^2}{c^2 r^3} + \frac{3\mu^2}{ac^2}\left(\frac{4}{\sqrt{1-e^2}} - 1\right)\frac{1}{r^2} \tag{3.2}$$

Hence, with this perturbation, Eq. (2.1) and Eq. (2.2) are recovered. This apparent improvement to the original model of Nobili & Roxburgh (1986) is not so at all as pointed out by Saha & Tremaine (1994). The point is that given an initial osculating a_0 the mean $\bar{a}$ generated by the evolution under Eq. (1.1) is different from the mean $\bar{a}_R$ generated by the evolution under Eq. (3.2). Then, model (3.2) will not reproduce the secular evolution of an object with the correct $\bar{a}$ but with a different mean semimajor axis given by $\bar{a}_R$. In consequence, the model introduces a secular drift in M and no evident progress is done in comparison with Eq. (3.1). This problem can be solved taking appropriate initial conditions.

3.2. *Our r-independent Model*

The corrections exposed above, though computationally better than Eq. (1.1) because the **v**-dependence is eliminated, still have the problem that near perihelion the perturbation can be high enough to introduce numerical errors in constant time-step integrators. In order to overcome this difficulty, we propose the following constant radial perturbation:

$$R = \frac{3\mu^2}{c^2 a^3 \sqrt{(1-e^2)^3}} \tag{3.3}$$

which generates the expected variation in the argument of perihelion given by Eq. (2.1). However, for mean anomaly we obtain a different expression:

$$<\dot{M}> = -\frac{9}{c^2}\sqrt{\frac{\mu^3}{a^5(1-e^2)^3}} \tag{3.4}$$

The discrepancy for the drift on M with respect to Eq. (2.2) is irrelevant in numerical simulations because, as we have explained, the predictions for M given by models cannot coincide with the true relativistic effect if the same set of initial conditions for all models are taken. The advantage of model (3.3) is that it maintains the precision of the numerical integration even for very small perihelion orbits while the original Eq. (1.1) and the other models need a strong reduction in the step size. For low eccentricity orbits, both Eq. (2.2) and Eq. (3.4) give the same result.

3.3. *On Initial Conditions*

Initial conditions should be used according to the theory they were determined. For example, the model of Saha & Tremaine (1992) can be used confidently for generating precise ephemeris only if the initial conditions were obtained adjusting observations to this model. That is the underlying idea of the very accurate method for computing ephemeris of low eccentricity orbits proposed by Vitagliano (1997). But, it is not possible to follow the exact evolution of the mean anomaly if there is no consistence between the initial conditions (which should also be very precise) and the model used. Then, any of the models we have analyzed are valid to follow the secular evolution of an orbit and, in particular, our constant radial perturbation model is computationally more convenient.

For illustrative purposes, in order to compare the variations of the orbital elements produced by the different models given by Eq. (1.1), Eq. (3.2) and Eq. (3.3), we integrated numerically our planetary system for one million years. Results corroborate what we have already stated by means of the Gauss equations: the secular relativistic evolution of all orbital elements is very well reproduced. And even with strong differences in M, of the order of 10^2 degrees, the orbital evolution of the different models is almost undistinguishable. For planet Mercury for example, at the end of the integration, differences of the order of 10^{-5} in e, 10^{-3} degrees on ω, 10^{-4} degrees on Ω, and even lower differences for i were obtained.

4. Conclusions

At least three models are able to reproduce the secular drift in ω but for high eccentricity orbits only the model by Saha & Tremaine (1992) can follow the exact secular drift in M and only if appropriate initial conditions are taken. For secular evolution studies of numerous populations by means of numerical integrations, if no precise ephemeris are required, our constant radial r-independent model is very convenient specially if low perihelion orbits are involved. For accurate ephemeris computations of real bodies, the original formula (1.1), or even the full relativistic N-body one should be used with a control of the integrator's precision in the case of low perihelion orbits.

Acknowledgements

We acknowledge support from PEDECIBA, CSIC and IAU. This work was done as part of the Project "Caracterización de las poblaciones de cuerpos menores del Sistema Solar" (ANII).

References

Anderson, J. D., Esposito, P. B., Martin, W., & Muhlemsn, D. O. 1975, *ApJ*, 200, 221
Benitez, F. & Gallardo, T., 2008, *Celest. Mech. and Dyn. Ast.*, 101, 289
Brumberg, V., 1991, *Essential Relativistic Celestial Mechanics.* (London: Adam Hilger)
Calura, M., Fortini, P., & Montanari, E. 1997, *Phys. Rev. D*, 56, 4782
Iorio, L. 2005a, *A&A*, 431, 385
Iorio, L. 2005b, *A&A*, 433, 385
Iorio, L. 2007, *Ap&SS*, 312, 331
Nobili, A. & Roxburgh, I. 1986, in: J. Kovalevsky & V. A. Brumberg (eds.), *Relativity in Celestial Mechanics and Astronomy*, p. 105
Quinn, T., Tremaine, S., & Duncan, M. 1991, *AJ*, 101, 2287
Rubincam, D. P. 1977, *Celest. Mech.*, 15, 21
Saha, P. & Tremaine, S. 1992, *AJ*, 104, 1633
Saha, P. & Tremaine, S. 1994, *AJ*, 108, 1962
Shahid-Saless, B. & Yeomans, D. 1994, *AJ*, 107, 1885
Soffel, M. 1989, *Relativity in Astrometry, Celestial Mechanics and Geodesy*, (Berlin: Springer)
Soffel, M., Klioner, S. A., Petit, G., *et al.* 2003, *AJ*, 126, 6, 2687
Vitagliano, A. 1997, *Celest. Mech. and Dyn. Ast.*, 66, 293

Part IV

Icy Satellites of the Outer Planets

Part IV

Icy Satellites of the Outer Planets

Icy Bodies of the Solar System
Proceedings IAU Symposium No. 263, 2009
J.A. Fernández, D. Lazzaro, D. Prialnik & R. Schulz, eds.

© International Astronomical Union 2010
doi:10.1017/S1743921310001602

Interior Models of Icy Satellites and Prospects of Investigation

Frank Sohl

Institute of Planetary Research, German Aerospace Center
Rutherfordstr.2, 12489, Berlin-Adlershof, Germany
email: Frank.Sohl@dlr.de

Abstract. The state of knowledge about the structure and composition of icy satellite interiors has been significantly extended by combining direct measurements from spacecraft, laboratory experiments, and theoretical modeling. Interior models of icy bodies will certainly benefit from future missions to the outer solar system, providing new and improved constraints on the surface chemistry, bulk composition and degree of internal differentiation, possible heterogeneities in radial mass distribution, the presence and extent of liquid reservoirs, and the amount of tidal heating for each target body. Here we summarize geophysical constraints on the interior structure and composition of selected Jovian and Saturnian icy satellites and investigate conditions under which potentially habitable liquid water reservoirs could be maintained. Future geophysical exploration which includes gravitational and magnetic field sounding from low-altitude orbit and close flyby, combined with altimetry data and in-situ monitoring of tidally-induced surface distortion and time-variable magnetic fields, would impose important constraints on the interiors of outer planet satellites.

Keywords. astrobiology, conduction, convection, equation of state, icy satellites, interiors, Io, Europa, Ganymede, Callisto, Enceladus, Rhea, Titan, Triton

1. Introduction

The internal structure and bulk composition of solar system bodies are key to understanding the origin and early evolution of the solar system. In general, solar system bodies are composed of rock, metal, ices, and gases. The terrestrial, i.e. Earth-like, bodies in the inner solar system are primarily composed of rock and metal since early condensation of silicate minerals and metals was initiated at relatively high temperature in the solar nebula. In turn, volatile-rich components in form of ices and gases are more abundant in the cold outer solar system because of their lower condensation temperatures. Interior structure models aim at inferring the bulk composition, masses of major chemical reservoirs, the depth to chemical discontinuities and mineral phase boundaries, variation with depth of temperature, pressure, density, and composition. In the absence of seismological data, the construction of depth-dependent models of planetary interiors must rely on high-pressure and -temperature laboratory experiments to deduce equations of state for the density, transport properties like viscosity and thermal conductivity, phase stability regions, and melting relations. Since there are usually fewer constraints than unknowns, even basic interior structure models that would involve only a few chemically homogeneous layers of constant density suffer from inherent non-uniqueness (e.g., Sohl & Schubert 2007).

2. Geophysical constraints

The state of knowledge about the structure and composition of icy satellite interiors has improved considerably in the late 70s and mid 80s, when the *Voyager* spacecraft flew by Jupiter, Saturn, Uranus, and Neptune, followed by the advent of the *Galileo*

F. Sohl

spacecraft in the Jovian system (1995-2003) and the *Cassini* spacecraft in the Saturnian system (since 2004). *Voyager* observations of the Jovian moons Io and Europa and, to a lesser extent, Ganymede suggested tidal heating to be a major heat source, possibly resulting in near-surface viscous deformation of warmed ice and cryovolcanic resurfacing (e.g., Johnson 2005).

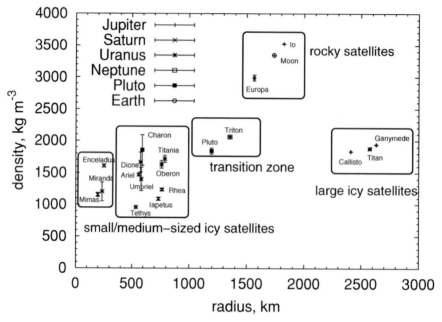

Figure 1. Radius-density relation for satellites and dwarf planets.

Models of the internal density distribution are then required to satisfy the mean density, derived from the total radius and mass, and the mean moment-of-inertia (MoI) factor, inferred from the mean radius and quadrupole moments of the gravitational field. Whereas the mean moment of inertia is a measure for the degree of internal differentiation or concentration of mass toward the center, the bulk chemical composition of a planet or satellite can be inferred from its mean *uncompressed* density. Additionally, self-sustained and/or induced magnetic fields, surface geology and composition, and the volatile inventory of a planet or satellite provide indirect information about the constitution of planetary and satellite interiors. Tectonic manifestations of endogenic activity preserved in the long-term surface record are particularly useful to infer the mode of internal heat transport (e.g., Schubert *et al.* 1986, Hussmann *et al.* 2007).

Most of the natural satellites are in synchronous rotation and subject to static and dynamic tidal forces exerted by their primaries. The non-spherical part of their gravity fields is predominated by rotational and/or tidal contributions to the quadrupolar and tesseral moments J_2 and $C_{2,2}$, measuring the polar oblateness and equatorial ellipticity of the gravity field, respectively. Superimposed are time-variable contributions which are induced by radial and librational tides along slightly eccentric orbits of a number of satellites. From the analysis of Doppler range and range rate observations acquired at around closest approach, the axial moments of inertia of only a few satellites have been inferred from gravitational perturbations on spacecraft trajectories by using the Radau-Darwin relation (e.g., Hussmann *et al.* 2007). However, the moment-of-inertia values derived in this way are almost entirely based on the assumption that the satellites are in hydrostatic equilibrium and the ratio $J_2/C_{2,2}$ taken constant at 10/3. Among those are the Galilean satellites Io, Europa, Ganymede, Callisto and the Saturnian moons Titan and Rhea. The shapes of most mid size Saturnian moons are consistent with hydrostatic equilibrium (Thomas *et al.* 2007), with the notable exception of Iapetus' shape that corresponds to a former rotational period of several hours (Castillo-Rogez *et al.* 2007). For Io, the volcanically most active body in the solar system, hydrostatic equilibrium was confirmed from different flyby geometries (near-polar and near-equatorial) of the *Galileo* spacecraft (Anderson *et al.* 1996a, Anderson *et al.* 2001a). Deviations from hydrostaticity are commonly attributed to uncompensated topography and/or internal dynamics.

3. Subsurface water oceans

Almost four decades have passed since extant liquid water oceans on icy moons were postulated (Lewis 1971, Consolmagno & Lewis 1978). The possible formation of liquid water layers below the outer ice shell of icy satellites is strongly dependent on their accretion history, initial thermal state, and degree of internal differentiation. The maintenance of liquid water layers at a depth of several tens of kilometres is closely related to the internal structure, chemical composition, and thermal state of icy satellite interiors subsequent to internal differentiation. Icy satellites are believed to operate in the stagnant-lid regime at present, i.e, the outer ice shell can be subdivided into an elastic conductive stagnant lid underlain by a viscoelastic convective sublayer in contact with the liquid water layer below. Controlling parameters for sub-surface water ocean formation are the competition between radiogenic heating of the silicate component, additional contributions due to, e.g., the dissipation of tidal energy, and the effectiveness of the heat transfer to the surface (Spohn & Schubert 2003). Furthermore, the melting temperature of ice I will be reduced by pressure increase with depth and the minimum melting temperature will be attained at the ice I/ice III transition. Moreover, impurities like salts and/or volatiles such as ammonia and methanol will lead to a significant melting point depression of the icy component, thereby causing even thicker and colder subsurface water oceans.

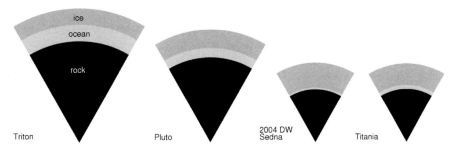

Figure 2. Interiors of ocean-bearing icy satellites and dwarf planets to scales.

In general, large icy bodies such as, e.g., the icy Galilean satellites, Titan and Triton, are more likely to harbour subsurface oceans because of slower cooling rates and more intense radiogenic heating caused by their larger rock mass fractions, as compared to smaller icy bodies. However, depending on the amount of antifreezes incorporated in the icy component during accretion, internal oceans cannot be ruled out for the largest of the medium-sized satellites of Saturn and Uranus, and the biggest trans-Neptunian objects (McKinnon *et al.* 2008), as illustrated in Fig. 2, provided those are differentiated into a rock core and a water ice/liquid shell (Hussmann *et al.* 2006). Water-rock interactions would affect oceanic composition and the mineralogy of rocks and oceanic sediments on ocean-bearing icy satellites like Europa and Triton where liquid reservoirs are likely in contact with silicate rock below. The most convincing argument for extant subsurface water oceans on icy satellites, however, results from the detection of induced magnetic fields in the *Galileo* magnetometer data collected near the icy Jovian satellites (Kivelson *et al.* 2004).

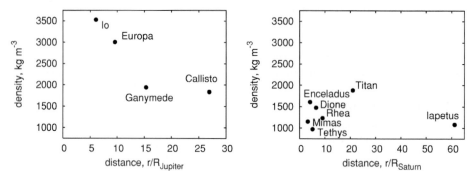

Figure 3. Mean density $\bar{\rho}$ of the (left) Galilean and (right) largest Saturnian satellites vs. distance from the primary r in units of Jupiter and Saturn radii R_{Jupiter} and R_{Saturn}, respectively.

4. Jovian satellites

The prominent density gradient with increasing distance from Jupiter (see left-hand side of Fig. 3) provides important constraints on the formation of the Jovian satellites (e.g., Coradini *et al.* 1995). Whereas the mean densities of Io and Europa indicate that their interiors are mainly composed of rock and metal, and, in the case of Europa, up to 10 wt.% water ice/liquid, Ganymede and Callisto contain water ice and rock-metal components in nearly equal amounts by mass. Further indirect constraints are provided by the surface geology, spectral properties, and chemistry of each individual satellite. Active volcanism, anomalous intrinsic luminosity, and high-temperature lava flow deposits at Io's surface emphasize the importance of tidal heating in the Jupiter system, essentially maintained by gravitational interaction due to the resonant orbits of Io, Europa, and Ganymede (Laplace resonance) (e.g., Peale 1999). Additionally, magnetic induction signals were observed at Europa, Callisto, and possibly Ganymede. The observed magnetic signatures suggest the presence of *globe-encircling*, electrically conducting reservoirs of liquid water below the surface (Kivelson *et al.* 2004). Those are interpreted in terms of briny subsurface water oceans that may contain even more water than all terrestrial oceans combined.

Table 1. Physical parameters of the Galilean satellites.

satellite	R [km]	GM [km^{3}s^{-2}]	$\bar{\rho}$ [kg m^{-3}]	a [km]	b [km]	c [km]
Io	1821.46	5959.91 ± 0.02	3529	1829.4	1819.3	1815.7
Europa	1562.09	3202.72 ± 0.02	3006	1564.13	1561.23	1560.93
Ganymede	2632.345	9887.83 ± 0.03	1940	2632.4	2632.29	2632.35
Callisto	2409.3	7179.29 ± 0.01	1837	2409.4	2409.2	2409.3

Note: The mean densities are calculated from $\bar{\rho} = 3M/(4\pi R^3)$. Values of mean radius R, mass GM (G is the gravitational constant) were taken from Seidelmann *et al.* 2007 and Schubert *et al.* 2004. a, b, c denote the Jupiter-facing, orbit-facing, and polar axis, respectively.

Ganymede, the Mercury-sized, largest satellite in the solar system, possesses a self-sustained magnetic dipole field with equatorial and polar field strengths at the surface of 750 and 1200 nT, respectively (Kivelson *et al.* 1996). Since the most likely source is dynamo action in a liquid Fe-FeS core, Ganymede's interior is believed to consist of an

Table 2. Gravity field parameters of the Galilean satellites.

satellite	R_{ref} [km]	$J_2 \times 10^{-6}$	$C_{2,2} \times 10^{-6}$	k_{f}	$C/(MR^2)$
Io	1821.6 ± 0.5	1859.5 ± 2.7	558.8 ± 0.8	1.3043 ± 0.0019	0.37824 ± 0.00022
Europa	1565.0 ± 8.0	435.5 ± 8.2	131.5 ± 2.5	1.048 ± 0.020	0.346 ± 0.005
Ganymede	2631.2 ± 1.7	127.53 ± 2.9	38.26 ± 0.87	0.804 ± 0.018	0.3115 ± 0.0028
Callisto	2410.3 ± 1.5	32.7 ± 0.8	10.2 ± 0.3	1.103 ± 0.035	0.3549 ± 0.0042

Note: R_{ref} is the reference radius associated with J_2 and $C_{2,2}$; the corresponding GM-values are given in Table 1. J_2 $(= -C_{2,0})$ and $C_{2,2}$ are degree-two gravity field coefficients, $C/M/R^2$ is the axial MoI factor, and k_{f} is the static fluid potential Love numbers calculated from $k_{\mathrm{f}} = 4C_{2,2}/q_r$, where $q_r = \omega^2 R_{\mathrm{ref}}^3/(GM)$ is the smallness parameter for the equilibrium figure of a synchronously rotating satellite with ω the mean angular frequency of rotation. Values were taken from Schubert *et al.* 2004.

iron-rich core surrounded by a silicate rock mantle overlain by an ice shell that may contain a subsurface water ocean sandwiched between a high-pressure water ice layer and an outermost ice I layer (Schubert *et al.* 2004). In fact, Ganymedes MoI factor of 0.3115 ± 0.0028 is the smallest measured value for any solid body in the solar system and indicates a strong concentration of mass towards the center (Anderson *et al.* 1996b, Sohl *et al.* 2002).

Callistos radius is about 200 km smaller than that of Ganymede and its mass is 70% that of Ganymede. The satellite's old, heavily cratered surface suggests that endogenic resurfacing has never happened since accretion was completed. Provided hydrostatic equilibrium is attained, the *Galileo* gravity data suggest that the satellite's axial MoI factor is equal to 0.3549 ± 0.0042. However, this value is not compatible with a fully differentiated interior and suggests partial or weak internal differentiation (Anderson *et al.* 2001b), augmented by a density increase with depth due to pressure-induced water ice phase transitions (McKinnon 1997). Furthermore, the magnetic data suggest that an ocean is present at around 150 km depth (Khurana *et al.* 1998, Zimmer *et al.* 2000). These two interpretations of geophysical data seem contradictory since the presence of an ocean would lead to internal differentiation. In order to reconcile these two observations, it was proposed that Callisto may have undergone incomplete gradual unmixing of ice and rock, proceeding from underneath the cold and immobile lithosphere, and with rock concentration increasing with depth up to the close-packing limit (Nagel *et al.* 2004).

5. Saturnian satellites

In contrast to the Jovian satellites, the lack of a prominent density gradient with increasing distance from Saturn (see right-hand side of Fig. 3) suggests an average bulk composition for the Saturnian satellites. Saturns largest moon Titan is intermediate between the Jovian satellites Ganymede and Callisto with respect to its radius and mean density. Whereas Tethys' low mean density suggests the presence of porous ice, the densities of Titan and Enceladus, Saturn's exceptionally active inner moon, indicate that both interiors are composed of ice and rock-metal in nearly equal amounts by mass. The range of interior structure models satisfying the degree-two gravity field of Rhea is attributed to the possible existence of a high-pressure ice layer and the extent of unmixing of ice and rock (Castillo-Rogez 2006). The axial MoI factor inferred from Doppler data acquired during a close *Cassini* flyby suggests that Rhea is only weakly differentiated (Iess *et al.* 2007), or even an almost homogeneous mixture of rock-metal and water ice compounds (Anderson & Schubert 2007).

From remote-sensing *Cassini* observations, it is obvious that Titan and Enceladus have been subject to intense endogenic activity in the course of their evolutions. Whether or

Table 3. Physical parameters of the largest Saturnian satellites.

satellite	R [km]	GM [km^{3}s^{-2}]	$\bar{\rho}$ [kg m^{-3}]	a [km]	b [km]	c [km]
Mimas	198.2 ± 0.5	2.5023 ± 0.0020	1150 ± 9	207.4 ± 0.7	196.8 ± 0.6	190.6 ± 0.3
Enceladus	252.1 ± 0.2	7.2096 ± 0.0067	1608 ± 5	256.6 ± 0.6	251.4 ± 0.2	248.3 ± 0.2
Tethys	533.0 ± 1.4	41.2097 ± 0.0063	973 ± 8	540.4 ± 0.8	531.1 ± 2.6	527.5 ± 2.0
Dione	561.7 ± 0.9	73.1127 ± 0.0025	1476 ± 7	563.8 ± 0.9	561.0 ± 1.3	561.7 ± 0.9
Rhea	764.3 ± 2.2	153.9416 ± 0.0049	1233 ± 11	767.2 ± 2.2	762.5 ± 0.8	763.1 ± 1.1
Titan	2575.5 ± 2	8978.1356 ± 0.0039	1880 ± 4			
Iapetus	735.6 ± 3.0	120.5117 ± 0.0173	1083 ± 13	747.4 ± 3.1		712.4 ± 2.0

Note: The mean densities are calculated from $\bar{\rho} = 3M/(4\pi R^3)$. Values of mean radius R, mass GM (G is the gravitational constant) were taken from Thomas *et al.* 2007 and Jacobson *et al.* 2006. a, b, c denote the Saturn-facing, orbit-facing, and polar axis, respectively.

not Titan's deep interior is further differentiated like Ganymede's into an iron core and a rock mantle above is more speculative, as there is no observational clue on the possible existence of a self-sustained magnetic field (Backes *et al.* 2005). Based on gravity data collected during several *Cassini* spacecraft encounters, it cannot be safely excluded that Titan's interior is only partly differentiated, more similar to that of Callisto. Titans surface shows a number of cryovolcanic units and tectonic features that can be related to endogenic activity, as revealed by imaging during the descent of the Huygens probe (Tomasko *et al.* 2005) and Cassini remote sensing Porco *et al.* 2005, Lopes *et al.* 2007. The detection of ^{40}Ar by the Huygens probe (Niemann *et al.* 2005) suggests methane replenishment of Titan's atmosphere by degassing of the interior (Atreya *et al.* 2006). The latter may involve episodes of methane clathrate dissociation and cryovolcanic activity coupled to the satellite's thermal-orbital evolution (Tobie *et al.* 2006). Titan's orbital eccentricity is remarkably high, suggesting a relatively recent origin and/or moderate tidal heating over time (Sohl *et al.* 1995, Tobie *et al.* 2005). Finally, Titan is likely to harbour a cold, extended internal liquid reservoir, similar to those first proposed for the large icy satellites of Jupiter, but more enriched in ammonia (Grasset *et al.* 2000, Sohl *et al.* 2003, Tobie *et al.* 2005, Grindrod *et al.* 2008).

The detection of plumes of water-vapour and ice grains (Dougherty *et al.* 2006) populating Saturn's E-ring and lineated thermal anomalies (Spencer *et al.* 2006) shows that the heavily tectonized south polar region of Enceladus is cryovolcanically active (Porco *et al.* 2006). In particular, the presence of non-condensable volatile species in the jet-like plumes, like molecular nitrogen, carbon dioxide, and methane, suggests an aqueous internal environment at elevated temperatures, facilitating aqueous, catalytic chemical reactions (Matson *et al.* 2007). Possible venting mechanisms are sudden decompression of near-surface reservoirs of liquid water (Porco *et al.* 2006), chlathrate decomposition (Kieffer *et al.* 2006), and cryovolcanic processes. The recent detection of a salt-rich and basic-pH population of E-ring ice grains from in-situ compositional analysis suggests that the plumes originate from a liquid reservoir in contact with silicate rock below (Postberg *et al.* 2009). Taken together, this strongly suggests that Enceladuss interior is differentiated into a rock-metal core overlain by a water-ice liquid shell (Schubert *et al.* 2007). However, the concentration of geologic and thermal activity toward the south-polar region and the energy source required to initiate and maintain the activity are not well understood. It is possible that those are associated with a low-degree mode of internal convection (Grott *et al.* 2007) and diapir-induced reorientation (Nimmo & Pappalardo 2006) early in Enceladus' history. Tidal heating above a liquid water reservoir confined to beneath the south-polar region (Tobie *et al.* 2008) would help explain Enceladus' south pole hot spot and associated circular topographic depression (Collins & Goodman 2007).

6. Summary and outlook

The existence of potentially habitable liquid water reservoirs on icy satellites is dependent on the radiogenic heating of the rock component, additional contributions such as the dissipation of tidal energy, the efficiency of heat transfer to the surface, and the presence of substances that depress the freezing point of liquid water. Gravitational and magnetic field sounding from low-altitude orbit and close flyby, combined with altimetry data and in-situ monitoring of tidally-induced surface distortion and time-variable magnetic field, would impose important constraints on the interiors of outer planet satellites. In particular, the hydrostatic assumption – central to the construction of interior structure models – needs to be carefully evaluated by separate determination of the static components of the low-degree gravity field coefficients from independent orbits (polar) and flybys (inclined and equatorial). These coefficients are required to be determined at a sufficiently high accuracy to distinguish between tidally-induced contributions and high-order static gravity anomalies. Future recovery of static *and* time-variable parts of satellite gravity fields would provide entirely new information on the gravitational signature of intrinsic density anomalies and regional topographic features as well as on the existence and radial extent of liquid subsurface water reservoirs on icy satellites. Global shapes and rotational states and orientations in space hint at the thickness and rigidity of the overlying ice crust and would be obtained from combinations of global altimetry, imaging, and limb profiling. In particular, the correction of non-hydrostatic effects requires combined collection of gravity and altimetry data. Magnetometer measurements conducted from orbiting spacecraft and surface probes would help distinguish between intrinsic and induced contributions to the observed magnetic field, thereby providing complementary information on the depth to liquid water reservoirs and their electrical conductivities. This taken together would improve our general understanding of the origin and early evolution of outer planet satellites.

Future exploration of the outer solar system should benefit from truly international cooperation. Current spacecraft mission proposals, jointly put forward by ESA and NASA, would facilitate synergistic observations shared between several platforms and mainly targeted at discovering potentially habitable, liquid reservoirs in the outer solar system. These missions would involve orbiting spacecraft around Europa and Ganymede (Blanc *et al.* 2009), to be launched around 2020, and possibly followed by a Titan-orbiting mission, including a long-lived aerial platform and a short-lived lake lander (Coustenis *et al.* 2009). Albeit more challenging in terms of mission duration and distance, outstanding scientific gain at long sight must be expected from focused missions to the Uranian system, Neptune and Triton, and beyond.

Acknowledgements

H. Hussmann is thanked for valuable comments and corrections. The author is very grateful to the organizers of IAU Symposium 263 for kindly inviting the presentation of this paper. This research is supported by the Helmholtz Alliance "Planetary Evolution and Life".

References

Anderson, J. D., *et al.* 1996a, *Science*, 272, 709
Anderson, J. D., *et al.* 1996b, *Nature*, 384, 541
Anderson, J. D., *et al.* 2001a, *J. Geophys. Res.*, 106, 32,963
Anderson, J. D., *et al.* 2001b, *Icarus*, 153, 15
Anderson, J. D. & Schubert, G. 2007, *Geophys. Res. Lett.*, 34, L02202
Atreya, S. K., *et al.* 2006, *Planet. Space Sci.*, 54, 1174
Backes, H., *et al.* 2005, *Science*, 308, 992

Blanc, M., *et al.* 2009, *Exp. Astron.*, 23, 849

Castillo-Rogez, J. 2006, *J. Geophys. Res.*, 111, E11005

Castillo-Rogez, J., *et al.* 2007, *Icarus*, 190, 179

Collins, G. C. & Goodman, J. C. 2007, *Icarus*, 189, 72

Consolmagno, G. J. & Lewis, J. S. 1978, *Icarus*, 34, 280

Coradini, A., *et al.* 1995, *Surv. Geophys.*, 16, 533

Coustenis, A., *et al.* 2009, *Exp. Astron.*, 23, 893

Dougherty, M., *et al.* 2006, *Science*, 311, 1406

Grasset, O., *et al.* 2000, *Planet. Space Sci.*, 48, 617

Grindrod, P. M., *et al.* 2008, *Icarus*, 197, 137

Grott, M., *et al.* 2007, *Icarus*, 191, 203

Hussmann, H., Sohl, F., & Spohn, T. 2006, *Icarus*, 185, 258

Hussmann, H., Sotin, C., & Lunine, J. I. 2007, in T. Spohn (ed.), *Treatise on Geophysics 10* (Amsterdam: Elsevier), p. 509

Iess, L., *et al.* 2007, *Icarus*, 190, 585

Jacobson, R. A., *et al.* 2006, *Astron. J.*, 132, 2520

Johnson, T. V. 2005, *Space Sci. Rev.*, 116, 401

Kivelson, M. G., *et al.* 1996, *Nature*, 384, 537

Khurana, K., *et al.* 1996, *Nature*, 395, 777

Kieffer, S. W., *et al.* 2006, *Science*, 314, 1764

Kivelson, M. G., *et al.* 2004, in F. Bagenal, T. Dowling, & W. McKinnon (eds.), *Jupiter: The Planet, Satellites and Magnetosphere* (Cambridge: Cambridge University Press), p. 513

Lewis, J. S. 1971, *Icarus*, 15, 174

Lopes, R., *et al.* 2007, *Icarus*, 186, 395

Matson, D. L., *et al.* 2007, *Icarus*, 187, 569

McKinnon, W. B. 1997, *Icarus*, 130, 540

McKinnon, W. B., *et al.* 2008, in M. A. Barucci, H. Boehnhardt, D. P. Cruikshank, & A. Morbidelli (eds.), *The Solar System Beyond Neptune* (Tucson: University of Arizona Press), p. 213

Nagel, K., *et al.* 2004, *Icarus*, 169, 402

Niemann, H. B., *et al.* 2005, *Nature*, 438, 779

Nimmo, F. & Pappalardo, R. T. 2006, *Nature*, 441, 614

Peale, S. J. 1999, *Ann. Rev. Astron. Astrophys.*, 37, 533

Porco, C. C., *et al.* 2005, *Nature*, 434, 159

Porco, C. C., *et al.* 2006, *Science*, 311, 1393

Postberg, F., *et al.* 2009, *Nature*, 459, 1098

Schubert, G., *et al.* 1986, in J. A. Burns & M. S. Matthews (eds.) *Satellites* (Tucson: University Arizona Press), p. 224

Schubert, G., *et al.* 2004, in F. Bagenal, T. Dowling, & W. McKinnon (eds.) *Jupiter: The Planet, Satellites and Magnetosphere* (Cambridge: Cambridge University Press), p. 281

Schubert, G., *et al.* 2007, *Icarus*, 188, 345

Seidelmann, P. K., *et al.* 2007, *Celest. Mech. Dynam. Astron.*, 98, 155

Sohl, F. & Schubert, G. 2007, in T. Spohn (ed.), *Treatise on Geophysics 10* (Amsterdam: Elsevier), p. 27

Sohl, F., *et al.* 1995, *Icarus*, 115, 278

Sohl, F., *et al.* 2002, *Icarus*, 157, 104

Sohl, F., *et al.* 2003, *J. Geophys. Res.*, 108, E12, 5130

Spencer, J., *et al.* 2006, *Science*, 311, 1401

Spohn, T. & Schubert, G. 2003, *Icarus*, 161, 456

Thomas, P. C., *et al.* 2007, *Icarus*, 190, 573

Tobie, G., *et al.* 2005, *Icarus*, 175, 496

Tobie, G., *et al.* 2006, *Nature*, 440, 61

Tobie, G., *et al.* 2008, *Icarus*, 196, 642

Tomasko, M. G., *et al.* 2005, *Nature*, 438, 765

Zimmer, C., *et al.* 2000, *Icarus*, 147, 329

Icy Bodies of the Solar System
Proceedings IAU Symposium No. 263, 2009
J.A. Fernández, D. Lazzaro, D. Prialnik & R. Schulz, eds.

© International Astronomical Union 2010
doi:10.1017/S1743921310001614

Long-term evolution of small icy bodies of the Solar System

Dina Prialnik

Dept. of Geophysics & Planetary Sciences, Tel Aviv University
Ramat Aviv, Tel Aviv 69978, Israel
email: dina@planet.tau.ac.il

Abstract. Detailed evolutionary calculations spanning 4.6×10^9 yr are presented for (a) a model representing main-belt comet 133P/Elst-Pizarro, considering different initial mixtures of ices and dust, and (b) a Kuiper Belt object heated by radioactive decay, growing in size from an initial radius of 10 km to a final 250 km.

It is shown that for the main-belt comet only crystalline H_2O ice may survive in the interior of the nucleus, and may be found at depths ranging from ∼50 to 150 m. Other volatiles will be completely lost. For the large Kuiper Belt object, evaporation and flow of water and vapor gradually remove the water from the core and the final (present) structure is differentiated, with a rocky, highly porous core of 80 km radius. Outside the core, due to refreezing of water vapor, a compact, ice-rich layer forms, a few tens of km thick. The amorphous ice is preserved in an outer layer about 20 km thick.

Keywords. comets: general, comets: individual (133p/Elst-Pizarro), Kuiper Belt

1. Introduction

A new class of comets has been identified in recent years – known as "main-belt comets" (MBCs) – the prototype being 133P/Elst-Pizarro (EP), which was first classified as a main-belt asteroid. Cometary activity for EP was first detected in 1996 as a dust tail, and at the time it was thought to be the result of an impact. The dust could be either a trail of debris (Toth 2000), or the manifestation of volatile-driven activity caused by the exposure of deep-lying ice (Boehnhardt *et al.* 1998). The latter interpretation gained support when the comet became active again one orbital period later, in 2002 (Hsieh *et al.* 2004, Ferrin 2006). The third episode of activity detected near perihelion another orbital period later (Jewitt *et al.* 2007b) left little doubt that mass ejection was triggered by volatile sublimation. Two more objects belonging to the MBC class soon joined EP: 176P/LINEAR and P/2005 U1 (Read), which showed vigorous activity (Hsieh *et al.* 2009a,b) and recently, a fourth object was detected, P/2008 R1 (Jewitt *et al.* 2009).

The orbits of these objects are stable; hypotheses regarding their origin are divided between formation in place (Haghighipour 2008) and capture from distant regions of the Solar System into the outer asteroid belt during the late heavy-bombardment era (Levison *et al.* 2008, based on the *Nice model*). Either way, MBCs must have spent between 3.9 and 4.6 Byr in the main belt. A plausible scenario that would explain both the prolonged nature of the dust emission and its recurrence at successive perihelia is sublimation of ices exposed by a recent impact, which implies that MBCs have ices buried below the surface. The question whether or not may ices survive 4.6 Byr of evolution in an orbit that is relatively close to the Sun requires detailed evolutionary calculations. Simplified analytical estimates (Schorghofer 2008) indicate that this may be possible, but such estimates do not consider gas flow through the porous nucleus, nor sublimation in the deep porous interior.

A different question that demands long-term evolutionary calculations is whether and to what extent are comets, or rather, their distant progenitors – Kuiper Belt objects (KBOs) – pristine bodies that hold clues to the formation of the solar system. Simplified models show that radiogenic heating by short-lived radionuclides, such as ^{26}Al, provides sufficient energy for melting the ice (e.g., Merk and Prialnik 2006, Podolak and Prialnik 2006, Jewitt et al. 2007a and McKinnon et al. 2007).

In this paper long-term evolutionary calculations for a MBC (Prialnik & Rosenberg 2009) and a KBO (Prialnik & Merk 2008) – meant to shed light on these questions – are presented and compared.

2. Long-term evolution – 4.6 Byr – of a Main-Belt comet

The code used for numerical modeling is a 1-D code described in detail by Prialnik (1992) and Sarid et al. (2005); discussion of the input physics may be found in Prialnik et al. (2004). The orbital period of EP is ~ 5.5 yr and in the long-term calculations aim to cover 4.6 Byr of evolution. Clearly, orbital changes cannot be resolved, as this would require a prohibitive number of time steps. To circumvent this difficulty, but still keep track of the correct overall amount of energy absorbed by the comet, the elliptic orbit is replaced by a circular one, corresponding to the same total orbital insolation. Different initial configurations are considered, having in common a radius of 2.5 km, an albedo of 0.04, and dust and ice in equal mass fractions.

For the first model (Model A), the initial composition is a mixture of pure amorphous water ice (no occluded gases) and dust, representing a captured body, formed in more distant and cold regions of the Solar System. The ice crystallizes completely during the first 6×10^5 yr of evolution, with the crystallization front – initiated by absorption of solar energy – advancing from the surface all the way to the center, feeding mainly on the released latent heat. If gases were trapped in the amorphous ice, they would have escaped to the surface and out of the nucleus. While the crystallization front is receding toward the center, the crystalline ice starts sublimating at the surface, with the sublimation front very slowly advancing inwards, leaving behind an ice-depleted dust mantle. Some of the vapor flows inwards and refreezes ahead of the crystallization front. Thus the ice fraction profile peaks at some depth below the dust mantle, as shown in Fig. 1. This peak is also fed by vapor flowing outward from the deep interior, where the temperature has risen to about 125 K, allowing for bulk sublimation. Hence the ice is slowly depleted, but far more slowly than the rate of free sublimation would imply, impeded by the build-up of internal pressure. As a result of internal sublimation, after 4.6×10^9 yr of evolution, the ice density throughout most of the interior is significantly lower than the initial density, as seen in Fig. 1. Overall, the total ice content of the body at present is slightly less than half of the initial. The depth at which ice may be found at present, assuming no collapse of the dust mantle, is about 90 m.

If the body formed in place in the asteroid belt, it is more likely that the ice was crystalline to start with. A second model is evolved (Model B), identical to the first, except that the ice is initially crystalline throughout. In this case the internal temperature attained is slightly lower, and less ice sublimates both in the interior and below the surface, so that ice may be found at present at a depth of ~ 45 m (see Fig. 1). The difference between the amorphous and crystalline ice bodies is due to the additional heat source in the former, provided by the exothermic crystallization.

In order to assess the effect of initial density (and thus ice content), a third model (Model C) is calculated, identical to the second, except that the bulk initial density is

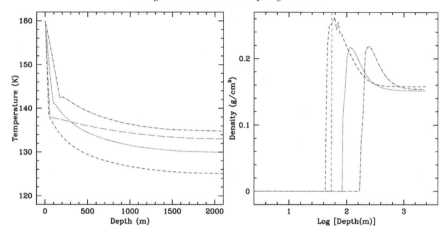

Figure 1. Comparison of the present structure of the 4 models: A (blue dot), B (red short dash), C (green long dash) and D (magenta dash-dot), for temperature profile (left) and H_2O density (right).

about 2.5 times larger and the porosity – correspondingly lower. This causes the internal vapor pressure to rise and hence the temperatures attained are higher. The relatively high pressure is maintained due to the lower porosity that hinders the flow of gas. In fact, the outflow of vapor is so weak, that only a very small fraction of the ice sublimates in the interior. The H_2O-depleted layer is about 50 m thick, similar to the previous case, but the internal density of ice remains almost unchanged.

The next model (Model D) includes a mixture of 4 volatile ices – CO_2, HCN, NH_3 and C_2H_2 – besides crystalline water ice and dust. Water ice amounts to only 0.45 of the mass, the remaining 0.05 being divided equally among the other volatiles. The significant result is that after about 10^8 yr, no volatile beside H_2O is left, not even a trace. As for the H_2O ice, the behavior is similar to the other cases, but sublimation is strongest in this case, with the ice boundary receding down to ~ 150 m. Over 3 times more H_2O ice is lost as compared to Model B (of the same density), in addition to the other volatiles. Energetically, this does not pose a problem since the energy required for sublimating the entire ice content of the body is only about 10^{-7} of the absorbed solar energy during the lifetime of the solar system.

3. Long-term evolution – 4.6 Byr – of a Kuiper-Belt object

The generic KBO considered has a radius of 250 km and the main energy source in this case is radioactive decay of ^{26}Al and ^{40}K. The evolution starts with a 100 Myr-long accretion episode, during which the body grows in size from an initial radius of 10 km to the final radius, amounting to a volume (mass) increase by a factor of 1.56×10^4. However, since the half lifetime of ^{26}Al (0.72 Myr) is far shorter than the accretion phase, most of the radiogenic heating by this short-lived nuclide occurs while the body has only increased its volume by only a factor of 10. The central temperature reaches its peak after less than 2 Myr of growth and then gradually declines.

The radioactive energy released by the decay of the longer lived radioisotope ^{40}K (half lifetime of 1.28 Byr), is supplied at too low a rate to compete with conductive cooling. Eventually, however, the heat diffusion timescale of the growing body exceeds the radioactive heating timescale, and the central temperature starts rising again, when about one third of the mass has been accreted and the radius has attained ~ 175 km. It

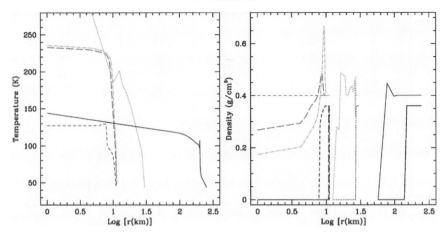

Figure 2. Profiles of temperature (left) and ice density (right) at various times during evolution: early evolution, at times: 1×10^5 yr (short-dashed line, red), 2×10^5 yr (long-dashed line, magenta), 3×10^5 yr (dash-dotted line, cyan), and 4×10^6 yr (dotted line, green); and present time, 4.6×10^9 yr (solid line, blue). Black (on the right) represents amorphous ice.

reaches another, lower, peak after the body has reached its final size. The decay of ^{40}K is then followed by a slow decline in central temperature to the present value that still exceeds the equilibrium temperature at the body's average heliocentric distance.

The evolution of the composition is shown in Figure 2 by profiles of the ice density as function of radial distance at different times. When the central temperature – initially 70 K throughout – exceeds ~ 120 K, the amorphous ice starts crystallizing and a crystallization front advances from the center outwards. As the temperature rises further, the crystalline ice in the core begins to sublimate and the vapor flows outwards. Upon reaching colder layers, most of it refreezes. Slowly, the core becomes depleted of ice, while further out the ice content increases, as exhibited by the early profiles of Fig. 2. Under the assumed conditions, this already happens before the melting temperature is attained. A considerable amount of water is produced only when the permeability to flow is significantly reduced. In all cases, however, the core becames eventually ice depleted.

Although by the end of the accretion phase, most of the ice is still amorphous, it slowly crystallizes during the first 1.5×10^9 years of evolution. The heat released by the exothermic transition delays cooling of the body. But at the end of 4.6 Byr, the KBO is cold, with the temperature ranging from slightly above 60 K at the center to ~ 40 K at the surface (the average equilibrium temperature). The layered structure has a very porous rocky core, ~ 80 km in radius (assuming that the material strength is high enough to prevent collapse), completely depleted of ice. Despite the fact that the core radius reaches out to almost 1/3 of the object's size, the core's volume constitutes only 3.3% of the total volume, and an even smaller fraction of its mass (the core porosity being higher than the average porosity).

Surrounding the core there is a relatively thin layer enriched in ice; its volume is roughly the same as the core's. The higher ice content is a result of refreezing of the water and vapor flowing out of the hot core into the cooler outer layers during the early stages of evolution. The bulk of the configuration has ice and dust in roughly the initial proportions, but the initially amorphous ice is now crystallized. Surrounding it is a thin layer, of amorphous and crystalline ice, with the ratio between them gradually increasing outwards. Finally, the surface ~ 20 km thick layer, comprising 22% of the volume, preserves the initial composition.

4. Conclusions

In conclusion, H_2O ice may be retained in the interior of a main belt body, despite its proximity to the sun and the long evolution time in its orbit. The ice is expected to be found at a depth of $\sim 50 - 150$ m, depending on initial structure and composition, but no other volatile ices would be able to survive. Therefore, any activity resulting from exposure of the ice-rich layer by an impact, will be driven by water sublimation and should be compatible with the H_2O sublimation pattern. This does not exclude, however, the presence of other volatile species, which may remain trapped in a fraction of the amorphous ice that does not crystallize, as indicated by laboratory experiments (Notesco & Bar-Nun 1997). These volatiles will be ejected together with the water molecules.

Regarding KBOs, pristine material may be retained in a relatively thick outer layer, out of which comets may be broken off by collisions (Davis & Farinella 1997). Beneath this layer, the ice is crystallized, while a central core is porous and completely ice-depleted. This layered structure is the typical outcome of long-term evolution under the initial conditions assumed, while changes in physical parameters merely affect to some extent the relative thicknesses of the various layers.

Finally, it is interesting to note that the two types of icy bodies considered here, although very different in nature, location in the Solar System, and major heat source, have in common an ice-depleted zone that amounts to no more than 10% of ther volume, and – at the boundary of this zone – an ice-enriched layer resulting from vapor migration and refreezing.

References

Boehnhardt, H., Sekanina, Z., Fiedler, A., Rauer, H., Schulz, R., & Tozzi, G. 1998, *Highlights of Astronomy*, 11, 233

Davis, D. R. & Farinella, P. 1997, *Icarus*, 125, 50

Ferrín, I. 2006, *Icarus*, 185, 523

Haghighipour, N. 2008, *LPI*, 1405, 8287

Hsieh, H. H., Jewitt, D. C., & Fernández, Y. R. 2004, *ApJ*, 127, 2997

Hsieh, H. H., Jewitt, D., & Fernández, Y. R. 2009, *ApJ Lett.*, 694, L111

Hsieh, H. H., Jewitt, D., & Ishiguro, M. 2009, *ApJ*, 137, 157

Jewitt, D., Chizmadia, L., Grimm, R., & Prialnik, D., 2007, in: B. Reipurth, D. Jewitt & K. Keil (eds.), *Protostars and Planets V* (Tucson: Univ. Arizona Press), p. 863

Jewitt, D., Lacerda, P., & Peixinho, N. 2007, *IAUC* 8847, 1

Jewitt, D., Yang, B., & Haghighipour, N. 2009, *ApJ*, 137, 4313

Levison, H. F., Bottke, W. F., Nesvorný, D., Morbidelli, A., & Gounelle, M. 2008, *LPI* 1405, 8156

McKinnon W. B., Prialnik, D., Stern, A. S., & Coradini, A. 2007, in: M. A. Barucci, H. Boehn-hardt, D. Cruikshank & A. Morbidelli (eds.) *The Solar System beyond Neptune* (Tucson: Univ. Arizona Press), p. 213

Notesco, G. & Bar-Nun, A. 1997, *Icarus*, 126, 336

Merk, R. & Prialnik, D. 2006, *Icarus*, 183, 283

Podolak, M. & Prialnik, D. 2006, in: P. J. Thomas, R. D. Hicks, C. F. Chyba & C. P. McKay (eds.), *Comets and the Origin and Evolution of Life*, (Berlin: Spriger), p. 303

Prialnik, D. 1992, *ApJ*, 388, 196

Prialnik, D., Benkhoff, J., & Podolak, M. 2004, in: M. C. Festou, H. U. Keller, H. A. Weaver & M. Festou (eds.), *Comets II*, (Tucson: Univ. Arizona Press), p. 359

Prialnik, D. & Merk, R. 2008, *Icarus*, 197, 211

Prialnik, D. & Rosenberg, E. D. 2009, *MNRAS*, 399, L79

Sarid, G., Prialnik, D., Meech, K. J., Pittichovà, J., & Farnham, T. L. 2005, *PASP*, 117, 796

Schorghofer, N. 2008, *ApJ*, 682, 697

Toth, I. 2000, *A&A*, 360, 375

Icy Bodies of the Solar System
Proceedings IAU Symposium No. 263, 2009
J.A. Fernàndez, D. Lazzaro, D. Prialnik & R. Schulz, eds.

© International Astronomical Union 2010
doi:10.1017/S1743921310001626

The surface composition of Enceladus: clues from the Ultraviolet

Amanda R. Hendrix and Candice J. Hansen

Jet Propulsion Laboratory, California Institute of Technology, Pasadena, USA

Abstract. The reflectance of Saturn's moon Enceladus has been measured at far ultraviolet (FUV) wavelengths (115–190 nm) by Cassini's UltraViolet Imaging Spectrograph (UVIS). At visible and near infrared (VNIR) wavelengths Enceladus' reflectance spectrum is very bright, consistent with a surface composed primarily of H_2O ice. At FUV wavelengths, however, Enceladus is surprisingly dark – darker than would be expected for pure water ice. We find that the low FUV reflectance of Enceladus can be explained by the presence of a small amount of NH_3 and a small amount of a tholin in addition to H_2O ice on the surface.

Keywords. planets and satellites: general; planets and satellites: Enceladus

1. Introduction

The canonical method for studying surface composition is the analysis of spectral reflectance measurements at visible-near infrared (VNIR) wavelengths. Using this method, the primary surface component of the Saturnian satellites has been found to be water ice (e.g., McCord *et al.* 1971; Fink *et al.* 1976; Morrison *et al.* 1976; Cruikshank 1980; Clark *et al.* 1984), possibly in mixtures with other constituents (e.g., Emery *et al.* 2005; Verbiscer *et al.* 2006; Clark *et al.* 2008). Most of the published spectral data of Enceladus focus on the near-IR: the Emery *et al.* (2005) and Cruikshank *et al.* (2005) spectra extend as short as 800 nm, as do the Verbiscer *et al.* (2006) Enceladus spectra. Grundy *et al.* (1999) present Enceladus spectra at wavelengths as short as 1200 nm. Published Enceladus spectra from the Cassini Visible-Infrared Mapping Spectrometer (VIMS) extend as short as 1000 nm (Brown *et al.* 2006). At visible wavelengths, the only published spectral information on Enceladus is photometric data from Voyager (Buratti & Veverka 1984; Buratti 1984) and Hubble Space Telescope (HST) (Verbiscer *et al.* 2005). Buratti (1984) pointed out that, at visible wavelengths, the brighter Saturnian satellites are less spectrally red, suggesting that the visibly darker satellites contain more contaminants that are reddish; in particular, Buratti *et al.* (1990) note that Enceladus' "high albedo and flat spectrum between 350 and 590 nm imply that its surface layer is depleted in opaque materials." The visible spectrum (350–600 nm) of Enceladus is flat to bluish, consistent with an almost pure water ice surface (Buratti 1984; Buratti & Veverka 1984). Toward shorter UV wavelengths, the spectrum of Enceladus has not been previously studied. Here we investigate the reflectance spectrum of Enceladus as measured by the Cassini Ultraviolet Imaging Spectrograph (UVIS) (Esposito *et al.* 2004).

The Enceladus observation discussed here was obtained on 27 May 2007, at an altitude of approximately 620,000 km. The observation lasted 70 minutes, during which the solar phase angle decreased from 3.1^o to 1.0^o and increased back to 3.1^o. The average phase angle during the observation was 1.93^o. The sub-spacecraft latitude was 13^oS during the observation; the sub-spacecraft longitude transitioned from 26^oW to 33^oW. The size of the Enceladus disk during this observation was 0.8 mrad; the size of a UVIS pixel during this observation was 1.0×1.5 mrad, so Enceladus was sub-pixel in size and UVIS measured a disk-average spectrum.

We calculate the disk-average reflectance by taking into account the size of the body in the UVIS slit and dividing by the solar spectrum as measured by SORCE SOLSTICE (McClintock *et al.* 2000), scaled to 9.2 AU, Enceladus' heliocentric distance on the day of the observation. We use the Hapke (2002) model to represent the reflectance spectra of individual species given their optical constants, and we use spectral mixing models with varying amounts of different species to try to replicate the spectral magnitude and shape observed at Enceladus, following the methods outlined by Hendrix & Hansen (2008).

Laboratory measurements of water ice in the ultraviolet have had varying results. Several measurements of the optical constants n and k have been made: Warren (1984) provides a compilation of the measurements of hexagonal ice in the 44–2270 nm wavelength range and Warren & Brandt (2008) recently updated the compilation. Absorption by ice in the near- and mid-UV ($\sim$ 200–400 nm) is very weak (Warren & Brandt 2008) but the location of minimum absorption is unclear. There is a lack of optical constant information in the ultraviolet, and as a result, the Warren (1984) and Warren & Brandt (2008) compilations include an interpolation between 161 nm and 180 nm, the Warren (1984) compilation also interpolates between 185 nm and 400 nm (Fig. 3), and the Warren & Brandt (2008) compilation extrapolates between 180 and 200 nm. The Warren & Brandt (2008) compilation uses a different data set in the 390–600 nm range than the Warren (1984) compilation, and leaves a gap in the 200–390 nm range. The Warren & Brandt (2008) data focus mainly on temperatures applicable at Earth; thus, these data are not appropriate for Enceladus-like (e.g., 80 K) temperatures. Nevertheless, we use the Warren & Brandt (2008)) optical constants for water ice in our models, as no other data are currently available. (As pointed out by Warren (1984), Shibaguchi *et al.* (1977) measured only small variations in optical density with temperature (83 K–160 K) in the 120–150 nm region; any temperature dependence in the optical constants of water ice at $\lambda > 150$ nm is unknown.)

2. The far-UV spectrum of Enceladus does not agree with pure H$_2$O ice

In Fig. 1a is shown the far-UV (FUV) reflectance spectrum of Enceladus. The UVIS spectrum is dark at wavelengths < 165 nm; the onset in brightness due to water ice is detected at 165 nm; at wavelengths $\sim$170–$\sim$185 nm, the spectrum is relatively flat, producing a "ledge" in the overall spectrum; the spectrum appears to begin increasing again (possibly the shoulder of an absorption edge) at $\sim$185 nm. We combine the FUV spectrum with published VNIR data to create a composite spectrum (Fig. 1b). Clearly, Enceladus exhibits a strong, abrupt decrease in reflectance in the middle-UV (MUV) region, as suggested by the single MUV data point of Verbiscer *et al.* (2006), and because the FUV spectrum is much darker than the visible spectrum. Water ice exhibits a strong dropoff in reflectance in the UV - but the H$_2$O absorption edge occurs in the FUV (at $\sim$165 nm, depending on grain size as shown in Fig. 1a), and is not responsible for the decrease in brightness of Enceladus at $\lambda < 336$ nm, nor for the darkness of the Enceladus spectrum at $\lambda = 170$–185 nm. Thus, pure water ice models (Fig. 1) are *much too bright to fit the UVIS Enceladus spectrum*.

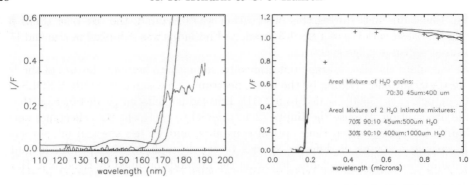

Figure 1. Enceladus spectra compared with water ice spectral models. (left) UVIS Enceladus disk-average reflectance spectrum at solar phase $= 2^o$. (right) Enceladus data from UVIS and Verbiscer *et al.* (2005) (plus signs), also at 2° solar phase angle. (red) areal mixture of 2 grain sizes; (blue) areal mixture of 2 intimate mixtures of 2 grain sizes. Both models are too bright to fit the UVIS data in the 160–190 nm range and do not exhibit the dropoff in brightness that evidently occurs in the near-UV (~ 300 nm).

3. Spectral mixture models

We have investigated candidate species that could be present at Enceladus and could meet the criteria of being bright and featureless in the visible, largely featureless in the near-IR, featureless in the 120–190 nm region, and having a UV absorption edge in the 190–280 nm, with perhaps a weaker absorption in the 280–400 nm region. We tried different mixtures of H_2O, CO_2, tholins and NH_3. Other candidate species are naturally possible, but published optical constant data in the far-UV are not available. The UV spectrum of carbon dioxide frost is not consistent with the Enceladus spectrum, so we do not consider CO_2 to be important in our considerations. Understanding that including more than $\sim 2\%$ of either a tholin or NH_3 adversely affects the spectral fits in the visible (as well as in the near-IR), we constrained the amount of those species to 1% in intimate mixtures with H_2O ice. We find that mixture models with 1% each of ice tholin and NH_3 result in spectra that are consistent with the Enceladus spectrum: they do not match exactly, but they reproduce the overall spectral features. We also find that if we model the NH_3–H_2O intimate mixture as being present on only a portion of the body (e.g., the south polar region), with pure water ice everywhere else on the surface, that we get poor fits to the Enceladus spectrum; areal mixtures with more than $\sim 20\%$ pure H_2O ice are simply too bright to agree with the UVIS spectrum; suggesting a more global coverage

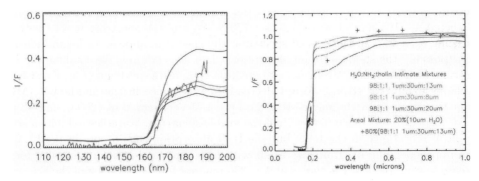

Figure 2. Enceladus data (as in Fig. 1) compared with spectral intimate mixture models using H_2O ice, ice tholin and NH_3 with varying grain sizes.

by the intimate mixture, possibly due to E-ring grain interaction. Satisfactory fits are obtained when assuming global coverage of the NH_3–H_2O-tholin mixture.

4. Implications

It is now known that Enceladus hosts water vapor jets with sources at the south polar hot spot (Hansen *et al.* 2006; Porco *et al.* 2006; Spencer *et al.* 2006; Dougherty *et al.* 2006). The plume gases include >90% H_2O (Hansen *et al.* 2006; Waite *et al.* 2006) in addition to small amounts of NH_3 (0.8%) and other trace species (Waite *et al.* 2006; Waite *et al.* 2009), including several types of hydrocarbons. UVIS likely detects the plume NH_3 that has been redeposited on the surface on Enceladus; however, it appears that the NH_3 is not present only in the south polar region, but globally, suggesting a different transport process such as E-ring interactions.

Though ammonia is not expected to be stable to photoloysis and radiolysis at the surface of Enceladus, ammonia hydrate may be stable, and we cannot rule out that it is not pure ammonia but ammonia hydrate (or a related species) that causes the strong UV dropoff. The presence of vaporous NH_3 in the Enceladus plume (Waite *et al.* 2009) suggests that the NH_3 could condense along with H_2O in the E-ring), which would replenish NH_3 on the surface of Enceladus via E-ring grain coating (Buratti 1988; Buratti *et al.* 1990), even in regions away from the south polar region. Redeposited plume material, along with coating by E-ring grains, contributes to the brightness of Enceladus' surface. Verbiscer *et al.* (2007) suggest that Enceladus is so bright because of this re-coating, an idea first suggested by Buratti (1988) and supported by Hamilton & Burns (1994).

Acknowledgements

This research was carried out at the Jet Propulsion Laboratory, California Institute of Technology, under a contract with the National Aeronautics and Space Administration.

References

Brown, R. H. *et al.* 2006, *Science*, 311, 1425

Buratti, B. 1984, *Icarus*, 59, 392

Buratti, B. J., Mosher, J. A., & Johnson, T. V. 1990, *Icarus*, 87, 339

Buratti, B. J. & Veverka, J. 1984, *Icarus*, 58, 254

Clark, R. N., Brown, R. H., Owensby, P. D., & Steele, A. 1984, *Icarus*, 58, 265

Clark, R. N., Curchin, J. M., Jaumann, R., Cruikshank, D. P., Brown, R. H., Hoefen, T. M., Stephan, K., Moore, J. M., Buratti, B. J., Baines, K. H., Nicholson, P. D., & Nelson, R. M. 2008, *Icarus*, 193, 372

Cruikshank, D. P. 1980, *Icarus*, 41, 246

Cruikshank, D. P., Owen, T. C., Dalle Ore, C., Geballe, T. R., Roush, T. L., de Bergh, C. Sandford, S. A., Poulet, F., Benedix, G. K., & Emery, J. P. 2005, *Icarus*, 175, 268

Dougherty, M. K., Khurana, K. K., Neubauer, F. M., Russell, C. T., Saur, J., Leisner, J. S., & Burton, M. E. 2006, *Science*, 311, 1406

Emery, J. P., Burr, D. M., Cruikshank, D. P., Brown, R. H., & Dalton, J. B. 2005, *A&A*, 435, 353

Esposito, L. W. *et al.* 2004. *Space Sci. Rev.*, 115, 299

Fink, U., Larson, H. P., Gautier III., T. N., & Treffers, R. R. 1976, *ApJ*, 207, L63

Grundy, W. M., Buie, M. W., Stansberry, J. A., & Spencer, J. R. 1999, *Icarus*, 142, 536

Hamilton, D. P. & Burns, J. A. 1994, *Science*, 264, 550

Hansen, C. J., Esposito, L., Stewart, A. I. F., Colwell, J., Hendrix, A., Pryor, W., Shemansky, D., & West, R. 2006, *Science*, 311, 1422

Hapke, B. W. 2002, *Icarus* 157, 523

Hendrix, A. R. & Hansen, C. J. 2008, *Icarus*, 193, 323

McClintock, W. E., Rottman, G. J., & Woods, T. N. 2000, *Earth Observing System V*, Proceedings of the SPIE, 4135, 225

McCord, T. B., Johnson, T. V., & Elias, J. H. 1971, *ApJ*, 165, 413

Morrison, D., Cruikshank, D. P., Pilcher, C. B., & Rieke, G. H. 1976, *ApJ*, 207, L213

Porco, C. C. *et al.* 2006, *Science*, 311, 1393

Shibaguchi, T., Onuki, H., & Onaka, R. 1977, *J. Phys. Soc. Jpn.*, 42, 152

Spencer, J. R., Pearl, J. C., Segura, M., Flasar, F. M., Mamoutkine, A., Romani, P., Buratti, B. J., Hendrix, A. R., Spilker, L. J., & Lopes, R. M. C. 2006, *Science*, 311, 1401

Verbiscer, A. J., French, R. G., & McGhee, C. A. 2005, *Icarus*, 173, 66

Verbiscer, A. J., Peterson, D. E., Skrutskie, M. F., Cushing, M., Helfenstein, P., Nelson, M. J., Smith, J. D., & Wilson, J. C. 2006, *Icarus*, 182, 211

Verbiscer, A., French, R., Showalter, M., & Helfenstein, P. 2007, *Science*, 315, 815

Waite, J. H., Combi, M. R., Ip, W.-H., Cravens, T. E., McNutt Jr., R. L., Kasprzak, W., Yelle, R., Luhmann, J., Niemann, H., Gell, D., Magee, B., Fletcher, G., Lunine, J., & Tseng W.-L. 2006, *Science*, 311, 1419

Waite, Jr., J. H., Lewis, W. S., Magee, B. A., Lunine, J. I., McKinnon, W. B., Glein, C. R., Mousis, O., Young, D. T., Brockwell, T., Westlake, J., Nguyen, M.-J., Teolis, B. D., Niemann, H. B., McNutt Jr., R. L., Perry, M., & Ip, W.-H. 2009, *Nature*, 460, 487

Warren, S. G. 1984, *Appl. Optics*, 23, 1206

Warren, S. G. & Brandt, R. E. 2008, *J. Geophys. Res.*, 113, D14220, doi: 10.1029/2007JD009744

Icy Bodies of the Solar System
Proceedings IAU Symposium No. 263, 2009
J.A. Fernández, D. Lazzaro, D. Prialnik & R. Schulz, eds.

© International Astronomical Union 2010
doi:10.1017/S1743921310001638

Effect of the tensile strength on the stability against rotational breakup of icy bodies

Imre Toth[1] and Carey M. Lisse[2]

[1]Konkoly Observatory
Postbus 67, H-1525 Budapest, Hungary
email: tothi@konkoly.hu

[2]Johns Hopkins University Applied Physics Laboratory
11100 Johns Hopkins Road, Laurel, MD 20723-6099, USA
email: carey.lisse@jhuapl.edu

Abstract. Focusing on primitive icy minor bodies in the solar system like cometary nuclei, centaurs, transneptunian objects (TNOs), and main-belt comets (MBCs) we investigate the stability of these objects against rotational breakup by comparing their location in (radius – rotational period) space with respect to separation lines of the stable and breakup zones in this plane. We estimate the bulk tensile strength according to new structural and elasto-mechanical models of grain-aggregates, using these tensile strengths to compute separation lines. We note that the process of grain coagulation and growth is highly uncertain in the field of solar system formation and we simply don't know how to grow interstellar grains to aggregates larger than about 1 mm but we apply in our calculations the recently available elasto-mechanical models of grain-aggregates. Accoring to this study most of the observed comets, centaurs, TNOs, and MBCs are stable against rotational breakup, with a few notable exceptions. E.g., we suggest that the rotational fission is a likely scenario for the Haumea-family in the Kuiper belt.

Keywords. solar system: general, solar system: formation, comets: general, comets: individual (133P/Elst-Pizarro), Centaurs, Kuiper Belt, Haumea-family, asteroids, Main-Belt Comets

1. Introduction and overview

The physical characteristics of primitive minor bodies in the solar system like bulk interior properties and mechanisms and ingredients of cometary emission activity, are very complex and have not been well understood yet (e.g., Weissman & Lowry 2008; Belton 2009; Biele *et al.* 2009; Jewitt 2009). For example, we have little knowledge of mechanical properties like the tensile strength of the interior material of primitive small bodies like comets, related asteroids, and dwarf planets. We note that there are different strength definitions throughout the literature (Holsapple 2009). Here we are concerned with the influence of the static tensile strength of the material on the stability against rotational breakup of primitive minor bodies.

Up to now the following methods have been used to derive the tensile strength of primitive small bodies of the solar system: (i) examining the observed rotational stability, (ii) modeling the splits due to tidal forces (e.g., tidal breakup of sungrazing comets, and D/Shoemaker-Levy 9 disrupted by Jupiter), (iii) modeling results from the Deep Impact experiment(to determine the dynamic and estimate the static tensile strengths), (iv) prediction from theory. The best estimate of tensile strength of cometary nuclei ranges from 10 Pa to 100 kPa (Jewitt 1992; Boehnhardt 2004; Weissman *et al.* 2004, and references therein).

Formerly it was common wisdom that cometary interiors consist of materials there is extremely porous structure and low strength (see review by Weissman *et al.* 2004).

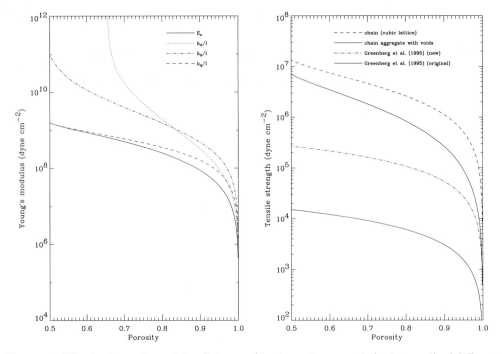

Figure 1. Effective Young's modulus (left panel) and tensile strength (right panel) of different aggregate models are displayed. Effective Young's modulus of an aggregate (solid line) and the deformation coefficients of rolling, sliding, and stretch are shown as dotted line, dash-dotted line and dashed line, respectively. Tensile strength of cubic lattice aggregate chain (dashed line), chain aggregate with voids (solid line), Greenberg *et al.* (1995) model with new value of the intermolecular energy (Sirono & Greenberg 2000), (dash-dot line), and the original aggregate model described by Greenberg *et al.* (1995). (dash-dot-dot line) are shown. Note that both the Young's modulus and tensile strength are rapidly decreasing for large porosities ($\gtrsim 0.7$).

However, hitherto we have less knowledge on the bulk interior physical properties of small primitive bodies, like elastic parameters. Despite the in-situ nucleus Deep Impact space experiment, we still have limited knowledge of the bulk interior physical properties. While Richardson *et al.* (2007) placed limits of 10^4 Pa on the effective bulk tensile strength of the comet 9P/Tempel-1 nucleus assuming a gravity dominated cratering mechanism, Holsapple & Housen (2007) concluded that "the observations are unlikely to answer the questions about the mechanical nature of the 9P/Tempel 1 surface". Even after the in-situ flybys the comets remain the poorest understood solar system objects. And according to Biele *et al.* (Biele *et al.* 2009), it appears doubtful that the tensile strength is as low as originally published for the nucleus of comet 9P/Tempel 1.

One approach to estimate the tensile strength of a rotating body is the examination of its rotational breakup stability or instability. For this purpose the early studies considered the stability of a rotating sphere or a homogeneous prolate ellipsoid rotating about its shortest axis, balancing the centrifugal and gravitational forces at the tips of the longest axis (Jewitt & Meech 1988; Luu & Jewitt 1992; Weissman *et al.* 2004).

This simplified model is a first guess to indicate the rotational instability because it describes material shedding from the elongated body surface at the tips, not a complete disintegration of the body as a whole. More complex numerical models like N-body simulations and hydrocodes now exist to follow rotational breakup for both rubble piles and continuum structures with and without internal cohesive forces (Weissman *et al.* 2003; Walsh *et al.* 2008; Richardson *et al.* 2008, 2009). But in order to easily compare models with observations there are simpler descriptions of the rotational breakup stability. One of these is based on the Drucker-Prager model and was successfully applied by Holsapple (2007) for both small rotators and large transneptunian dwarf planets like 136198 Haumea (2003 EL61) and 20000 Varuna (2000 WR106).

Another recent approach was developed by Davidsson (1999, 2001) for rotating spheres and biaxial ellipsoids. Applying the Davidsson models Toth & Lisse (2006) explored the regions of stability, fragmentation, and destruction for cometary bodies versus rotational breakup in the radius – rotational period plane. In this work we are continue this line of investigation, and present new results taking into account the elasto-mechanical parameters of new grain-aggregate models. We then compare the location of well observed objects like comets, centaurs, transneptunian objects, and a new class of primitive minor bodies, the main belt comets to the predicted regions of stability and disintegration.

2. Outline of the method and analysis

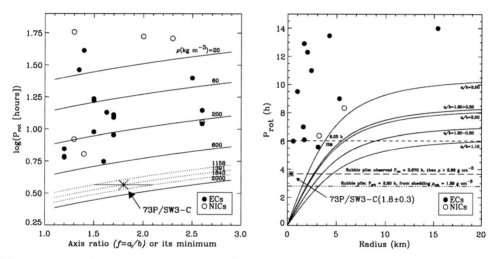

Figure 2. The fragment C of comet 73P/Schwassmann-Wachmann 3 is a fast rotator and material shedding from its surface possible (left panel) but the body itself is safe against the rotational breakup (right panel). Set of curves correspond to bulk density of the boady (left panel) and the curves of axial ratios of a rotating prolate body are shown (right panel). Ecliptic comets (filled dots) and nearly-isotropic comets (open circles) are also displayed.

Bulk material parameters are a consequence of planetesimal or cometesimal formation history. According to the formation scenario for planetesimals and cometesimals, the sub-kilometer to kilome-ter-sized planetary and minor body precursors form due to nonelastic collisions between dust agglomerates in combination with adhesive surface forces (Greenberg *et al.* 1995; Sirono & Greenberg 2000; Weidenschilling 2004; Weissman *et al.* 2004, and references therein). The process of grain coagulation determines the resulted aggregate structure, bulk density, porosity, effective Young's modulus, and tensile strength of grain-aggregates. But note that the process of grain coagulation and growth is highly uncertain in the field of solar system formation and we simply don't know how to grow ISM grains to aggregates larger than about 1 mm. Collisions in the midplane of larger aggregates should be destructive by all measures and see the recent work of Chiang & Yudin (2009).

Here we apply the newer models of grain-aggregate model mechanical properties for the interior material properties of bulk nuclear material. This problem is certainly not solved by Sirono & Greenberg (2000) but we apply it because i) it is a recently available elasto-mechanical model of primitive small body-forming grain-aggregates, and ii) this model predicts a stronger material which is suspected by some recent studies (Holsapple 2007; Biela et al. 2009).

Aggregate material tensile strength. We calculate the effective Young's modulus and derive the tensile strength grain-aggregate models following Sirono & Greenberg (2000). With the use of elastic constants for rolling, sliding, and stretching displacements of a grain, based on to the formulas obtained by Dominik & Tielens (1997), the effective Young's modulus of the aggregate is computed.

We note that most recently the reality of the grain-aggregate models are also tested in laboratory and confirmed by new calculations and comparison with some observations (Blum et al. 2006). In our approach we calculate the tensile strength for pure chain aggregates and cubic lattice structured aggregates. The derived effective Young's modulus and tensile strength of different aggregate models are displayed in Fig. 1. Note that both the Young's modulus and tensile strength are rapidly decreasing for large porosities (porosity $\gtrsim 0.7$) for all aggregate models considered here.

Model. To qualify the stability against rotational disintegration of small bodies, very useful, user friendly, and realistic models were developed by Davidsson (1999, 2001) for both spheres and biaxial ellipsoids, which are realistic enough to apply for minor bodies and compare them with observed sizes and rotational periods. Despite that the models are based on simplified assumptions (internal homogeneity and incompressibility) they are very good approaches to the problem. Using his formalism, the model problem reduces to the question of the appropriate choice of the material strength and bulk density parameters, taking into account the typical material parameters for primitive minor bodies. Toth and Lisse (2006) addressed the question of the internal material dependence of the rotational breakup, namely how the critical spin periods are influenced by the tensile strength for various grain-aggregate structured models. By testing different plausible physical models of the bulk material parameters of the primitive bodies the plane is divided into three segments: allowed, damaged, and forbidden regions and we investigate the location of selected primitive small bodies with respect to these zones separated by lines corresponding to the elastic parameters of grain-aggregates.

Observational data Recent surveys of the physical properties of cometary nuclei, centaurs, transneptunian objects, and main-belt comets are used to compare the expected regions of stability to the observed populations of these primitive minor bodies. In the size determinations the typical relative uncertainty is roughly order of ~10%, and the accuracy of rotational periods is on the order of 0.25 hr.

3. Results and discussion

Comets, centaurs, and transneptunian objects. Split comets are very important to understand the deep interior structure of the cometary nuclei via observing and physical characterization of the fragments (Boehnhardt 2004). We examined the rotational breakup stability of the fragment C of split comet 73P/Schwassmann-Wachmann 3 (73P/SW3-C) investigating its location in the radius – rotational period diagram. The size and period were determined from the Hubble Space Telescope observations made in April of 2006 when the comet was in favorable visibility conditions close to Earth (Toth et al. 2006, 2008). We applied two approaches: the balance between gravitational and centrifugal forces at the tip of longest body axis of a rotating prolate ellipsoid and the model developed by Davidsson (2001). Figure 2 displays the results obtained by these

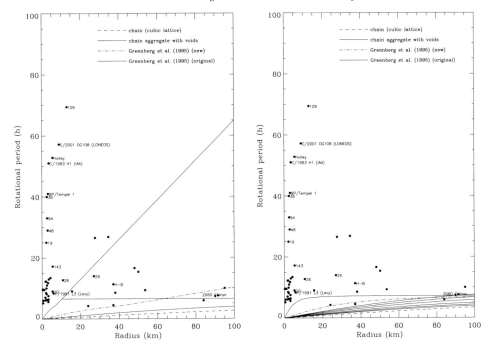

Figure 3. Comets and centaur Chiron in the radius – rotational period plane are shown (dots). Left panel: separation lines of allowed, damaged, and forbidden zones are computed for spheres according to the model given by Davidsson (1999). Right panel: set of separation lines corresponding to axial ratios (ranging from 1.01 to 3.0) of allowed/forbidden zones are computed for prolate ellipsoids according to the model given by Davidsson (2001) but only for the grain-aggregate model with voids (Sirono & Greenberg 2000).

methods. We conclude that material shedding from the surface of this fragment is possible (left panel of Fig. 2) but despite its rapid rotation the fragment is stable against the complete rotational fission (right panel of Fig. 2).

Regarding a larger sample of comets and centaurs, for which both size and rotational period data are known the range of constituent material parameters of various grain-aggregate models we found that (i) the observed comets are in the allowed region, (ii) comet C/1995 O1 (Hale-Bopp) resides in the damaged region where the body is fractured and only held together gravitationally, (iii) Comet C/1996 B2 (Hyakutake) observed to emit fragments close to its perihelion and perigee in 1996, may be near the boundary of the damaged region, (iv) split comet C/1999 S4 (LINEAR) was solidly in the rotationally allowed region, making its disintegration in July 2000 due to centrifugal forces unlikely.

In contrast to the comets, the centaurs do not cluster in the allowed region, with the majority falling instead into the rotationally damaged region region for the weakest grain-aggregate models (Greenberg *et al.* 1995), but for stronger materials they are stable against the rotational breakup. These bodies thus seem to have different bulk physical properties than cometary nuclei.

Transneptunian objects are displayed in Figure 4. The large majority of them prove stable against rotational breakup, assuming any material property model. Only (150642) 2001 CZ31 and 135108 Haumea (2003 EL61) are definitely in the forbidden zone according to the model criteria (Davidsson 2001). Haumea is particularly intriguing, because it is a very large TNO with two small satellite companions and is the largest member of the first

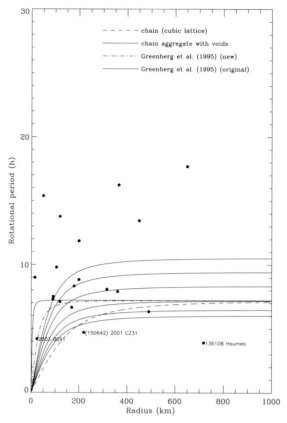

Figure 4. Transneptunian objects in the radius – rotational period plane are shown (dots). Set of separation curves (solid lines) corresponding to axial ratios (ranging from 1.01 to 3.0) of allowed/forbidden zones are computed for prolate ellipsoids according to the model given by Davidsson (2001) but only for the grain-aggregate model with voids (Sirono & Greenberg 2000). Separation lines for other grain-aggregate models are computed for the nearly-spherical axial ratio of 1.01.

collisional family in the Kuiper belt. The new solid body model developed by Holsapple (2007) is able to stabilize Haumea against rotational breakup assuming size- and depth-dependent tensile strength of a stronger icy material. On the other hand according to the most recent analysis by Schlichting & Sari (2009) the scenario of collisional origin of Haumea's family is doubtful. They argue that the velocity dispersion of fragments is too small to support a collision formation model. According to the close proximity to the rotational breakup limit of this large TNO in the frames of models by Davidsson (2001) and Holsapple (2007), we suggest that the mechanism of rotational fission is a likely formation scenario for this dynamical family.

Main-Belt Comets The existence of a population of comets residing in the outer region of the main asteroid belt was discovered by Hsieh & Jewitt (2006). The MBCs occupy dynamically asteroidal orbits decoupled from Jupiter in the main asteroid belt but show cometary activity with gas-driven mass loss. The existence of the MBCs lends new support to the idea that main belt objects could be a major source of terrestrial water. If so, the physical characterization of the MBCs is extremely important. Four MBCs are currently known: 133P/Elst-Pizarro (7968 = 1996 N2), 176P/LINEAR (also 118401 = 1999 RE70), P/2005 U1 (Read), and P/2008 R1 (Garradd). In order to examine their

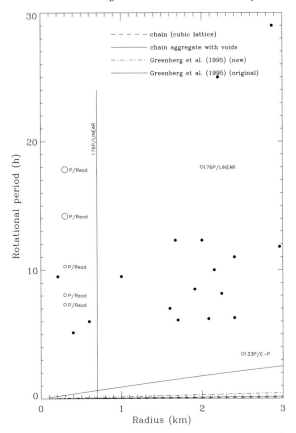

Figure 5. Main-Belt Comets with known size and rotational period in the radius – rotational period plane (open circles). The rotational period of P/2005 U1 (Read) has not been determined yet unambiguously therefore all of the proposed alternatives are shown along the vertical line. Separation lines of allowed/forbidden zones are computed for spheres according to the model given by Davidsson (1999). 133P/Elst-Pizarro is close to the rotational breakup limit for the weak grain-aggregate material (Greenberg *et al.* 1995). Ecliptic comets falling in the diagram are also indicated for a comparison (filled dots).

stability against rotational disintegration we adopted the observed rotational period and size data from Hsieh *et al.* (2004) for 133P, Hsieh & Jewitt (2006), and Hsieh *et al.* (2009a) for 176P, and Hsieh *et al.* (2009b) for P/2005 U1. For P/2005 U1 (Read) the rotational period has not been determined yet unambiguously and the proposed periods are 14.20, 17.82, 7.32, 8.06, 10.20 hr (Hsieh *et al.* 2009b), all of which are shown in Fig. 5. The size and rotational period of P/2008 R1 (Garradd) are still unknown.

Justification of the validity of the application of grain-aggregate models for MBCs is based on (i) their cometary nature, (ii) the MBCs orbit in the Themis-zone in the cold icy reservoir of the outer main belt and owing to the presumably collisional origin of MBCs from a larger icy parent body, the dusty-icy aggregate structure can be assumed, (iii) the example of the D/Shoemaker-Levy 9 as a probable escaped Hilda-zone asteroid (or a temporarily resident comet).

In the radius – rotational period diagram we observe that all of the MBCs are stable against the rotational disintegration including 133P (Fig. 5). Elst-Pizarro is a fast rotating small body and it is very close to the rotational breakup limit but this object is also stable against rotational splitting even for the weakest material of the grain-aggregate model (Greenberg et al 1995). The two other MBCs are far from the forbidden zones for all of the aggregate material models.

The idea that the activity of the MBCs is driven by sublimation of volatiles as in the case of "classical" bona-fide ecliptic and nearly-isotropic comets (Hsieh *et al.* 2004; Licandro 2009; Prialnik 2009), is widely accepted.

However, the rapid rotation of 133P is intriguing and challenging: aside from the mechanism of the solar insolation triggered sublimation, after the discovery of the recurrent activity of the prototype MBC 133P, Hsieh et al. (2004) discussed the possible consequences of its rapid rotation as a possible scenario of the non gas-driven mass loss for this object. According to this scenario if 133P was an asteroid losing mass from its surface because it rotates on the verge of centripetal instability. The natural consequence of spin-up to instability is a temporary shedding of mass followed by settling into a state just below instability.

The problem with the fast rotation to explain the long-lasting (many months duration) cometary activity is that the material shedding is only an episodic short-term event. However, the fast rotation can influence the gas-driven activity in the interior or at the surface of a small body as it was proposed for some comets and centaurs (Toth & Lisse 2006). The details of this process are poorly known, although the fast rotation induced shear fracture may open fissures beneath the body surface, which allows more efficient gas and dust outflow from the interior and near-surface regions of the body.

4. Summary and conclusions

Our recent analysis of the rotational breakup stability of primitive small bodies of the solar system led to the following conclusions:

(*a*) We applied more complex models of grain-aggregates to explore the elasto-mechanical properties of the interior structure of primitive minor bodies of the solar system. We calculated the effective Young's modulus and static tensile strength of grain-aggregates and we found a strong decrease of the tensile strength for porosities greater than ∼0.7-0.8. But there are caveats in the understanding the elasto-mechanical properties of primitive small bodies. Recently available models of primitive icy bodies predict both weak (Greenberg *et al.* 1995) and strong (Sirono & Greenberg 2000) bulk tensile strengths. Further studies are needed in order to (i) better understand the coagulation and grain growing processes which form primitive small icy bodies and determine their bulk tensile strengths, and (ii) decide whether these bodies are separated according to weaker or stronger bulk tensile strengths.

(*b*) Using the resulting tensile strengths we constructed the segments of the rotational disintegration classes (allowed, damaged, and forbidden zones) separated by limit curve lines in the effective body radius – rotational period plane.

(*c*) Most of the comets, centaurs, and small transneptunian objects are stable against the rotational breakup even for the weakest grain-aggregate model (Greenberg *et al.* 1995). For stronger grain-aggregate (or rocky, stronger icy) models they are surely stable against the rotational disintegration.

(*d*) There are a few comets, and many centaurs and TNOs in the damaged region, suggesting that comets are fundamentally weaker objects. Some large fast rotators are close to the breakup limit (Davidsson 1999, 2001) unless they have stronger internal strength.

(*e*) Main-belt comets are stable - but 133P/Elst-Pizarro is very close to the breakup limit for the weakest aggregate model but this object is safe for stronger aggregates or rocky materials.

(*f*) We suggest that Haumea-family could be formed via rotational fission.

Acknowledgements

I.T. acknowledges the financial support from Konkoly Observatory Budapest, and from the CNRS and CNES during the data analysis and interpretation of the Hubble Space Telescope imaging data at OAMP-LAM, France. C.M.L. acknowledges support from the NASA Deep Impact Extended project for this work.

References

Belton, M. J. S. 2009, IAU Symposium No. 263 *Icy Bodies of the Solar System*, Invited.
Biele, J., Ulamec, S., Richter, L., Kührt, E., Knollenberg, Möhlmann, D., & Rosetta Philae Team 2009, in: *Deep Impact as a World Observatory Event: Synergies in Space, Time and Wavelength*, ESO Astrophysics Symposia Volume, (Berlin, Heidelberg: Springer), p. 285
Blum, J., Schräpler, R., Davidsson, B. J. R., & Trigo-Rodriguez, J. M. 2006, *ApJ*, 652, 1768
Boehnhardt, H. 2004, in: M. C. Festou, H. U. Keller & H. A. Weaver (eds.), *Comets II*, Space Science Series, (Tucson: University of Arizona Press), p. 301
Chiang, E. & Yudin, A. 2009, *astro-ph arXiv:0909.2652, 2009 September 14*
Davidsson, B. J. R. 1999, *Icarus*, 142, 525
Davidsson, B. J. R. 2001, *Icarus*, 149, 375
Dominik, C. & Tielens, A. G. G. M. 1997, *ApJ*, 480, 647
Greenberg, J. M., Mizutani, H., & Yamamoto, T. 1995, *A&A*, 295, L35
Holsapple, K. A. 2007, *Icarus*, 187, 500
Holsapple, K. A. 2009, *Planet. Space Sci.*, 57, 127
Holsapple, K. A. & Housen, K. R. 2007, *Icarus*, 187, 345.
Hsieh, H. H., Jewitt, D., & Fernández, Y. R. 2004, *AJ*, 127, 2997
Hsieh, H. H. & Jewitt, D. 2006, *Science*, 312, 561
Hsieh, H. H., Jewitt, D., & Fernández, Y. R. 2009, *ApJ*, 694, L111
Hsieh, H. H., Jewitt, D., & Ishiguro, M. 2009, *AJ*, 137, 157
Jewitt, D. 1992, in: A. Brahic, D.-C. Gerard & J. Surdej (eds.), *Proc. of the 30th Liége International Astrophysical Colloquium*, (Liége: University of Liége Press), p. 85
Jewitt, D. 2009, IAU Symposium No. 263: *Icy Bodies of the Solar System*, Invited.
Jewitt, D. C. & Meech, K. J. 1988, *ApJ*, 328, 974
Jewitt, D., Yang, B., & Haghighipour, N. 2009, *AnJ*, 137, 4313
Luu, J. X. & Jewitt, D. C. 1992, *AnJ*, 104, 2243
Licandro, J. 2009, IAU Symposium No. 263: *Icy Bodies of the Solar System*, this confence.
Prialnik, D. 2009, in: D. Lazzaro, D. Prialnik, R. Schulz, & J. A. Fernández (eds.), *Icy Bodies of the Solar System*, IAU Symposium 263, (Cambridge: Cambridge University Press), this volume.
Richardson, J. E., Melosh, H. J., Lisse, C. M., & Carcich, B. 2007, *Icarus*, 190, 357
Richardson, D. C., Schwartz, S. R., Michel, P., & Walsh, K. J. 2008, *BAAS*, 40, 498, abstr. 55.02
Richardson, D. C., Michel, P., Walsh, K. J., & Flynn, K. W. 2009, *Planet. Space Sci.*, 57, 183
Schlichting, H. E. & Sari, R. 2009, *ApJ*, 800, 1242
Sirono, S.-I. & Greenberg, J. M. 2000, *Icarus*, 145, 230
Toth, I. & Lisse, C. M. 2006, *Icarus*, 181, 162
Toth, I., Lamy, P., Weaver, H. A., A'Hearn, M. F., Kaasalainen, M., & Lowry, S. C. 2006, *BAAS*, 38, 489, abstr. [06.01]
Toth, I., Lamy, P. L., Weaver, H. A., Noll, K. S., & Mutchler, M. J. 2008, *BAAS*, 40, 394, abstr. [05.08]

Walsh, K. J., Richardson, D. C., & Michel, P. 2008, *BAAS*, 40, 498, abstr. 55.03

Weidenschilling, S. J. 2004, in: M. C. Festou, H. U. Keller & H. A. Weaver (eds.), *Comets II*, Space Science Series, (Tucson: University of Arizona Press), p. 97

Weissman, P. R., Richardson, D. C., & Bottke, W. F. 2003, *BAAS*, 35, 1012, abstr. 47.06

Weissman, P. R., Asphaug, E., & Lowry, S. C. (2004), in: M. C. Festou, H. U. Keller & H. A. Weaver (eds.), *Comets II*, Space Science Series, (Tucson: University of Arizona Press), p. 337

Weissman, P. R. & Lowry, S. C. 2008, *Meteoritics & Planet. Sci.*, 43, 1033

Icy Bodies of the Solar System
Proceedings IAU Symposium No. 263, 2009
J.A. Fernández, D. Lazzaro, D. Prialnik & R. Schulz, eds.
© International Astronomical Union 2010
doi:10.1017/S174392131000164X

Ground-based observations of Phoebe (S9) and its rotational dynamics

Ekaterina Yu. Aleshkina, Alexandr V. Devyatkin, and Denis L. Gorshanov

Main (Pulkovo) Astronomical Observatory of the Russian Academy of Sciences,
Pulkovskoye ave., 65-1, 196140 Saint-Petersburg, RUSSIA
email: aek@gao.spb.ru

Abstract. Analysis of CCD observations of Phoebe, the 9th satellite of Saturn (visual magnitude of about 16.5), with a mirror astrograph ZA-320*M* at Pulkovo Observatory in Saint-Petersburg are presented. Photometric observations are performed both in the integral band of the telescope and in bands *BVR* of the Johnson system. Reference catalogues USNO-A2.0 (for *R* - filter and integral observations) and Ticho-2 (for *V* and *B* - filters) were used. Rotational light-curve data for Phoebe taken over a short time span (2 - 8 hours) for several nights are presented. Numerical investigation of the evolution of Phoebe's rotational dynamics is carried out. The probability of Phoebe's capture in resonant states that are distinct from 1:1 is estimated.

Keywords. planets and satellites: Phoebe, techniques: photometric, solar system: formation.

1. Introduction

In the Solar System, potential candidates that could still be in a chaotic or nonsynchronous state are satellites having distant or eccentric orbits – the so called irregular satellites. There is a probability of their capture in resonant states that are distinct from 1:1. Phoebe, Saturn's 9th satellite, is the only irregular satellite with known inertial parameters and fast nonsynchronous rotation. Its retrograde, eccentric and inclined orbit indicates that it could be an object captured from a heliocentric orbit. Numerical investigation of the evolution of Phoebe's rotational dynamics and theoretical estimation of its tidal despinning time are carried out in this study.

Since Phoebe has a fast and nonsynchronous rotation, which is sufficiently reliable (Melnikov (2002), Devyatkin *et al.* (2004)), its observations are important for revealing probable changes of its rotation characteristics. More then 250 photometric and about 150 astrometric observations of Phoebe were obtained during 2007 - 2008 (Aleshkina *et al.* (2009)).

Our observational program is performed with an automatically-operated 0.32-m mirror astrograph ZA-320*M* at Pulkovo observatory in Saint-Petersburg (Latitude: 059 46 15.000 N, Longitude: 030 19 45.000 E, Altitude: 75.00 m), having the following characteristics: $D = 320$ mm, $F = 3200$ mm, $M = 65"/$mm, limit magnitude 19^m, and a CCD detector FLI (1024×1024 pixels, 28' $\times$ 28 ').

2. Phoebe - 9th satellite of Saturn

Phoebe is an irregular satellite of Saturn with known inertial parameters, fast rotation and an eccentric, inclined orbit. It is supposed to be an ice-rich body coated with a thin layer of dark material (Cruikshank *et al.* (2008)). Phoebe's orbital, inertial and physical characteristics are presented in Tables 1 and 2. Values are given in these tables for:

Table 1. Orbital parameters of Phoebe (according to Jacobson (2006)).

r (10^6 km)	12.947780
e	0.1635
n (deg/day)	0.6541824
T_orb (day)	550.31
i (day)	175.986

Table 2. Inertial and physical parameters of Phoebe.

Parameter	Value	Ref.	Parameter	Value	Ref.
					Simonelli
$P_{rot}(h)$	9.2735 ± 0.0006	Bauer *et al.*(2004)	albedo	0.081 ± 0.002	*et al.* (1999)
$R(10^5 cm)$	106.6 ± 1.1	Jacobson *et al.* (2005)	A/C	0.93623	[1]
$Gm(km^3/s^2)$	0.5531 ± 0.0006	Jacobson *et al.* (2005)	B/C	0.94455	[1]
$\rho(g/cm^3)$	1.633 ± 0.049	Jacobson *et al.* (2005)	$g(cm/s^2)$	4.87	[2]
$a_e/b_e/c_e, (km)$	108.6/107.7/101.5	Johnson *et al.* (2000)	$\mu(dyn/cm^2)$	10^{11}	[3]

Notes:

[1] The inertial parameters $A/C, B/C$ are calculated by means of formulas for a triaxial ellipsoidal satellite of homogeneous density $A/C = (b_e^2 + c_e^2)/(a_e^2 + b_e^2), B/C = (a_e^2 + c_e^2)/(a_e^2 + b_e^2)$.

[2] The value of gravitational acceleration g is calculated as $g = Gm/R^2$.

[3] Values of rigidity μ for planetary satellites are practically unknown; the theory gives $\mu = 5 \times 10^{11} dyn/cm^2$ for rock with $\rho = 2$ g/cm^3 and $\mu = 3.5 \times 10^{10} dyn/cm^2$ for ice with $\rho = 1$ g/cm^3 (Dobrovolskis (1995))

r, e, i, n, T_{orb} – semi-major axis, eccentricity, inclination of orbit, mean motion and orbital period, respectively, and a_e, b_e, c_e – semi-axes of ellipsoid of inertia, P_{rot}, R, Gm, ρ – rotation period, mean radius, gravitation constant of Phoebe and its mean density, respectively.

3. Photometric observations of Phoebe

Photometric observations were performed both in the integral band of the telescope (300–900 nanometers) and in bands *V* and *R* of the Johnson system. Reference catalogues USNO-A 2.0 (for *R* filter and integral observations) and Ticho-2 (for *V* filter) were used. In 2007 – 2008 we obtained about 250 frames of Phoebe. All our observations of 2007 – 2008 are available on www.ad-astra.len.su/Phoebe07-08.html.

Results of differential photometry over a short time span (2-8 hours) for 11 observational nights are presented in Table 3. The brightness of stars in the integral band of the instrument was calculated using the values of B and R for these stars from a USNO-A 2.0 catalog. It is the only catalog containing both astrometric and photometric information for stars falling within the instrument's field of view, although it has low accuracy (0.15^m) of photometric data. A considerable number (in general about 30-50) of star images suitable for measurement made it possible to slightly compensate for the low accuracy of the catalogue. The brightness of stars in the *V* band was taken from Ticho-2. The intrinsic accuracy indicated for each value is a standard deviation obtained by averaging the Phoebe brightness values relative to each of the measured stars on a frame. Mean values of accuracy are equal to $0.07^m, 0.05^m$ and 0.3^m for the integral, *R* and *V* bands, respectively. The average value of color index for Phoebe (Fig.1b) for the performed dense observational series 2008.03.24 (in the instrumental system) is $< R - V >= 1.6$.

Table 3. Statistics of photometric observations of Phoebe 2007 - 2008 (number of frames).

Date	Time span (h)	Integral band	R - filter	V - filter	B - filter
2007-03-26	5.25	20	–	–	–
2007-03-30	3	–	16	5	–
2007-03-31	1.5	–	5	–	–
2007-04-01	3.5	22	–	–	–
2007-04-16	3.5	18	–	–	–
2008-01-28	1.5	1	5	1	1
2008-02-25	2	5	6	6	6
2008-03-21	5	1	2	5	–
2008-03-24	8	18	21	18	–
2008-03-28	2	8	7	–	–
2008-03-31	6.5	29	13	–	–

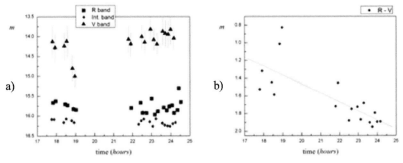

Figure 1. Photometric observations of Phoebe. (a) Light-curves over a short time span obtained on 24.03.2008. (b) The average value of the color index $< R - V >$ (in the instrumental system).

4. Rotational dynamics of Phoebe

A numerical investigation of the evolution of Phoebe's rotational dynamics was carried out, using the same dynamical model as in the previous paper Aleshkina (2009). Because irregular satellites are potential candidates for still being in chaotic or nonsynchronous states, the probability for Phoebe being captured in resonant states that are distinct from 1:1 and its stability were estimated.

Theoretical estimates for the probability of Phoebe being captured in the 3:2 resonant state for two models of tidal interaction Murray & Dermott (2006) yield 0.7 and 0.5, respectively. Results of our numerical experiments show that for different initial conditions (spin angle $0.5 < \psi_0 < 1.77$ and angular velocity $6 < \omega_0/n < 14$) the probability of Phoebe's capture in spin-orbit resonant states 5:2, 2:1, 3:2 is high. At the same time, before capture, there are long periods of time (about a hundred thousand years) when Phoebe is in nonsynchronous resonant states (for instance, Fig. 2 shows the capture in resonance 3:2 and a chaotic zone in resonant state 2:1). Estimation of resonance stability was obtained according to criteria derived in Murray & Dermott (2006). For modern orbital parameters of Phoebe all of its possible resonant states are stable.

We also estimated the tidal despinning time of Phoebe τ_D (Table 4) both theoretically (according to Dobrovolskis (1995)) and by numerical experiments.

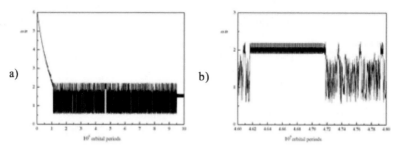

Figure 2. Capture in resonance 3:2. (a) Initial conditions $\psi_0 = 1.27$, $\omega_0/n = 6$, (b) Chaotic zone between 4.5 and 5 millions orbital periods in (a).

Table 4. Tidal despinning time of Phoebe.

	τ_D (year)
Numerical experiment	1.5×10^{13}
Theoretical estimation	10^{15}
Peale (1977)	1.7×10^{14}

References

Aleshkina, E. 2009, *Solar System Research*, 43, 71

Aleshkina, E., Devyatkin, A., & Gorshanov, D. 2009, *Izv.GAO RAN*, 219, (in press)

Bauer, J. M. *et al.* 2004, *ApJ*, 610, L57

Cruikshank, D. P. *et al.* 2008, *Icarus*, 193, 334

Devyatkin, A. V. *et al.* 2004, *Izv. GAO RAN*, 217, 229

Dobrovolskis, A. R. 1995, *Icarus*, 118, 181

Jacobson, R. A. 2006, *SAT252 - JPL satellite ephemeris*

Jacobson, R. A. *et al.* 2005, *BAAS*, 37, 729

Johnson, T. V., Castillo-Rogez, J. C., Matson, D. L., & Thomas, P. C. 2000, *40th Lunar and Planetary Science Conference*, (The Woodlands: Lunar and Planetary Science XL), id.2334

Melnikov, A. V. 2002, *IAA Transactions*, 8, 131

Murray, C. & Dermott, S. 2006, *Solar System Dynamics* (Cambridge: Cambridge Univ. Press)

Peale, S. J. 1977, in: J. A. (ed.), *Planetary satellites*, (Tucson: Univ. of Arizona Press), p. 87

Simonelli, D. P. *et al.* 1999, *Icarus*, 138, 249

Icy Bodies of the Solar System
Proceedings IAU Symposium No. 263, 2009
J.A. Fernández, D. Lazzaro, D. Prialnik & R. Schulz, eds.
© International Astronomical Union 2010
doi:10.1017/S1743921310001651

Adenine synthesis at Titan atmosphere analog by soft X-rays

Sergio Pilling[1,2], Diana P. P. Andrade[1,2], Alvaro C. Neto[3], Roberto Rittner[3] and Arnaldo N. de Brito[4]

[1]Pontifícia Universidade Católica do Rio de Janeiro (PUC-Rio), 22453-900, Rio de Janeiro, RJ, Brazil

[2]Instituto de Pesquisa & Desenvolvimento (IP&D), Universidade do Vale do Paraíba (UNIVAP), 12244-000, São Jose dos Campos, SP, Brazil

[3]Universidade Estadual de Campinas (UNICAMP), 13084-971, Campinas, SP, Brazil

[4]Laboratório Nacional de Luz Síncrotron (LNLS), 13083-970, Campinas, SP, Brazil

Abstract. In this work, we investigate the possible effects produced by soft X-rays (and secondary electrons) on Titan aerosol analogs in an attempt to simulate some prebiotic photochemistry. The experiments have been performed inside a high vacuum chamber coupled to the soft X-ray spectroscopy beamline at the Brazilian Synchrotron Light Source (LNLS). In-situ sample analysis were performed by a Fourier transform infrared spectrometer. The infrared spectra have presented several organic molecules, including nitriles and aromatic CN compounds. After the irradiation, the brownish-orange organic residue was analyzed ex-situ by gas chromatographic technique revealing the presence of adenine ($C_5H_5N_5$), one of the constituents of the DNA molecule.

Keywords. astrochemistry, astrobiology, molecular processes, methods: laboratory, planets and satellites: Titan, Sun: X-rays

1. Introduction

Titan, the largest satellite of Saturn, has an atmosphere chiefly made up of N_2 and CH_4, and including many simple organic compounds. This atmosphere also partly consists of hazes and aerosols particles which shroud the surface of this satellite, giving it a reddish appearance. As a consequence of its high surface atmospheric pressure (~ 1.5 bar) the incoming solar UV and soft X-ray photons are mostly absorbed allowing virtually no energetic photons to reach the surface. However, during the last 4.5 gigayears, the photolysed atmospheric molecules and aerosol particles have been deposited over the Titan surface composed by water-rich ice (80-90 K) delivered by comets. This process may have produced in some regions a ten meter size, or even higher, layers of organic polymer Griffith *et al.* (2003).

2. Experimental methodology and results

In this work investigate the chemical effects induced by soft X-rays in the Titan aerosol analog. The experiments have been performed inside a high vacuum chamber coupled to the soft X-ray spectroscopy (SXS) beamline at the Brazilian Synchrotron Light Source (LNLS), Campinas, Brazil. Briefly, a gas mixture simulating the titan atmosphere (95% N_2, 5% CH_4) was continuously deposited onto a polished NaCl substrate previously cooled at 13-14 K and exposed to synchrotron radiation (maximum flux between 0.5-3 keV) up to 73 hs.

A small fraction of water and CO_2 was also continuously deposited on the frozen substrate simulating thus, a possible heavy cometary delivery at Titan. The total energy deposited on the sample was about $\sim 10^{12}$ erg. In-situ sample analysis were performed by a Fourier transform infrared spectrometer (FTIR) during the irradiation and during the sample slowly heating to room temperature. The IR analysis has shown several organic molecules created and trapped in the ice, including the reactive cyanate ion ONC^-, nitriles, and possibly amides and esters (Figure 1a).

After the irradiation, the brownish-orange organic residue (tholin) were analyzed ex-situ by chromatographic (GC-MS) technique (Figure 1b) revealing the presence of adenine ($C_5H_5N_5$), one of the constituents of DNA molecule. The complete description of the experimental setup and results can be found elsewhere Pilling *et al.* (2009).

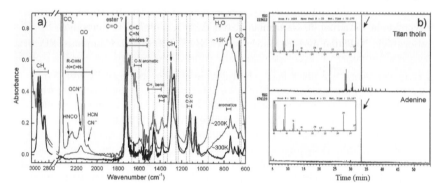

Figure 1. (a) Comparison between FTIR spectra of 73 h irradiated sample at 15, 200, and 300 K. (b) Total-ion current chromatogram of processed the Titan tholin and of the adenine standard. Details can be found elsewhere (Pilling *et al.* (2009))

3. Conclusions

In this work, we present the chemical alteration produced by the interaction of soft X-rays (and secondary electrons) on Titan aerosol analogs. The experiments simulate roughly 7×10^6 years of solar soft X-ray exposure on Titan atmosphere. Thermal heating of frozen tholin drastically changes its chemistry, resulting in an organic residue rich in C-C and C-N aromatic structures.

On Titan, the processed aerosols will be deposited along the time at the surface or at the bottom of lakes/rivers, leaving with them newly formed organic species. Gas chromatography analysis of the organic residue at room temperature has shown that among several nitrogen compounds, adenine, is one of the most abundant species produced due to irradiation by soft X-rays. This confirms previous studies suggesting that the organic chemistry in the Titan atmosphere and on the surface should be complex, being rich in prebiotic molecules such as adenine and amino acids (or its precursors species). Molecules such as these on the early Earth have found a place that allows life (as we know) to flourish, a place with liquid water.

References

Griffith, C. A., Owen, T., Geballe, T. R., Rayner, J., & Rannou, P. 2003 *Science*, 300, 628
Pilling, S., Andrade, D. P. P., Neto, A. C., Rittner, R., & de Brito, A. N. 2009 *J. Phys. Chem. A* 2009, In press, DOI:10.1021/jp902824v
Lappi, S. E. & Franzen, S. 2004 *Spectrochim. Acta, Part A*, 60, 357

Icy bodies of the Solar System
Proceedings IAU Symposium No. 263, 2009
J.A. Fernández, D. Lazzaro, D. Prialnik & R. Schulz, eds.
ⓒ International Astronomical Union 2010
doi:10.1017/S1743921310001663

Water masers in the Kronian system

**Sergei V. Pogrebenko[1], Leonid I. Gurvits[1], Moshe Elitzur[2],
Cristiano B. Cosmovici[3], Ian M. Avruch[4], Salvatore Pluchino[5],
Stelio Montebugnoli[5], Emma Salerno[5], Giuseppe Maccaferri[5],
Ari Mujunen[6], Jouko Ritakari[6], Guifre Molera[6], Jan Wagner[6],
Minttu Uunila[6], Giuseppe Cimo[1], Francesco Schilliro[7],
and Marco Bartolini[5]**

[1] Joint Institute for VLBI in Europe,
Postbus 2, NL-7990AA, Dwingeloo, The Netherlands
emails: pogrebenko@jive.nl, lgurvits@jive.nl, cimo@jive.nl

[2] University of Kentucky, Department of Physics and Astronomy
600 Rose Street, Lexinton, KY, 40506-0055 USA
email: moshe@pa.uky.edu

[3] Istituto Nazionale di Astrofisica (INAF) - Istituto di Fisica dello Spazio Interplanetario
(IFSI), Via del Fosso del Cavaliere, I-00133, Rome, Italy
email: cosmo@ifsi-roma.inaf.it

[4] Rijksuniversiteit Groningen, Postbus 72, NL-9700 AB Groningen, Nederlanden
email: iavruch@gmail.com

[5] Istituto Nazionale di Astrofisica (INAF), Istituto di Radio Astronomia (IRA), Stazione
Radioastronomica di Medicina , Via Fiorentina, Medicina (BO), Italy
emails: s.pluchino@ira.inaf.it, s.montebugnoli@ira.inaf.it, esalerno@ira.inaf.it,
g.maccaferri@ira.inaf.it, mbartolini@ira.inaf.it

[6] Aalto University of Science and Technology, Metsähovi Radio Observatory, Metsähovintie
114, FIN-02540 Kylmälä, Finland
emails: amn@kurp.hut.fi, jr@kurp.hut.fi, gofrito@kurp.hut.fi, jwagner@kurp.hut.fi,
minttu@kurp.hut.fi

[7] Istituto Nazionale di Astrofisica (INAF), Istituto di Radio Astronomia (IRA), Stazione
Radioastronomica di Noto, Casella Postale 141, I-96017 Noto (SI), Italy
email: f.schilliro@noto.ira.inaf.it

Abstract. The presence of water has been considered for a long time as a key condition for life in planetary environments. The Cassini mission discovered water vapour in the Kronian system by detecting absorption of UV emission from a background star (Hansen *et al.* 2006). Prompted by this discovery, we started an observational campaign for search of another manifestation of the water vapour in the Kronian system, its maser emission at the frequency of 22 GHz (1.35 cm wavelength). Observations with the 32 m Medicina radio telescope (INAF-IRA, Italy) started in 2006 using Mk5A data recording and the JIVE-Huygens software correlator. Later on, an on-line spectrometer was used at Medicina. The 14 m Metsähovi radio telescope (TKK-MRO, Finland) joined the observational campaign in 2008 using a locally developed data capture unit and software spectrometer. More than 300 hours of observations were collected in 2006-2008 campaign with the two radio telescopes. The data were analysed at JIVE using the Doppler tracking technique to compensate the observed spectra for the radial Doppler shift for various bodies in the Kronian system (Pogrebenko *et al.* 2009). Here we report the observational results for Hyperion, Titan, Enceladus and Atlas, and their physical interpretation. Encouraged by these results we started a campaign of follow up observations including other radio telescopes.

Keywords. Planets and satellites, masers, molecular data.

1. Introduction

The Cassini Ultraviolete Imaging Spectrometer (UVIS) detected water vapour in a plume emanating from Enceladus via absorption of UV emission from a background star (Hansen *et al.* 2006). This discovery confirmed Enceladus as a supplier of water into the Saturns ring system and triggered search for other manifestations of water in the Kronian system. One of these manifestations is the maser emission of water molecules at the frequency of 22 GHz (1.35 cm wavelength). Possibilities of natural masers associated with different bodies of the Solar System were first discussed by Mumma (1993) and the first detection of the 22 GHz water maser emission from the planetary system was made during the collision of the Shoemaker-Levy comet with Jupiter (Cosmovici *et al.* 1996). To achieve the maser amplification optical depth of $\tau > 1$, with collisional pumping for H_2O molecules at a kinetic temperature around 200 K, several requirements should be met (Elitzur 1992, Elitzur & Fuqua 1989, Elitzur *et al.* 1992). If the water molecules are the dominant component (Waite *et al.* 2009) in the masing cloud, water-water collisions can ignite the maser emission if the water number density is $n_{H2O} = 10^9$ cm^{-3}. The presence of free electrons with energies in the range of 0.1 - 0.2 eV and number density $n_e = 10^3$ - 10^5 cm^{-3} relaxes the requirements (Elitzur & Fuqua 1989), to a cloud size of 300 km and the water molecules number density of $n_{H2O} = 10^7$ cm^{-3}, or the column number density of $n_{H2O} = 3 \times 10^{14}$ cm^{-2}. This low energy electron density, required for effective pumping of the maser, is consistent with in situ measurements made by the Cassini mission (Schippers *et al.* 2008), while the water molecules column density is consistent with the data from Hansen *et al.* (2006) and Hansen *et al.* (2008)

2. Observations, data processing and results

The observational campaign was conducted in 2006 - 2008, using the 32 m Medicina (INAF-IRA, Italy) and 14 m Metsähovi (AUST-MRO, Finland) radio telescopes. We observed with the bandwidth of 8 MHz, which corresponds to the radial Doppler velocity coverage of 100 km/s at the observational frequency of 22 GHz. The system temperature of the antennas during the observations was in the range of 140-250 K for the Medicina telescope and 90-150 K for the Metsähovi. The Metsähovi radio telescope observed with both right and left circular polarizations, while the Medicina telescope used the right circular polarization only. For spectrum acquisition we used the real-time online spectrometer at the Medicina station (Montebugnoli *et al.* 1996) and an off line software spectrometer at Metsähovi, both with the instrumental spectral resolution of 1 kHz or 13.5 m/s in velocity terms at 22 GHz, while the expected maser line width was in the range of 100-1000 m/s. The antenna beams of 2 arcmin for Medicina and 3.5 arcmin for Metsähovi covered Saturn, most of its inner satellites and rings, partly covering the E-ring and outer satellites at certain orbital phases. We also used a targeted pointing of antennas on individual Kronian satellites. The system temperature range, antenna's efficiency and the polarization factor made the sensitivity of these two antennas comparable in spite of the difference of their diameters.

The data analysis was carried out at JIVE, using a whole data base of observed spectra, more than 300 hours with a 3-10 minutes of integration time per spectrum. The observed spectra were Doppler corrected in frequency domain according to the predicted radial velocity of the satellites in the beam. We used the JPL Solar System Dynamic Group's online Horizons tool (Georgini 1996) to get the station-centric radial Doppler velocity for a number of Kronian satellites. The spectra for each observed satellite were integrated over the time of certain orbital phases of the satellites. We used 16 orbital phases within each satellite's orbital period. The total integration time per orbital phase bin was between 6 and 10 hours. Not all the orbital phases showed the detection, which is an indication of the narrow beaming of the maser emission.

After integrating the radial Doppler corrected spectra for 20 of major Kronian satellites for each of the 16 orbital phases, we found statistically significant detections for 4 satellites.

The 22 GHz water maser line was detected in association with orbital phases and Doppler shifts at a level of 0.3-0.5 K antenna temperature for Enceladus, Atlas, Titan and Hyperion, with a statistical confidence level of 4.2, 7.0, 3.8 and 4.0 sigma, respectively. As an example, a summary plot of our detection for Enceladus is shown in Figure 1.

3. Physical implications

The amount of experimental data available is insufficient to construct a detailed physical model of the masers detected to date. In different regions and bodies of the Kronian system, the cause or character of the maser pumping might vary. Following Elitzur *et al.* 1992, and assuming a spherical geometry and collisions with a neutral molecular agent or low energy electrons as the main pumping mechanism, and a water-vapour ambient temperature in the range of $120 - 200$ K, the required column density of water molecules of $5 \times 10^{14} - 5 \times 10^{15}$ cm^{-2} will provide a sufficient population inversion to achieve the amplification length with the optical depth of $\tau > 3$.

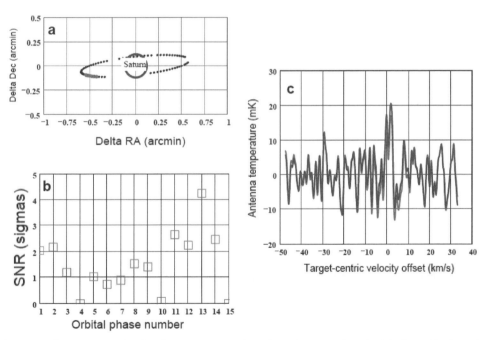

Figure 1. A summary plot for the water maser line detection associated with the orbital motion of Enceladus. Top left: the distribution of the Enceladus position with respect to Saturn during the 3 days in May 2008; black dots represent the positions from which the spectra were acquired, while red dots indicate the positions yielding the best SNR. Bottom left: the detection SNR vs the orbital phase, with the highest level of 4.2 sigma for the 13th orbital phase bin. Right: the spectrum, Doppler corrected and integrated over 4 hours of the orbital phase 13. The red line indicates a raw accumulated spectrum, while the blue line shows the base line ripple corrected one.

We note that such the column density is a small fraction of the peak column density of 1.5×10^{16} cm^{-2} measured by the Cassini UVIS (Hansen *et al.* 2008) for a characteristic length of 100 km. The elongated geometry of the masing cloud, with the major axis directed toward the observer can further relax the requirements on the volume density of water molecules. Other pumping mechanisms, such as interaction with the Kronian magnetosphere, solar-wind plasma and shocks, can also play a determining role.

We consider the results presented here as a strong case for in depth study of the water maser emission in the Kronian system. This study might provide a new insight on the physical conditions in the Kronian system and help to focus further in situ investigations.

Acknowledgements

We are grateful to C.J. Hansen, J.-P. Lebreton, B. Magee, M. Perry, D.E. Shemansky and J.H.Waite Jr. for advice and stimulating discussions, and E. Flamini for continuous interest and support. The work at JIVE was partially supported by the ESA-ESTEC Contract No. 18386, and at INAF-IRA by the Italian Space Agency (ASI) contract I/R/059/04, and by the NSF grant AST0507421 at UKY.

References

Cosmovici, C. B., Montebugnoli, S., Orfei, A., *et al.* 1996, *PSS*, 44, 735

Elitzur, M., 1992, *Astronomical masers*, (Kluwer Academic Publishers, Dordrecht)

Elitzur, M. & Fuqua, J. B. 1989, *ApJ*, 347, L35

Elitzur, M., Hollenbach, D. J., & McKee, C. G. 1989, *ApJ*, 346, 983

Elitzur, M., Hollenbach D. J., & McKee, C. G. 1992, *ApJ*, 394, 221

Giorgini, J. D., Yeomans, D. K., Chamberlin, A. B., *et al.* 1996, *Bull. Am. Astron. Soc.*, 28(3), 1158; also http://ssd.jpl.nasa.gov/horizons.cgi

Hansen, C. J., Esposito, L., Stewart, A. I. F., *et al.* 2006, *Science*, 311, 1422

Hansen, C. J., Esposito, L., Stewart, A. I. F., *et al.* 2008, *Nature*, 456, 477

Montebugnoli, S., Bortolotti, C., Buttaccio, S., *et al.* 1996, *Rev. Sci. Instrum*, 67(2), 365

Mumma, M. J. 1993, in: A. W. Clegg & G. E. Nedoluha (eds.), *Astrophysical Masers*, (Berlin: Springer), p. 455

Pogrebenko, S. V., Gurvits, L. I., Elitzur, M., *et al.*, 2009, *A&A*, 494, L1

Schippers, P., Blanc, M., Andre, N., *et al.* 2008, *J. Geophys. Res.*, 113, A07208

Waite, J. H. Jr., Lewis, W. S., Magee, B. A., *et al.* 2009, *Nature*, 460, 487

Icy bodies of the Solar System
Proceedings IAU Symposium No. 263, 2009
J.A. Fernández, D. Lazzaro, D. Prialnik & R. Schulz, eds.
© International Astronomical Union 2010
doi:10.1017/S1743921310001675

A cometary perspective of Enceladus

Daniel C. Boice and Raymond Goldstein

Space Science & Engineering Div., Southwest Research Institute,
6220 Culebra Road, San Antonio, TX 78238 USA
email: dboice@swri.edu
email: rgoldstein@swri.edu

Abstract. Icy plumes venting from Enceladus draw obvious comparisons to such features seen in comets. This paper outlines a consistent evolution from cometary activity to larger icy bodies in the outer solar system. The major differences are due to the systematic effects of increased gravity, including more spherical solid bodies (self-gravity), less porosity, the possible existence of liquid water due to internal sources of heat (Enceladus) versus possible cometary cyrovolcanism, internal inhomogeneities leading to jet-like features, and the possibility of a quasi-bound dusty gas atmosphere, as opposed to the extensive exospheres of comets. Similarities exist also, including gas and dust emission and the filamentary nature of jet-like features (caused by surface topography in comets), surface evolution by dust accumulation, heat and gas transport through the surface layers, among others. Initial results regarding the plume chemistry and comparisons to CAPS ion data show similarities too. Others have considered additional effects such as, charging of particles, micrometeorite impacts and complex interactions with the E-ring neutrals and plasma in great detail so these topics remain outside the scope of this paper.

Keywords. Planets and Satellites: Enceladus, Comets: Chiron, Comets: General

1. Introduction

The discovery of icy plumes emanating from Saturn's moon Enceladus by the Cassini spacecraft has raised questions about the cometary nature of this small satellite. The release of gas and dust from cometary nuclei is restricted to 'jets' or plumes also and this activity has been observed in comets at distances much further than Saturns orbit, including comet-like activity and a resolved coma of the Centaur Chiron. Enceladus and Chiron have sizes that are much larger than cometary nuclei but their atmospheres are still largely unbound, similar to the exospheres of comets. With Chiron, Enceladus may represent a transitional object in this respect, intermediate to the tightly bound, thin atmospheres typical of planets and large satellites and the greatly extended atmospheres in free expansion typical of cometary comae.

Measurements of the neutral and ion composition of the plumes reveal the presence of water group species, nitrogen-bearing molecules, and other species that have been found in comets (see Table 1). The nature of the volatile materials in Enceladus may also bear similarities with ideas of cometary ices (see Table 2). In other respects, the large size of Enceladus relative to comets and the presence of Saturn and its magnetosphere nearby, brings into question the validity of applying scaling laws to cometary results in order to understand the environment surrounding Enceladus. In addition, release mechanisms for the icy grains and gases at Enceladus, including liquid water mixtures below the cold, icy surface, are not thought to be applicable to comets. These issues and others are discussed as we offer a cometary perspective on our current understanding of Enceladus.

D. C. Boice & R. Goldstein

Table 1. Volatile composition of Enceladus and a typical comet.

Body	H_2O	CH_4	CO/N_2	CO_2
Enceladus	100	1.9	4.4	3.5
Comets	100	$0.14 - 1.5$	$3.5 - 14$	$3 - 6$

Table 2. Molecular production rates for ices at Enceladus (near perihelion[1]).

Parameter	CO	CO_2	H_2O (Crystalline)	H_2O (Amorphous)
$Z_{surface}$ [molecules/cm^{2}s]	10^{17}	$2 \cdot 10^{16}$	10^{10}	10^{16}
$n_{surface}$ [/cm^3]	$8.3 \cdot 10^{12}$	$1.3 \cdot 10^{12}$	$3.5 \cdot 10^{5}$	$3.1 \cdot 10^{11}$
$T_{surface}$ [K]	35	94	105	115
$v_{surface}$ [km/s]	0.12^2	0.16^2	0.21^2	0.35

Notes:
[1] $r_h = 9.0$ AU.
[2] Average speed is less than the escape velocity of Enceladus (0.24 km/s).

2. Observations and *in situ* Measurements

The Cassini spacecraft is opening a new chapter in our understanding of the icy moon Enceladus and the role of venting as the major source of neutral gasses and dust in its surrounding environment. Recent Cassini observations have established that the icy moon Enceladus is actively venting and ejecting water, other neutral molecules, and dust; indicating that it is the origin of the surrounding atmosphere and E-ring torus. The interactions between the subsurface gas source and the jet-like activity of the neutrals and dust are critical processes with significant implications for the evolution of the broader environment. Understanding these interactions would enable us to establish source properties during past and future Cassini flybys and investigate its time variability from encounter to encounter. Due to many similarities to cometary behavior, comparisons to cometary models may allow us to make better estimates of the gas and dust production rates at Enceladus and the likely composition of neutrals and ions in the venting region.

3. Enceladus and the Surrounding Environment

The Cassini spacecraft has performed several close flybys of Enceladus, revealing the moon's surface and environment in great detail and discovering a water-rich plume venting from its South Polar Region (SPR). The composition of the Enceladean plume as measured by the INMS instrument is similar to that seen at most comets (Table 1), containing mostly water vapor as well as minor components of CO (and possibly N_2), CH_4, CO_2, and simple and complex hydrocarbons, such as propane, ethane, and acetylene (Waite *et al.* 2006). This discovery, along with the presence of escaping internal heat and very few impact craters in the SPR, indicates that Enceladus is geologically active. The discovery of the plume supports the notion that material released from Enceladus is the source of the E-ring, composed of water ice grains that are primarily 0.3 to 3 mm in size (Nicholson *et al.* 1996). There are two mechanisms contributing to the ring

(Spahn *et al.* 2006). The most important source of particles comes from the cryovolcanic plume. For gas or dust to escape from a small icy body, the radial component of the gas velocity must exceed the gravitational attraction. For a satellite orbiting a planet we must also consider the Hill sphere. For Enceladus, the Hill sphere radius is 949 km, the escape velocity from the surface is 239 m/s and this speed is reduced to 205 m/s at the Hill sphere radius. While a majority of particles fall back to the surface, some escape and enter orbit around Saturn. The second mechanism comes from hypervelocity micrometeoroid impacts of Enceladus, raising dust particles from the surface, but this alone cannot explain the dust data (Spahn *et al.* 2006). This leaves gas entrainment of dust from the vents as a viable source. Dusty gas interaction is necessary to understand the complex interactions in the plumes and its surrounding environment. We estimate that the maximum grain size that can be lifted due to gas entrainment is about 10 - 100 μm. This doesn't appear to be sufficient to lift the icy grains seen in the E-ring.

Cassini images show fine structures within the plumes, revealing numerous filaments (perhaps due to numerous distinct vents) within a larger, faint component extending out nearly 500 km from the surface. Cassini CDA data are compatible with a dust source and the UVIS later observed gas jets coinciding with the dust jet-like features during recent flybys of Enceladus. Various investigations have estimated H_2O release rates necessary to maintain a steady state of about $2 \cdot 10^{27}$ H_2O molecules/s (Shemansky *et al.* 1993), possibly as high as $3.75 \cdot 10^{27}$ (Jurac *et al.* 2002) or 10^{28} H_2O molecules/s (Jurac & Richardson 2005). Potential H_2O sources identified by these investigators include sputtering and collisions; however, the rates of these processes are not sufficient to replace the lost neutrals (Jurac *et al.* 2002). Plume modeling near closest approach gives the total H_2O production rate from Enceladus of 1.5 to $4.5 \cdot 10^{26}$ molecules/s, and the total gas production rate 1.7 to $5.0 \cdot 10^{26}$ molecules/s, assuming the gas is 90% H_2O from INMS measurements. The bounds of the minimum and maximum estimates indicate a highly variable source rate with time scales of less than 1 hour over the range from 10^{26} to $3 \cdot 10^{27}$ molecules/s. The fissures from which the gas originates must be large enough to allow semicollisional gas flow below the surface (i.e., the size of the fissure must be larger than the mean free path between molecules). If the fissures are too small, molecules will stick to the cold surfaces, eventually sealing it off. Even the average temperature observed by the CIRS of 140K is far below the sublimation temperature so most molecules striking the surface would stick (Waite *et al.* 2006).

4. Source Mechanisms

Analysis of the outgassing suggests that it originates from a body of sub-surface liquid water, which along with the unique chemistry found in the plume has important astrobiology implications. Moons of gas giants can become trapped in orbital resonances that lead to forced libration or orbital eccentricity; proximity to the planet can then lead to tidal heating of the satellite's interior, offering a possible explanation for the activity. However, recent work has shown that Enceladus doesn't oscillate about the tidal equilibrium (Meyer & Wisdom 2008) as required by the tidal heating model of Ojakangas & Stevenson (1986), so other mechanisms must be responsible. The combined analysis of imaging, mass spectrometry, and magnetospheric data suggests that the observed south polar plume emanates from pressurized sub-surface chambers, similar to geysers on Earth (Porco *et al.* 2006). Since no ammonia was found in the vented material, which could act as an anti-freeze, the heated, pressurized chamber would consist of nearly pure liquid water with a temperature of at least 270K. Pure water would require more energy to melt, either from tidal or radiogenic sources, than an ammonia-water mixture. Another possible method for generating a plume is sublimation of warm surface ice. Temperatures found near the south pole by CIRS range from 85-90K to as high as 157K in small areas. This is too warm to be explained by solar heating, indicating heating from the interior of Enceladus (Spencer *et al.* 2006). Ice at these temperatures is warm enough to sublimate at a much faster rate than the background surface, thus generating a plume. This hypothesis is attractive since the sub-surface layer heating the surface water ice could be a mixture of ammonia and water at temperatures as low as 170K, and thus not as much energy is required to produce the plume activity. However, the abundance of particles in the south polar plume favors the cold geyser model, as opposed to ice sublimation (Porco *et al.* 2006). Alternatively, Kieffer *et al.* (2006) suggest that the plumes originate from clathrate hydrates; where carbon dioxide, methane, and nitrogen are released when exposed to the vacuum of space by the active, tiger stripe fractures. This hypothesis would not require the amount of heat needed to melt water ice as required by the cold geyser model, and would explain the lack of ammonia. An additional possibility is the presence of amorphous water ice below the Enceladean surface. The exothermic phase change from amorphous to crystalline ice occurs in the temperature regime at Enceladus and releases sufficient water gas to explain the observed release rate as our preliminary modeling indicates (see Table 2). These mechanisms are appropriate for comets when gravity and low porosity considerations are included.

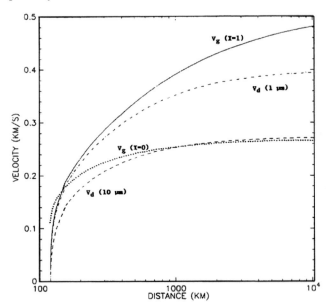

Figure 1. Gas and dust speeds for Chiron at perihelion. The dust acceleration region is within ~ 10 R_{Chiron}. For comparison, the dotted line is the gas speed for a dust-free model ($\chi = 0$).

5. The Dusty Gas Outflow

To test the cometary approach for Enceladus, a preliminary model of Chiron that treats the physics and chemistry of the comet coma in great detail (Schmidt *et al.* 1988) was prepared by Boice *et al.* (1991) and Boice & Huebner (1993). CO was assumed to be the only volatile. Being a diatomic molecule, CO is not an efficient emitter in the infrared so the radiative cooling term is negligible. In the simulations, gas and dust are rapidly accelerated upon leaving the nucleus as illustrated in Figure 1 (χ is the dust-to-gas mass ratio). For standard dust densities, small particles are more efficiently entrained with the gas flow than large particles, resulting in higher terminal speeds. The acceleration zone for all particles is approximately within 10 radii of the surface.

The inclusion of dust has two important effects on the gas flow. The first is an initial mass-loading of the gas, maintaining the gas velocity at subsonic values close to the surface of the nucleus. The second effect is a strong thermal coupling of the gas and dust near the nucleus as shown in Figure 2. Upon release, the dust heats rapidly to its radiative equilibrium value of 95K. Collisions of molecules with dust particles heat the CO gas (initially at 30K) to 85K within a Chiron radius. This results in a terminal gas velocity about 80% higher than that calculated from a pure gas model. Even with a modest amount of dust (dust-to-gas mass ratio of 0.1), the gas is significantly heated in the near-nucleus region. Recent progress of the model includes the addition of the gravity to Chiron and the incorporation of dust fragmentation and distributed coma sources of gas-phase species related to the dust.

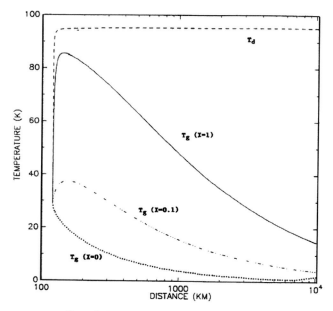

Figure 2. Temperature profiles of the gas and dust for Chiron at perihelion. Upon release from the nucleus, both sizes of dust quickly reach radiative equilibrium at 95K. The gas is strongly heated by the dust in the near-nucleus region for the $\chi = 1$ case. Even with a modest dust-to-gas mass ratio ($\chi = 0.1$), the gas is heated significantly above the dust-free model ($\chi = 0$).

6. Summary and Conclusion

This study represents a consistent evolution from our comet nucleus and coma experience to larger icy bodies in the outer solar system. The major difference is the systematic effects of increased gravity, including more spherical solid bodies (self-gravity), less porosity, the possible existence of liquid water due to internal sources of heat (Enceladus) versus possible cometary cyrovolcanism (Belton *et al.* 2008), internal inhomogeneities leading to jet-like features, and the possibility of a quasi-bound dusty gas atmosphere, as opposed to the extensive exospheres of comets (Johnson *et al.* 2008). Similarities exist also, including jet-like features with filaments (caused by surface topography in comets) gas and dust emission, surface evolution by dust accumulation, heat and gas transport through the surface layers, and others. Initial results regarding the plume chemistry and comparisons to CAPS ion data show similarities too (Boice & Goldstein 2008, Goldstein *et al.* 2008). Other effects that need to be considered in a realistic model include charging of particles, micrometeorite impacts and complex interactions with the E-ring neutrals and plasma. These topics remain outside the scope of this work as other researchers have considered them in great detail.

Acknowledgements

We acknowledge support from the SwRI Internal Research & Development Program, the NASA Cassini (CAPS) Mission, and the NSF Planetary Astronomy Program.

References

Belton, M. J. S., Feldman, P. D., A'Hearn, M. F., & Carcich, B. 2008, *Icarus*, 198, 189

Boice, D. C. & Goldstein, R. 2008, *Eos Trans. AGU*, 89(53), #P23B-1368

Boice, D. C., Stern, S. A., & Huebner, W. F. 1991, *LPSC XXII*, 22, 121

Boice, D. C. & Huebner, W. F. 1993, in: W. F. Huebner (ed.), *Workshop on the Activity of Distant Comets* (San Antonio: SwRI), p. 134

Goldstein, R., Boice, D. C., Young, D. T., Reisenfeld, D. B., & Smith, H. T. 2008, *Eos Trans. AGU*, 89(53), #P23B-1382

Johnson, R. E., Combi, M. R., Fox, J. L., Ip, W.-H., Leblanc, F., McGrath, M. A., Shematovich, V. I., Strobel, D. F., & Waite, J. H. 2008, *Space Science Reviews*, 139, 355

Jurac, S., McGrath, M. A., Johnson, R. E., Richardson, J. D., Vasyliunas, V. M., & Eviatar, A. 2002, *Geophys. Res. Lett.*, 29, 25

Jurac, S. & Richardson, J. D. 2005, *J. Geophys. Res.*, 110, A09220

Kieffer, S. W., Lu, X., Bethke, C. M., Spencer, J. R., Marshak, S., & Navrotsky, A. 2006, *Science*, 314, 1764

Meyer, J. & Wisdom, J. 2008, *Icarus*, 198, 178

Nicholson, P. D., Showalter, M. R., Dones, L., French, R. G., Larson, S. M., Lissauer, J. J., MeGhee, C. A., Seitzer, P., Sicardy, B., & Danielson, G. E. 1996, *Science*, 272, 509

Ojakangas, G. W. & Stevenson, D. J. 1986, *Icarus*, 66, 341

Porco, C. C., Helfenstein, P., Thomas, P. C., Ingersoll, A. P., Wisdom, J., West, R., Neukum, G., Denk, T., Wagner, R., Roatsch, T., Kieffer, S., Turtle, E., McEwen, A., Johnson, T. V., Rathbun, J., Veverka, J., Wilson, D., Perry, J., Spitale, J., Brahic, A., Burns, J. A., Del Genio, A. D., Dones, L., Murray, C. D., & Squyres, S. 2006, *Science*, 311, 1393

Schmidt, H. U., Wegmann, R., Huebner, W. F., & Boice, D. C. 1988, *Comp. Phys. Comm.*, 49, 17

Shemansky, D. E., Matheson, P., Hall, D. T., Hu, H.-Y., & Tripp, T. M. 1993, *Nature* 363, 329

Spahn, U., Schmidt, J,, Albers, N., Hörning, M., Makuch, M., Seiß, M., Kempf, S., Srama, R., Dikarev, V., Helfert, S., Moragas-Klostermeyer, G., Krivov, A.V., Sremčević, M., Tuzzolino, A. J., Economou, T., & Grün, E. 2006, *Science*, 311, 1416

Spencer, J. R., Pearl, J. C., Segura, M., Flasar, F. M., Mamoutkine, A., Romani, P., Buratti, B. J., Hendrix, A. R., Spilker, L. J., & Lopes, R. M. C. 2006, *Science*, 311, 1401

Waite, J. H., Combi, M. R., Ip, W.-H., Cravens, T. E., McNutt, R. L., Kasprzak, W., Yelle, R., Luhmann, J., Niemann, H., Gell, D., Magee, B., Fletcher, G., Lunine, J., & Tseng, W.-L. 2006, *Science*, 311, 1419

Icy Bodies of the Solar System
Proceedings IAU Symposium No. 263, 2009
J.A. Fernández, D. Lazzaro, D. Prialnik & R. Schulz, eds.
© International Astronomical Union 2010
doi:10.1017/S1743921310001687

On the Origin of Retrograde Orbit Satellites around Saturn and Jupiter

Yuehua Ma[1,2], Jiaqing Zheng[2,1] and Xiaohai Shen[3]

[1] Purple Mountain Observatory, Nanjing 210008, China,
email: yhma@pmo.ac.cn

[2] Turku University, Tuorla Observatory, 21500 Piikkiö, Finland
email: zheng@utu.fi

[3] Jiaozuo Teachers College, Henan, 454000, China

Abstract. Many Retrograde Orbit Satellites around Jupiter and Saturn have been found recently. Most of them are small with irregular shapes. They are farther from the planet than regular satellites. Their orbits have big eccetricities.

We tested their dynamical origin and found:

1. The small bodies can be captured by normal satellites and form retrograde orbits. But these orbits are not stable. Sooner or later, they would escape from planetary region or fall down into the planets.

2. Another way is that they have formed by collisions just after regular moons formed. We studied the mechanism and obtained good results.

Keywords. Solar System, planets and satellites: formation.

1. Introduction

Jupiter and Saturn have 63 (R = 48) and 61 (R = 29) moons, respectively (Figure 1, or http://ssd.jpl.nasa.gov). "R" is the number of retrograde orbits (dark colour in Figure).

We list retrograde irregular moons of Saturn on table 1 and put a photo of Phoebe - biggest irregular moon (Figure 1, right side) here. We also list prograde irregular moons on the table since they have irregular shapes and big i and e.

From the table and Figures we can see:

(1) Normal moons are very close to their planets and with regular orbits (small i, e). They have regular shapes (ball).

(2) Irregular moons are far from their planets with irregular shapes.

(3) Retrograde irregular moons mostly are far from their planet. They are located about half of the radius of planetary activity sphere and their orbits are near planetary orbital plane (few near 145°, most > 165°).

(4) Prograde irregular moons are closer than retrograde irregular moons, about 25% of radius of planetary activity sphere. Their inclinations are big, mostly about 45°.

Normal moons have formed by accretion. This is why they have regular shapes and regular orbits. Irregular moons cannot form by normal process as normal moons. We suppose that they were from outer solar system.

2. Test and problems of capture process

In an early work (Zheng 1994), we proved that short period comets were captured when they came from Oort cloud (or Kuiper belt) by close encounters with planets. The capture process of moons can be similar when small bodies come into planetary activity sphere (see Figure 2).

Table 1. Irregular moons of Saturn

Name	$a(km)$	$i(°)$	e	size(km)					
Kiviuq	11111000	45.71	0.334	16	Jarnsaxa	18811000	163.3	0.216	6
Ijiraq	11124000	46.44	0.316	12	Narvi	19007000	145.8	0.431	7
Phoebe	12944300	174.8	0.164	240	Bergelmir	1933800	158.5	0.142	6
Paaliaq	15200000	45.13	0.364	22	Suttungr	19459000	175.8	0.114	7
Skathi	15541000	152.6	0.270	8	Hati	19856000	165.8	0.372	6
Albiorix	16182000	33.98	0.478	32	Bestla	20129000	145.2	0.521	7
Bebhionn	17119000	35.01	0.469	6	Farbauti	20390000	156.4	0.206	5
Erriapo	17343000	34.62	0.474	10	Thrymr	20474000	176.0	0.470	7
Siarnaq	17531000	45.56	0.295	40	Aegir	20735000	166.7	0.252	6
Skoll	17665000	161.2	0.464	6	Kari	22118000	156.3	0.478	7
Tarvos	17983000	33.82	0.531	15	Fenrir	22453000	164.9	0.136	4
Tarqeq	18009000	46.09	0.160	7	Surtur	22707000	177.5	0.451	6
Greip	18206000	179.8	0.326	6	Ymir	23040000	173.1	0.335	18
Hyrrokkin	18437000	151.4	0.333	8	Loge	23065000	167.9	0.187	6
Mundilfari	18685000	167.3	0.210	7	Fornjot	25108000	170.4	0.206	6
Unnamed	Irregular	moons							
S/2004 S07	19800000	165.1	0.580	6	S/2006 S1	18981135	154.2	0.130	6
S/2004 S12	19650000	164.0	0.401	5	S/2006 S3	21132000	150.8	0.471	6
e S/2004 S13	18450000	167.4	0.273	6	S/2007 S2	16560000	176.7	0.218	6
e S/2004 S17	18600000	166.6	0.259	4	S/2007 S3	20518500	177.2	0.130	5

Notes: We put some prograde orbit moons on the table since they are also "Irregular" by shapes and big e.

Retrograde orbit moons can be captured by normal moons of the planets, but these orbits always cross the orbits of normal moons, then sooner or later, they would have new close encounters which could cause them to escape. These orbits are not stable. In other words, if retrograde orbit moons were captured by inner moons of Saturn, they can be the transfer source of short period comets, but they may not become into stable orbits as current retrograde moons we see now.

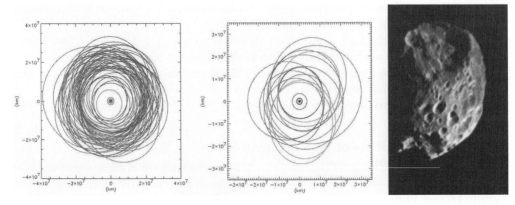

Figure 1.   Jupiter and Saturn's Known Satellites   Phoebe (by Cassini)

3. Collision process – formation of retrograde orbits

Now, we study retrograde moons of Saturn (similar for Jupiter). Since current orbits of retrograde moons are stable, they might form by another way at early time after Saturn formed. At that time, around Saturn's orbit, still there were some planetesimals, and around Saturn, after normal moons formed, many small bodies remained also. When these two kinds bodies met in Saturn's activity area, the relative motion between them were opposite: planetesimals were 'retrograde" and small bodies around Saturn were prograde, when they collided or combined with each other, some retrograde moons could form.

Let's consider a small body moves with a planet, the planet moves on a circular orbit and small body has eccentricity 0.1:$a_p = 1, e_p = 0, a_s = 1, e_s = 0.1$.

In Figure 3, left side is the orbits of planet and the small body in a fixed coordinate, right side is the orbits on a rotating coordinate. In a rotating coordinate, the small body moves around the planet looks like a satellite but moves on a retrograde direction.

The activity sphere of Saturn is

$$m_s^{0.4} * a_s = 0.0382 * 9.555 AU = 0.365 AU = 5.5 * 10^7 km$$

Most retrograde orbits are located about half of this value.

Suppose a prograde moon moves in a circular orbit with $a = 0.2AU$, the orbital velocity is

$$V_m = \sqrt{m_s/0.2} * V_e = 1.13 km/sec$$

.

A particle with $a = a_s$ and $e < 0.038$ moves on saturn's orbital plane would come into the activity sphere. Saturn's orbital velocity is $9.65 km/sec$, an orbit with $e = 0.02$ at perihelion ($q = a_s - 0.2AU$) would have a relative velocity (to Saturn) about $0.2 km/sec$. When this particle arrives into activity sphere at $r = 0.2AU$, its velocity is about $1.15 km/sec$. A little more than the circular orbital velocity but on a retrograde direction.

If this "retrograde" body collided by the prograde satellite (in different orientations), their relative velocity can be reduced, and one of them could become a retrograde satellite.

4. Collision probability

Suppose we have two particles with $r = 10km$. One particle moves in Saturn's activity sphere and another moves as a quasi-satellite around Saturn. When a quasi-satellite comes into Saturn's activity sphere, in one revolution, the collision probability is

$$p_{revo} = (2r)^3 / R_{act}^3 = 5 * 10^{-19}$$

where $R_{act} = 0.365 AU = 5.5 * 10^7 km$

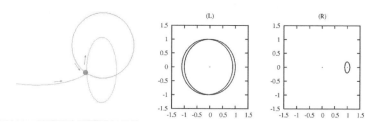

Figure 2. Sketch of a capture process   **Figure 3.** Quasi-satellite

Table 2. First results from simulation

Test	c	$a(10^7 km)$	$i(°)$	e
1	0.2	1.895	171.3	0.159
1	0.2	2.121	176.3	0.122
1	0.2	2.154	170.6	0.085
1	0.2	2.429	179.5	0.198
2	0.5	2.620	169.0	0.419
2	0.5	2.936	167.1	0.405

At the planetary and satellite system formation time, there are billions small bodies (Titan is about 1.3E8 greater than a $r = 10$km small body) .

The number of collisions should be multiplied by

$$N_{particle1} * N_{particle2} * N_{revolution}.$$

Let't put

$$N_{particle1} = 1 * 10^8, N_{particle2} = 1 * 10^9, N_{revolution} = T/30yr = 3 * 10^7/30 = 1 * 10^6,$$

Then

$$N_{collision} = p * N_{particle1} * N_{particle2} * N_{revolution} = 5 * 10^4$$

Of course this is not very accurate, but not very far from real process.

5. Simulation

In simulation, we set a planet as Saturn revolves around the Sun. Put 10^n $(n = 5)$ particles around Saturn with random orbital elements (small i, e in Saturn's activity sphere, put $10 \cdot 10^n$ particles (quasi-satellites) move in similar orbits as Saturn but with $0 < e < 0.038$. We check the condition for impacts (collision coefficient $c = 0.2 - 0.5$, $\Delta V_{new} = c \cdot \Delta V_{old}$ along the centers of two bodies) and obtain a few retrograde satellite orbits (They are all from quasi-satellites). We will simulate Large N in order to obtain more results.

Acknowledgements

We would like to acknowledge the support of the National Natural Science Foundation of China (No. 10573037, 10933004), the exchange program between Finnish Academy and NSFC and the Minor Planetary Foundation of Purple Mountain Observatory.

Reference

Zheng, J. Q. 1994, *ApJ Suppl.*, 108, 1994

Icy Bodies of the Solar System
Proceedings IAU Symposium No. 263, 2009
J.A. Fernández, D. Lazzaro, D. Prialnik & R. Schulz, eds.
© International Astronomical Union 2010
doi:10.1017/S1743921310001699

Long-term dynamics of Methone, Anthe and Pallene

Nelson Callegari Jr. and Tadashi Yokoyama

Instituto de Geociências e Ciências Exatas, Unesp - Univ Estadual Paulista,
Departamento de Estatística, Matemática Aplicada e Computação.
Av. 24-A, Rio Claro/SP/Brazil, CEP 13506-700.
email: `calleg@rc.unesp.br`, `tadashi@rc.unesp.br`

Abstract. We numerically investigate the long-term dynamics of the Saturn's small satellites Methone (S/2004 S1), Anthe (S/2007 S4) and Pallene (S/2004 S2). In our numerical integrations, these satellites are disturbed by non-spherical shape of Saturn and the six nearest regular satellites. The stability of the small bodies is studied here by analyzing long-term evolution of their orbital elements.

We show that long-term evolution of Pallene is dictated by a quasi secular resonance involving the ascending nodes (Ω) and longitudes of pericentric distances (ϖ) of Mimas (subscript 1) and Pallene (subscript 2), which critical argument is $\varpi_2 - \varpi_1 - \Omega_1 + \Omega_2$. Long-term orbital evolution of Methone and Anthe are probably chaotic since: i) their orbits randomly cross the orbit of Mimas in time scales of thousands years); ii) long-term numerical simulations involving both small satellites are strongly affected by small changes in the initial conditions.

Keywords. Celestial Mechanics; planets and satellites: individual (Saturn, Aegaeon, Methone, Anthe, Pallene); methods: numerical.

---

1. Introduction

Short-term dynamics (i.e., time scales of a few years), and determinations of orbital elements of Methone and Pallene, are reported in (Porco *et al.* 2005, Spitale *et al.* 2006, Porco *et al.* 2007, Jacobson *et al.* 2008). Anthe's orbit, the most recent small body detected in Saturnian system, is studied in Cooper *et al.* 2008.

In spite of some recent investigations (Porco *et al.* 2005, Callegari & Yokoyama 2008), long-term numerical integrations (i.e, in time scales of millennia), of the orbits of the three small satellites are not reported yet in the literature. Ephemeris of Methone, Anthe and Pallene are limited to 1,000 years (*Horizons data system*, http://ssd.jpl.nasa.gov). Moreover, inspection of different references listed above show slightly different values for some elements (in particular, the semi-major axis of the small satellites†).

In this work, we analyze the results of a great deal of numerical integrations over 60,000-years of the orbits of small satellites similar to Methone, Anthe and Pallene, which initial semi-major axes are varied within the range of different values published in literature.

We have considered the canonical set (Hamiltonian form) of equations of motion within the domain of general N-body problem (e.g. Ferraz-Mello *et al.* 2005):

† After this paper was nearly complete we noted that new ephemeris for Methone, Anthe and Pallene have been published in http://ssd.jpl.nasa.gov. The new values of semi-major axes of Methone ($\sim 194,684$ km), Anthe ($\sim 198,131$ km) and Pallene ($\sim 212,708$ km) are within the ranges of initial conditions covered in this work. We are now testing additional simulations considering these new data.

$$H = H_0 + H_1 + H_{\mathbf{J}} \tag{1.1}$$

$$H_0 = \sum_{i=1}^{N} \left(\frac{|\vec{p_i}|^2}{2\beta_i} - \frac{\mu_i \beta_i}{|\vec{r_i}|} \right), \tag{1.2}$$

$$H_1 = \sum_{0 < i < j} \left(-\frac{Gm_i m_j}{\Delta_{ij}} + \frac{\vec{p_i} \cdot \vec{p_j}}{M} \right), \quad j = 1, \ldots, N, \tag{1.3}$$

$$H_{\mathbf{J}} = -\sum_{i=1}^{N} \frac{\mu_i \beta_i}{|\vec{r_i}|} \left[-\sum_{l=2}^{\infty} J_l \left(\frac{R_e}{|\vec{r_i}|} \right)^l P_l(\sin \varphi_i) \right], \tag{1.4}$$

where $\vec{r_i}$ and $\vec{p_i}$ (canonical variables), are position vectors of the satellites relative to the center of the planet and momentum vectors relative to the center of mass of the system, respectively. $\mu_i = G(M + m_i)$, $\beta_i = \frac{M m_i}{M + m_i}$, $\Delta_{ij} = |\vec{r_i} - \vec{r_j}|$, where G is the gravitational constant, and M, m_i are the planet mass and the individual masses of satellites, respectively. The chosen units are the equatorial radius of the planet ($R_e = 60,268$ km), day and M.

Equation (1.2) defines the Keplerian motion of each satellite around the planet and Eq. (1.3) gives the mutual interaction among the satellites. Equation (1.4) represents the perturbation of the non-sphericity of the planet, $P_l(\sin \varphi_i)$ are the classical Legendre polynomials, φ_i are the latitudes of the orbits of the satellites referred to the *equator* of the planet. We consider $l = 2, 4$, where in Eq. (1.4) J_2, J_4 are the zonal oblateness coefficients.

Equations of the motion of the satellites, which are solved numerically using RA15 code (Everhart 1985), are

$$\frac{d\vec{r_i}}{dt} = \frac{\partial H}{\partial \vec{p_i}}, \quad \frac{d\vec{p_i}}{dt} = -\frac{\partial H}{\partial \vec{r_i}}, \quad i = 1, \ldots, N, \tag{1.5}$$

All outputs shown in this work are *planetocentric* ones. In the simulations which we show here, we have neglected the mutual perturbations between the small bodies, and numerical simulations include a small satellite and six inner regular satellites (Mimas, Enceladus, Tethys, Dione, Rhea and Titan) (i.e., $N = 7$ in all equations above). However, we have tested the effects of the mutual perturbations between the small satellites: the masses of these last bodies are very small and the simulations considering the whole system, clearly confirmed that their mutual effects are negligible.

Initial *osculating* planetocentric elements and parameters of *regular* satellites, and Saturn gravity field, are obtained from Jacobson *et al.* 2006 and *Horizons data system* (data January 1, 2007). Initial semi-major axis of the small bodies will be indicated in the figures given in next section. In our code, the mass of the small body is equal to the mass of a spherical body with density similar to the Mimas's and ∼3 km in diameter (e.g. Porco *et al.* 2005, Porco *et al.* 2007).

2. Results

In this section, we discuss the results of our numerical simulations on long-term evolution of orbits of Methone, Anthe and Pallene. Simulations are shown in the time span of 60,000 years, although several of them were continued until 100,000 years. More than

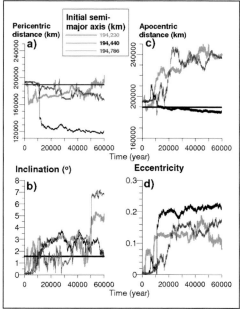

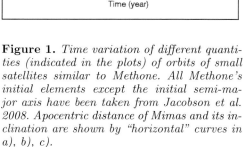

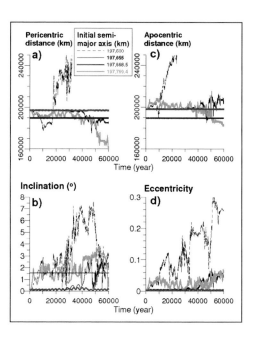

Figure 1. *Time variation of different quantities (indicated in the plots) of orbits of small satellites similar to Methone. All Methone's initial elements except the initial semi-major axis have been taken from Jacobson et al. 2008. Apocentric distance of Mimas and its inclination are shown by "horizontal" curves in a), b), c).*

Figure 2. *Similar to Fig. 1, but for clones of Anthe. Except for the semi-major axis, all Anthe's initial conditions have been obtained from osculating elements given in Cooper et al. 2008. The only exception is the strong-gray curves, where mean values of Cooper et al. 2008 are used instead of osculating elements.*

ten different simulations for each satellite have been done with our Fortran code utilizing different desktop computers, operational systems (Windows and Linux), and step-size in the integrations.

2.1. *Methone*

In Fig. 1(a), the black (almost horizontal) line is the apocentric distance of Mimas during the time integration of one simulation. The remaining curves are the pericentric distances for different clones of Methone. Fig. 1(b) shows that the orbital inclination of fictitious Methone varies quite randomly and in particular, during the 60,000 years, Mimas and Methone can share the same orbital plane several times. Therefore a collision seems to be a possibility in this scenario, unless a favorable mutual inclination of these satellites is always guaranteed.

The values of initial Methone's semi-major axis 194, 230 km and 194, 440 km (light-gray and black curves in Fig. 1, respectively), are given by Jacobson *et al.* 2008 and Spitale *et al.* 2006, respectively. Porco *et al.* 2007 lists the same value of Spitale *et al.* 2006 for the semi-major axis of Methone. Note the large divergence between their corresponding trajectories. The largest variation of eccentricity occurs when we adopt the value 194, 440 km (Fig. 1(d)). The high sensitivity of the evolution of the orbits with the initial conditions is probably an indication of the long-term chaotic motion of Methone. In Callegari & Yokoyama 2009 we are studying in details the dynamics of Methone with

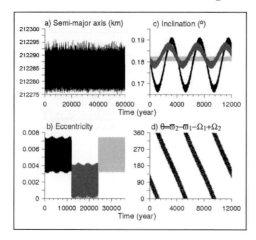

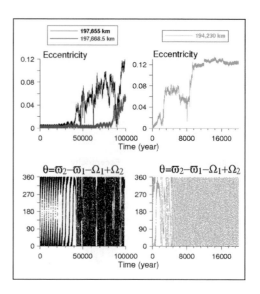

Figure 3. *Black curves: time variation of different quantities (indicated in the plots) of orbits of small satellites similar to Pallene. θ is the critical angle associated to quasi-resonance. Except the initial eccentricity in strong-gray curves (where initial eccentricity is null), all initial elements of Pallene have been taken from Jacobson et al. 2008. Light-gray curves: Mimas' mass is set as almost zero in the code. In b), the different portions of the eccentricity variation of the three simulations are shown.*

Figure 4. *Time variation of eccentricity and critical angle of some orbits with same colors given in Figure 1 (right column, Methone) and Figure 2 (left column, Anthe), in different time scales than those shown in Figs. 1, 2. Note that "stable" strong-gray curve in Fig. 2(d) suffers large variation on larger time scales.*

the technique of dynamical maps, where we show that Methone is close to the collision curve with Mimas, its main disturber.

Figure 1(c) shows that the apocentric distances attain in general larger values: in all cases, the quantity attains the Pallene's semi-major axis, ($\sim 212,280$) km; in some cases the orbit of Enceladus may be crossed ($\sim 238,408$ km).

2.2. *Anthe*

In Fig. 2(a), the listed Anthe's semi-major axes are slightly smaller than $197,770$ km given in Porco *et al.* 2007.

Curves in Fig. 2(a) show that pericentric distances of all but one fictitious Anthe are smaller than the apocentric distance of Mimas (black trajectories in Figs. 2(a,b)). The discussion above on the possibility of collisions between Methone and Mimas is also valid here. In the case of Anthe, if the initial conditions of the simulations are the osculating elements as given in Cooper *et al.* 2008, this satellite seems to be stable during the adopted time span (strong-gray curves in Fig. 2). However, for longer times as shown in Fig. 4, this scenario is not maintained.

It is remarkable the effect caused when a small change is applied in the initial semi-major axis of Anthe given in Cooper *et al.* 2008 ($a = 197,655$ km). For instance, for $a = 197,600$ km, Anthe's eccentricity attains large values (Fig. 2(d)), so that this satellite can penetrate Dione's orbit, crossing also the orbits of Tethys and Enceladus (see dashed curves are interrupted in order to keep the scale of y-axes).

The previous discussion on probable long-term (chaotic) instability of Methone is also a possibility in the case of Anthe, though its orbit is slightly far from collision curve with Mimas (Callegari & Yokoyama 2009).

2.3. *Quasi secular resonance between Mimas and Pallene*

Black curves in Fig. 3 show two typical results taken from several long-term numerical simulations involving small satellites similar to Pallene. Contrary to the cases discussed above, now the semi-major axes suffer small variations (on the order of 20 km), while eccentricity and inclination also vary between small extremes with a quasi-periodic modulation (not present in semi-major axis evolution).

The oscillations seen in Figs. 3(b,c) are due to proximity of Pallene's orbit of a type of quasi secular resonance associated to the critical angle $\theta = \varpi_2 - \varpi_1 - \Omega_1 + \Omega_2$, which circulates in retrograde sense with period ~ 4400 years (Fig. 3(d)).

Porco *et al.* 2005 pointed out that the non-null value of the current Pallene's eccentricity (~ 0.004, similar to the Enceladus'), could be explained by some secular resonance. Here we identify a possible candidate, but it is a *quasi-resonance*, and its effect is not strong enough to increase the eccentricity of the small satellite.

In fact, light-gray curve in Fig. 3(b) shows eccentricity of a small body in a simulation where Mimas' mass has been taken almost zero: though the quasi-periodic oscillations disappears, the interval of variation in eccentricity is almost the same as that seen in the real case, black line. (Annulling Mimas' effects, the variation of inclination however is very small; see Fig. 3(c)).

The variation in eccentricity is a natural outcome due to joint J_2 and Mimas' effects. (J_4 effects can be negligible; $J_2 = 0.0162906$ and $J_4 = -0.000936$.) Though not shown in Fig. 3, we have tested the individual effects of Enceladus and Titan (annulling their effects on the program): they are not responsible for the variation in eccentricity. Plot in strong-gray in Figs. 3(b,c) show that an initial null eccentricity evolves to a maximum near the current value. So, if initial Pallene's eccentricity were null, currently Pallene is near this maximum.

Some curves presented in Fig. 1 are shown again in Fig. 4. It can be observed that the jump of the eccentricity seems to occur when the critical angle θ alternates between circulation and libration. Alternatively we can say that transition of θ is due to the significative jump suffered by the eccentricity and inclinations which was caused by the close approach between the small satellite and Mimas. Since this close approach seems to cause more drastic effects that the θ-quasi resonance (which is of the order of $sin(i_1)sin(i_2)e_1e_2$), the alternation circulation-libration is only a consequence and not the reason of the increase of the small satellite's eccentricities and inclinations.

3. Conclusions

Complementing previous simulations (Callegari & Yokoyama 2008), we show here that the orbits of Methone and Anthe are probably chaotic and they can cross Mimas' orbit several times in a few thousand years. The possibility of collision of the satellites with Mimas must be taken into account in the evolutionary studies of the bodies (see Porco *et al.* 2007 and references therein). The orbit of Pallene is long-term stable in the time scale of the numerical simulations studied here. A quasi secular resonance involving Pallene and Mimas orbits was identified, but its effects are not important to long-term evolution of the small satellite.

Recently (Porco 2009), a new small body (S/2008 S1; Aegaeon) was found close to the G-ring. Preliminary 100,000-years numerical simulations with our code show that the orbit of S/2008 S1 is long-term stable, suffering only small variations in the elements ($\Delta a = 10$km, $\Delta e = 0.006$, $\Delta i = 0.004°$). It is worth noting that due the to the proximity to the main ring system and its very small size (radius of about 250 meters), long-term dynamics of S/2008 S1 is more complex than their neighbors.

Acknowledgements: Fapesp (06/58000-2, 06/61379-3, 08/52927-2) and CNPQ.

References

Callegari, N. Jr, & Yokoyama, T. 2008, *Bulletin of the American Astronomical Society*, 40, 479

Callegari, N. Jr, & Yokoyama, T. 2009, *Resonances in the system of Saturnian inner Satellites (Preprint)*

Cooper, N. J., Murray, C. D. *et al.* 2008, *Icarus*, 195, 765

Everhart, E. 1985, *IAU Coloquium*, 83, 185

Ferraz-Mello, S., Michtchenko, T. A., Beaugé, C., & Callegari, N. Jr. 2005, *Lecture Notes in Physics*, 683, 219

Jacobson, R. A. *et al.* 2006, *AJ*, 132, 2520

Jacobson, R. A., Spitale, J. *et al.* 2008, *AJ*, 135, 261

Porco, C. C. *et al.* 2005, *Science*, 307, 1226

Porco, C. C., Thomas, P. C., Weiss, J. W., & Richardson, D. C. 2007, *Science*, 318, 1602

Porco, C. C. 2009, *IAU Circ.*, 9023, 1

Spitale, J. N., Jacobson, R. A., Porco, C. C., & Owen, Jr. W. M. 2006, *AJ*, 132, 792

Icy Bodies of the Solar System
Proceedings IAU Symposium No. 263, 2009
J.A. Fernández, D. Lazzaro, D. Prialnik & R. Schulz, eds.
© International Astronomical Union 2010
doi:10.1017/S1743921310001705

How do the small planetary satellites rotate?

Alexander V. Melnikov and Ivan I. Shevchenko

Pulkovo Observatory of the Russian Academy of Sciences,
Pulkovskoje ave. 65, St.Petersburg 196140, Russia
email: melnikov@gao.spb.ru, iis@gao.spb.ru

Abstract. We investigate the problem of the typical rotation states of the small planetary satellites from the viewpoint of the dynamical stability of their rotation. We show that the majority of the discovered satellites with unknown rotation periods cannot rotate synchronously, because no stable synchronous 1:1 spin-orbit state exists for them. They rotate either much faster than synchronously (those tidally unevolved) or, what is much less probable, chaotically (tidally evolved objects or captured slow rotators).

Keywords. Planets and satellites: general.

The majority of planetary satellites with known rotation states rotates synchronously (like the Moon, facing one side towards a planet), i.e., they move in synchronous spin-orbit resonance 1:1. The data of the NASA reference guide (NASA website data) combined with additional data (Maris *et al.* 2001; Maris, Carraro & Parisi 2007; Grav, Holman & Kavelaars 2003) implies that, of the 33 satellites with known rotation periods, 25 rotate synchronously.

For the tidally evolved satellites, this observational fact is theoretically expected. The planar rotation (i.e., the rotation with the spin axis orthogonal to the orbital plane) in synchronous 1:1 resonance with the orbital motion is the most likely final mode of the long-term tidal evolution of the rotational motion of planetary satellites (Goldreich & Peale 1966; Peale 1977). In this final mode, the rotational axis of a satellite coincides with the axis of the maximum moment of inertia of the satellite and is orthogonal to the orbital plane.

Another qualitative kind of rotation known from observations is fast regular rotation. There are seven satellites that are known to rotate so (Maris *et al.* 2001; Maris, Carraro & Parisi 2007; Grav, Holman & Kavelaars 2003; Bauer *et al.* 2004; NASA website data): Himalia (J6), Elara (J7), Phoebe (S9), Caliban (U16), Sycorax (U17), Prospero (U18), and Nereid (N2); all of them are irregular satellites. These satellites, apparently, are tidally unevolved.

A third observationally discovered qualitative kind of rotation is chaotic tumbling. Wisdom, Peale & Mignard (1984) and Wisdom (1987) demonstrated theoretically that a planetary satellite of irregular shape in an elliptic orbit could rotate in a chaotic, unpredictable way. They found that a unique (at that time) probable candidate for the chaotic rotation, due to a pronounced shape asymmetry and significant orbital eccentricity, was Hyperion (S7). Besides, it has a small enough theoretical timescale of tidal deceleration of rotation from a primordial rotation state. Later on, a direct modelling of its observed light curves (Klavetter 1989; Black, Nicholson & Thomas 1995; Devyatkin *et al.* 2002) confirmed the chaotic character of Hyperion's rotation. Recent direct imaging from the *CASSINI* spacecraft supports these conclusions (Thomas *et al.* 2007).

It was found in a theoretical research (Kouprianov & Shevchenko 2005) that two other Saturnian satellites, Prometheus (S16) and Pandora (S17), could also rotate chaotically (see also Melnikov & Shevchenko 2008). Contrary to the case of Hyperion, possible chaos in rotation of these two satellites is due to fine-tuning of the dynamical and physical parameters rather than to a large extent of a chaotic zone in the rotational phase space.

We see that the satellites spinning fast or tumbling chaotically are a definite minority among the satellites with known rotation states. However, the observed dominance of synchronous behaviour might be a selection effect, exaggerating the abundance of the mode typical for big satellites. This is most probable. Peale (1977) showed on the basis of tidal despinning timescale arguments that the majority of the irregular satellites are expected to reside close to their initial (fast) rotation states.

A lot of new satellites has been discovered during last years. Now the total number of satellites exceeds 160 (see NASA website data). The rotation states for the majority of them are not known. In what follows, we investigate the problem of typical rotation states among all known satellites.

We consider the motion of a satellite with respect to its mass centre under the following assumptions. The satellite is a nonspherical rigid body moving in a fixed elliptic orbit about a planet. We consider the planet to be a fixed gravitating point. The shape of the satellite is described by a triaxial ellipsoid with the principal semiaxes $a > b > c$ and the corresponding principal central moments of inertia $A < B < C$. The dynamics of the relative motion in the planar problem (i.e., when the satellite rotates/librates in the orbital plane) are determined by the two parameters: $\omega_0 = \sqrt{3(B - A)/C}$, characterizing the dynamical asymmetry of the satellite, and e, the eccentricity of its orbit. Under the given assumptions, the planar rotational/librational motion of a satellite in the gravitational field of the planet is described by the Beletsky equation (Beletsky 1965):

$$(1 + e \cos f)\frac{\mathrm{d}^2\theta}{\mathrm{d}f^2} - 2e \sin f \frac{\mathrm{d}\theta}{\mathrm{d}f} + \omega_0^2 \sin \theta \cos \theta = 2e \sin f,$$

where f is the true anomaly, θ is the angle between the axis of the minimum principal central moment of inertia of the satellite and the "planet – satellite" radius vector.

As follows from an analysis of the Beletsky equation (see Melnikov & Shevchenko (2000) and references therein), for a satellite in an eccentric orbit, at definite values of the inertial parameters, synchronous resonance can have two centres in spin-orbit phase space; in other words, two different synchronous resonances, stable in the planar rotation problem, can exist. Consider a section, defined at the orbit pericentre, of the spin-orbit phase space. At $\omega_0 = 0$, there exists a sole centre of synchronous resonance with coordinates $\theta = 0 \bmod \pi$, $\mathrm{d}\theta/\mathrm{d}t = 1$. If the eccentricity is non-zero, upon increasing the value of ω_0, the resonance centre moves down the $\mathrm{d}\theta/\mathrm{d}t$ axis, and at a definite value of ω_0 (e. g., for $e = 0.1$ this value is $\simeq 1.26$) another synchronous resonance appears. Following Melnikov & Shevchenko (2000), we call the former synchronous resonance (emerging at zero value of ω_0) the *alpha* mode, and the latter one – the *beta* mode of synchronous resonance. Upon increasing the ω_0 parameter, the alpha and beta modes coexist over some limited interval of ω_0 (the extent of this interval depends on the orbital eccentricity), and in the section there are two distinct resonance centres situated at one and the same value of the satellite's orientation angle. Such a phenomenon takes place for Amalthea (J5) (Melnikov & Shevchenko 1998; Melnikov & Shevchenko 2000). On further increasing the ω_0 parameter, at some value of ω_0 the alpha resonance disappears, i. e., it becomes unstable in the planar problem, and only the beta resonance remains.

The "ω_0–e" stability diagram is presented in Fig. 1. Theoretical boundaries of the zones of existence (i.e., stability in the planar problem) of synchronous resonances are drawn in accordance with Melnikov (2001). Regions marked by "Ia" and "Ib" are the domains of sole existence of alpha resonance, "II" is the domain of sole existence of beta resonance, "III" is the domain of coexistence of alpha and beta resonances, "IV" is the domain of coexistence of alpha and period-doubling bifurcation modes of alpha resonance, "V" is the domain of non-existence of any 1:1 synchronous resonance, "VI" is the domain of sole existence of period-doubling bifurcation modes of alpha resonance.

The solid circles in Fig. 1 represent the satellites with known ω_0. The open circles represent the satellites with the ω_0 parameter determined by means of an approximation of the observed dependence of ω_0 on the satellite size r, accomplished following an approach by Melnikov & Shevchenko (2007). In total, the data on sizes and orbital eccentricities are available for 145 satellites (Karkoschka 2003; Sheppard & Jewitt 2003; Sheppard, Jewitt & Kleyna 2005; Sheppard, Jewitt & Kleyna 2006; Porco *et al.* 2007; Thomas *et al.* 2007; JPL website data); so, there are 145 "observational points" in the stability diagram "ω_0–e" in Fig. 1. The horizontal bars indicate three-sigma errors in estimating ω_0. They are all set to be equal to the limiting maximum value 0.21.

From the constructed diagram we find that 73 objects are situated in domain V, and in domain Ib 12 objects are situated higher than Hyperion (a sole solid circle in domain Ib), while in domain Ib there are 15 objects in total. Synchronous state of rotation does not exist in domain V. For the majority of satellites in domain Ib (namely, for those that are situated higher than Hyperion) synchronous rotation is highly probable to be attitude unstable as in the case of Hyperion. So, 73 satellites in domain V and 12 satellites in domain Ib rotate either regularly and much faster than synchronously (those tidally unevolved) or chaotically (those tidally evolved). Summing up the objects, we see that a major part (at least 85 objects) of all satellites with unknown rotation states (132 objects), i.e., at least 64%, cannot rotate synchronously.

In summary, though the majority of planetary satellites with known rotation states rotates synchronously (facing one side towards the planet, like the Moon), a significant part (at least 64%) of all satellites with unknown rotation states cannot rotate synchronously. The reason is that no stable synchronous 1:1 spin-orbit state exists for these bodies, as our analysis of the satellites location on the "ω_0–e" stability diagram demonstrates. They

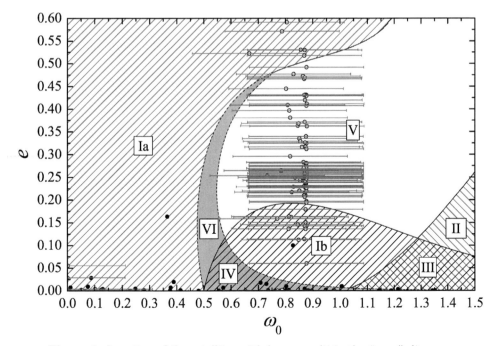

Figure 1. Location of the satellites with known radii in the "ω_0–e" diagram.

rotate either regularly and much faster than synchronously (those tidally unevolved) or chaotically (tidally evolved objects or captured slow rotators).

With the advent of new observational tools, more and more satellites are being discovered. Since they are all small, they are all irregularly shaped (Kouprianov & Shevchenko 2006). Besides, the newly discovered objects typically move in strongly eccentric orbits (see JPL website data; Sheppard & Jewitt 2003). Therefore these new small satellites are expected to be located mostly in domain V of the "ω_0–e" stability diagram. Consequently, either fast regular rotation (most probable) or chaotic tumbling (much less probable), but not the ordinary synchronous 1:1 spin-orbit state, can be a typical rotation state for the newly discovered planetary satellites.

References

Bauer, J. M., Buratti, B. J., Simonelli, D. P., & Owen, W. M. Jr. 2004, *ApJ*, 610, L57

Beletsky, V. V. 1965, *The Motion of an Artificial Satellite about its Mass Center* (Moscow: Nauka Publishers)

Black, G. J., Nicholson, P. D., & Thomas, P. C. 1995, *Icarus*, 117, 149

Devyatkin, A. V., Gorshanov, D. L., Gritsuk, A. N., Melnikov, A. V., Sidorov, M. Yu., & Shevchenko, I. I. 2002, *Sol. Sys. Res.*, 36, 248

Goldreich, P. & Peale, S. 1966, *AJ*, 71, 425

Grav, T., Holman, M. J., & Kavelaars, J. J. 2003, *ApJ*, 591, L71

Karkoschka, E. 2003, *Icarus*, 162, 400

Klavetter, J. J. 1989, *AJ*, 98, 1855

Kouprianov, V. V. & Shevchenko, I. I. 2005, *Icarus*, 176, 224

Kouprianov, V. V. & Shevchenko, I. I. 2006, *Sol. Sys.,Res.* 40, 393

Maris, M., Carraro, G., Cremonese, G., & Fulle, M. 2001, *AJ*, 121, 2800

Maris, M., Carraro, G., & Parisi, M. G. 2007, *A&A*, 472, 311

Melnikov, A. V. 2001, *Cosmic Res.*, 39, 68

Melnikov, A. V. & Shevchenko, I. I. 1998, *Sol. Sys. Res.*, 32, 480

Melnikov, A. V. & Shevchenko, I. I. 2000, *Sol. Sys. Res.*, 34, 434

Melnikov, A. V. & Shevchenko, I. I. 2007, *Sol. Sys. Res.*, 41, 483

Melnikov, A. V. & Shevchenko, I. I. 2008, *Celest. Mech. Dyn. Astron.*, 101, 31

Peale, S. J. 1977, in: J. A. Burns (ed.), *Planetary Satellites*, (Tucson: Univ. of Arizona Press), p. 87

Porco, C. C., Thomas, P. C., Weiss, J. W., & Richardson, D. C. 2007, *Science*, 318, 1602

Sheppard, S. S. & Jewitt, D. C. 2003, *Nature*, 423, 261

Sheppard, S. S., Jewitt, D., & Kleyna, J. 2005, *AJ*, 129, 518

Sheppard, S. S., Jewitt, D., & Kleyna, J. 2006, *AJ*, 132, 171

Thomas, P. C., Armstrong, J. W., Asmar, S. W., Burns, J. A., Denk, T., Giese, B., Helfenstein, P., & Iess, L. 2007, *Nature*, 448, 50

Wisdom, J. 1987, *AJ*, 94, 1350

Wisdom, J., Peale, S. J., & Mignard, F. 1984, *Icarus*, 58, 137

JPL website data, http://ssd.jpl.nasa.gov/ (Site Manager: D.K.Yeomans.)

NASA website data, http://solarsystem.nasa.gov/planets/

Part V

Icy Dwarf Planets and TNOs

Part V

Icy Dwarf Planets and TNOs

Icy Bodies of the Solar System
Proceedings IAU Symposium No. 263, 2009
J.A. Fernández, D. Lazzaro, D. Prialnik & R. Schulz, eds.
© International Astronomical Union 2010
doi:10.1017/S1743921310001717

Physical and dynamical characteristics of icy "dwarf planets" (plutoids)

Gonzalo Tancredi[1,2]

[1]Departamento Astronomía, Facultad de Ciencias, Montevideo, Uruguay
email: gonzalo@fisica.edu.uy

[2]Observatorio Astronómico Los Molinos, Ministerio de Educación y Cultura, Uruguay

Abstract. The geophysical and dynamical criteria introduced in the "Definition of a Planet in the Solar System" adopted by the International Astronomical Union are reviewed. The classification scheme approved by the IAU reflects dynamical and geophysical differences among planets, "dwarf planets" and "small Solar System bodies". We present, in the form of a decision tree, the set of questions to be considered in order to classify an object as an icy "dwarf planet" (a plutoid). We find that there are 15 very probable plutoids; plus possibly 9 more, which require a reliable estimate of their sizes. Finally, the most relevant physical and dynamical characteristics of the set of icy "dwarf planets" have been reviewed; e.g. the albedo, the lightcurve amplitude, the location in the different dynamical populations, the size distributions, and the discovery rate.

Keywords. solar system: general, "dwarf planets", plutoids, TNOs, Kuiper Belt

1. Introduction

In 2006 the XXVIth General Assembly of the International Astronomical Union adopted the Resolution 5: "Definition of a Planet in the Solar System". In this resolution 3 categories of objects orbiting around the Sun were distinguished: planets, "dwarf planets" and "small Solar System bodies"†.

There was also a Resolution 6 which established that "Pluto ... is recognized as the prototype of a new category of Trans-Neptunian Objects" and "an IAU process will be established to select a name for this category". On June, 2008, the Executive Committee of the IAU had decided on the term *plutoid* as a name for "dwarf planets" like Pluto.

Up to know 4 icy objects (plutoids) and one rocky object have been officially classified as "dwarf planets" by the IAU; i.e.: Eris, Pluto, Makemake, Haumea and Ceres. Nevertheless, there might exist many other objects which satisfy the criteria adopted in the Resolution 5 for "dwarf planets".

In the following sections, we review the scientific grounds of the resolution (Section 2), we present a list of potential icy "dwarf planets" (plutoids) (Section 3), and we discuss the main characteristics of this population of objects (Section 4).

We have adopted the following convention to define the transneptunian region: transneptunian objects (TNOs) have a semimajor axis greater than Neptune's one $(a > a_N)$.

---

† The final text of the Resolution can be found in:
http://www.iau.org/static/resolutions/Resolution_GA26-5-6.pdf

2. The criteria to distinguish among the categories of Solar System objects

2.1. *The dynamical criterion*

According to the Resolution 5, the difference between planets and "dwarf planets" is that the former ones "have cleared the neighborhood around its orbit", while the second ones not. The problem of clearing the accretion zone were analyzed by Stern & Levison (2002) and Soter (2006).

The likelihood that in a timescale τ a small body would suffer an encounter with a planet that leads to a deflection of an angle θ is given by (Stern & Levison 2002):

$$\Lambda = \frac{\mu^2}{a_P^{3/2}} \left[\frac{\tau\sqrt{GM_\odot}(1+2\Gamma)}{2U^3\pi^2\sin(i)|U_x|\Gamma^2} \right] \tag{2.1}$$

where μ is the ratio between the planet's (M) and the solar mass ($M_\odot$); G is the gravitational constant; a_P is the semimajor axis of the planet (AU); i is the mutual inclination of the small body's orbit respect to the planet; U is the relative velocity of the small body and the planet in units of the of the planet's orbital velocity; U_x is the radial component of U; and $\Gamma \equiv \tan(\theta/2)$.

We assign values for the parameters that appear within the brackets in eq. (2.1) typical of a population of dynamically warm objects interacting with a planet. We assume a mean $e \approx \sin(i) \approx 0.1$, which corresponds to a Tisserand parameter $T \approx 2.98$ and $U = \sqrt{3-T} \approx 0.14$. U_x is calculated from the assumption of isotropical decomposition between the 3 components; i.e. $U_x = U/\sqrt{3}$. The clearing of the accretion zone would occur if the deflection angle is large, let's say $\theta \approx 1$; i.e. $\Gamma \approx 1/2$.

Since the adopted planet definition applies to objects in the Solar System, we consider the age of the Solar System as the relevant timescale for the clearing process; i.e. $\tau \approx 4.5 \times 10^9$ yr. Note that Stern & Levison (2002) and Soter (2006) use instead the age of the Universe for the relevant timescale.

According to the previous considerations, the condition for clearing the neighborhood around its orbit can be translated into the following condition for the mass:

$$\mu > \sqrt{\frac{a_P^{3/2}}{5\times 10^{14}}} \tag{2.2}$$

In Fig. 1 we plot the masses of the planets and several massive asteroids and TNOs as a function of the semimajor axis in logarithmic scale. A couple of vertical dashed lines are included in the plots that corresponds to the possible inner and outer limits of the so-called "snow line", the heliocentric distance where the water condensates. A thick full-line is also drawn which corresponds to the condition of eq. (2.2). There is a huge gap of 3-4 order of magnitude in mass between planets and "dwarf planets" in the inner or outer region of the Solar System. The condition for clearing the neighborhood stated above clearly separates the two type of objects.

2.2. *The geophysical criterion*

A geophysical criterion separates planets and "dwarf planets" from the group of "small Solar System bodies". The former ones "have sufficient mass for its self-gravity to overcome rigid body forces so that it assumes a hydrostatic equilibrium (nearly round) shape". The geophysical criterion introduced in the definition distinguishes between objects that had suffered (or not) important internal transformations due to the action of the self-gravitation. It separates between two extreme cases: objects that are just an

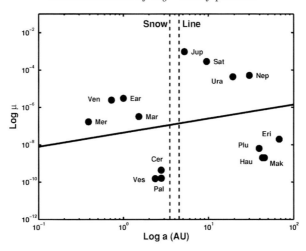

Figure 1. The ratio (μ) between the planet's and the solar mass versus the semimajor axis of the planet in AU (a) in logarithmic scale for the planets and several massive asteroids and TNOs. See the text for the explanation of the additional lines.

agglomeration of planetesimals with little mutual cohesion (the "small Solar System bodies"); and objects where the material has been largely metamorphosed due to the high pressure and temperatures produced under the weight of the outer layers (planets and "dwarf planets"). The former ones adopt an irregular shape, while the later ones tend to acquire a figure in hydrostatic equilibrium. Tancredi & Favre (2008) (hereafter Paper I) has revised the information about Solar System objects in hydrostatic equilibrium either from the theoretical and the observational perspective.

The definition takes into account two concepts that we discuss below: i) the figures of equilibrium and ii) the overcome of rigid body forces by the self-gravity.

For an strengthless isolated object in rotation, the equilibrium figures are a set of ellipsoidal shapes depending on the angular momentum. Chandrasekhar (1987) introduced two dimensionless parameter to characterize the problem: one associated with the angular momentum ($\Gamma = L/(GM^3 R)^{1/2}$), where L is the angular momentum, G is the gravitational constant, M the mass and R the radius; and the other one is associated with the angular velocity ($\Omega = \omega^2/(\pi G \rho)$), where ω is the angular velocity and ρ is the density. A non-rotating body acquires a spherical shape, while a body with a low angular momentum acquires the figure of an oblate ellipsoid (a MacLaurin spheroid two equal axes larger than the third axis). For higher angular momentum ($\Gamma > 0.303$), the body acquires a triaxial ellipsoidal figure (a Jacobi ellipsoid), up to a critical value of $\Gamma = 0.39$ where the body becomes unstable against further increase of the angular momentum (see Fig. 1a in Paper I). In the range $0.303 < \Gamma < 0.39$, although the angular momentum increases, the angular velocity is being reduced while the figure becomes more elongated. The ratios between the axis for the Jacobi ellipsoids go from $b/a = 1$ and $c/a = 0.583$ at the transition between MacLaurin to Jacobi ellipsoids, to $b/a = 0.432$ and $c/a = 0.345$ at the edge of instability. In the same range the dimensionless angular velocity is being reduced from $\Omega < 0.374$ to 0.284 (see Fig. 1b in Paper I). Therefore, for bodies with densities over 1 gcm^{-3}, the full rotational period is constrained to values less than 7.15h.

Assuming a strengthless isolated object, we then have a given relation between the rotational period, the shape (represented by the axis ratios of an ellipsoid) and the density for figures in hydrostatic equilibrium.

Different criteria have been used in the literature to estimate the condition when a self-gravitating body overcomes the material strength. All the criteria reduce to the following expression that relates the critical diameter (D) for a self-gravitating body to overcome the material strength, the density (ρ) and the material strength (S) (Paper I):

$$D\rho/2 = \alpha\sqrt{\frac{3}{2\pi G}}\sqrt{S} \qquad (2.3)$$

where α is a parameter that depends on the chosen criteria, but it has typical values in the range $1 < \alpha < 5^{1/2}$ (Note that in eq. 2.3 we correct a typo that appeared in eq. 1 of Paper I; a division by 2 of the diameter was missing in that eq.; nevertheless the computations based on this eq. and the plots were correct). The material strength depends on the constituents and the temperature. Typical values for mixtures of ice and soil are in the range of a few tens to a few hundreds MPa (Petrovic 2003).

From the observational perspective it is possible to put some constraints from the analysis of the shapes of the icy bodies visited by spacecrafts; e.g. the icy satellites of the outer planets. In Paper I it was found that the Saturnian satellite Mimas and the Neptunian satellite Proteus are among the smallest objects with a figure close to equilibrium. It was noticed a possible dependence of the critical diameter with the temperature, as it was expected. They concluded that in the TNO region the transition to an equilibrium figure should occur at a value of $D\rho \sim 600$ km gcm^{-3}. Assuming a value of $\rho = 1.3$ gcm^{-3} (typical of the Uranian and Neptunian satellites of this size), the critical size for TNOs is $D \sim 450$ km. This value is in correspondence to the theoretical critical limit presented above for a low-strength material of $S \sim 1 - 10$ MPa and $\rho \sim 1 - 2$ gcm^{-3}.

Therefore, we will use the value $D \sim 450$ km as the limit between "dwarf planets" and "small Solar System bodies" in the transneptunian region.

3. The list of icy "dwarf planets"

Unless the case of Pluto and Eris, there are not direct estimates of the size of the TNOs. A model dependent estimate of the size comes from measurements of thermal emission using IR space telescopes. Using the Spitzer Space Telescope Stansberry *et al.* (2008) obtain size estimates of a large fraction of the largest TNOs. Combined with accurate determinations of the absolute magnitude in V (H), it is possible to compute the geometrical albedo (p_V). In Fig. 2 we plot p_V against the diameter (D in km) from the data collected by Stansberry *et al.* (2008).

Finally, a rough idea of the size can be obtained from the total absolute magnitude and an assumed value for the geometrical albedo. There is a wide range of albedo estimates for TNOs, from values of $0.6 - 0.8$ for the largest objects down to values of 0.03 (see Fig. 2). Assuming the constraint $p_V \leqslant 1$, objects with an absolute magnitude brighter than $H < 2.4$ are certainly larger than 450 km. For the TNOs without size estimates, we assume a conservative value $p_V = 0.1$ to left behind as less dwarf candidates as possible. A diameter of 450 km would correspond to $H < 4.9$ for this given albedo.

From the list of TNOs listed by the Minor Planet Center by July 22, 2009 (†) and the list of objects observed by Stansberry *et al.* (2008), we extract 46 objects with an estimated size larger than 450 km. This list is an updated version of the one presented in Paper I. This preliminary list of icy dwarf candidates is presented in Table 1. The objects are listed in increasing order of absolute magnitude H.

† The latest lists of TNOs and Scattered Disk Objects is given in: http://www.cfa.harvard.edu/iau/lists/MPLists.html

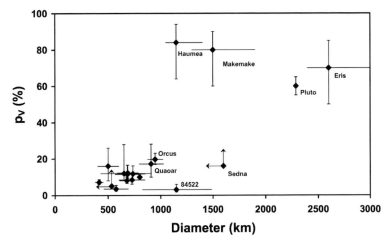

Figure 2. Plot of the geometrical albedo (p_V) against the diameter (D in km) from the data collected by Stansberry *et al.* (2008). We include the error bars as stated by the authors.

In order to finally classify these objects as "dwarf planets", we have to get some information about their shapes. In Paper I, it was proposed to analyze the rotational lightcurve, i.e. the variation of the observed brightness as a function of the rotational phase angle. The brightness is proportional to the projected shape in the sky and the surface albedo. A sphere or MacLaurin spheroid with uniform albedo distribution produces a flat lightcurve; while a Jacobi ellipsoid produces a lightcurve with two identical maximum peaks. The existence of albedo spots could introduce weird patterns, but the vast experience in asteroidal lightcurves has shown that the albedo contribution is generally less important than the shape (Magnusson 1991).

The viewing geometry also affect the shape of the lightcurve of an ellipsoidal figure. The maximum amplitude is obtained when the observer is in the plane of the object's equator; and it is reduced to zero if the object is pole on. For the following analysis, since we do not have any information of the viewing geometry, we will assume that the observed amplitudes correspond to the maximum amplitude for the object. Unfortunately, this situation can not change in the near future because, due to the slow movement of the TNOs, the viewing geometry may take decades to show a significant variation.

In the case that the lightcurve amplitude is small ($\Delta m < 0.15$), we can considered the object as a small departure from a sphere or MacLaurin spheroid with small albedo spots. But if the amplitude is larger than the previous value, we have to analyze whether the observed lightcurve is compatible with the lightcurve produced by a Jacobi ellipsoid of a plausible density range.

For an ellipsoidal figure, the square of the projected shape in the sky as a function of time can be described as a Fourier series of order 2 with a null term of order 1 (see Barucci *et al.* (1989)) and eq. 2 in Paper I). We make the assumption that the brightness is directly proportional to the projected shape. Therefore, if we develop the square of the brightness in a Fourier series up to order two, the ratio between the quadratic sums of the coefficients of order one and two (defined as the β parameter) is an indicator of the closeness to an ellipsoidal shape. Values of $\beta \sim 0$ would correspond to a perfect ellipsoid.

After modeling synthetic lightcurves, it was found that important departures from an ellipsoidal shape comparable to the ones observed in the irregular satellites can be detected if the value of $\beta \geqslant 0.25$ (see Paper I). Therefore, lightcurves with $\beta \geqslant 0.25$ can be discarded as produced by a smooth ellipsoid. If $\beta < 0.25$, the shape deduced from the

Table 1. List of icy "dwarf planets" candidates.

Number	Name	Provisional Designation	Abs. Mag. H_V		Diameter (km)	Dwarf	Case
136199	Eris	2003 UB313	-1.1		2600	Yes	I
134340	Pluto		-0.7		2390	Yes	I
136472	Makemake	2005 FY9	0		1500	Yes	II
136108	Haumea	2003 EL61	0.5		1150	Yes	III
90377	Sedna	2003 VB12	1.8		1600	Yes	II
		2007 OR10	1.9	#	1752		
90482	Orcus	2004 DW	2.5		946	Yes	II
50000	Quaoar	2002 LM60	2.6		908	Yes	II
		2005 QU182	3.1	#	1008		
202421		2005 UQ513	3.4		878		
55636		2002 TX300	3.49		800	Yes	II
174567		2003 MW12	3.6	#	801	Yes?	II
		2007 UK126	3.6	#	801		
55565		2002 AW197	3.61		735	Yes	II
		2003 AZ84	3.71		686	Yes	II
55637		2002 UX25	3.8		681	???	V
		2006 QH181	3.8	#	730		
28978	Ixion	2001 KX76	3.84		650	Yes	II
145452		2005 RN43	3.9	#	697	Yes?	II
20000	Varuna	2000 WR106	3.99		500	Yes	III
		2002 MS4	4		726		
145453		2005 RR43	4	#	666	Yes?	II
		2004 XA192	4	#	666		
84522		2002 TC302	4.1		1150		
120178		2003 OP32	4.1	#	636	Yes?	III
90568		2004 GV9	4.2		677	Yes	II
84922		2003 VS2	4.2	#	607	No	IV
42301		2001 UR163	4.2	#	607	Yes?	II
120347		2004 SB60	4.2	#	607	Yes?	II
		2003 UZ413	4.3	#	580		
119951		2002 KX14	4.4	#	554		
145451		2005 RM43	4.4	#	554	Yes?	II
		2004 NT33	4.4	#	554		
120348		2004 TY364	4.5	#	529	No	IV
		2004 XR190	4.5	#	529		
144897		2004 UX10	4.5	#	529	Yes?	II
-19308		1996 TO66	4.5	#	529		
		2004 PR107	4.6	#	505		
26375		1999 DE9	4.7	#	482	Yes?	II
145480		2005 TB190	4.7	#	482		
		2007 JH43	4.7	#	482		
		2003 QX113	4.7	#	482		
175113		2004 PF115	4.7	#	482		
24835		1995 SM55	4.8	#	461	No	IV
38628	Huya	2000 EB173	5.23		533	Yes	II
15874		1996 TL66	5.46		575	Yes	II

lightcurve can be approximated to an ellipsoid; but, is it an ellipsoid of the Jacobi-family? The Jacobi ellipsoids have a given set of relations between the axis ratios, the rotational period and the density, as it was stated in Subsection 2.2 (see Fig. 1b in Paper I). From the lightcurve, we obtain the rotational period and a possible range of ratios of the two major axis (b/a), depending on the aspect angle at the time of the observation (the angle between the rotation axes and the visual). Based in the equations for the Jacobi ellipsoids

(Chandrasekhar 1987, and Paper I), a range of densities compatible with the data can be found. Since, all the icy satellites with equilibrium-shape bodies and the large TNOs have densities $\rho > 1$ gcm^{-3}, we accept candidates for an icy "dwarf planet" with a Jacobi shape, if there are estimates of the density based in the previous calculations with values larger than 1 gcm^{-3}.

The previous set of criteria to qualify a candidate as an icy "dwarf planet" was presented in detail in Paper I, but here we have compiled them in the decision tree presented in Fig. 3.

We apply this decision tree to the list of "dwarf planet" candidates listed in Table 1. A column is added to answer the question: is the object a "dwarf planet"? The following answers are considered to this question:

- *Yes* - accepted as a "dwarf planet"
- *Yes?* - possibly acceptable case, those are objects that show very small amplitudes, but we do not have enough information to support the size estimate
- *No* - rejected as a "dwarf planet"
- *???* - the observational evidence is conflicting: the object seems to be larger than 450 km based on the IR data, but there are important differences between the lightcurve data among different observers.
- blank space - for objects without a lightcurve or any other kind of information to decide whether they can be considered as "dwarf planets" or not.

In addition we include in Table 1 another column to list under which of the Cases presented in the decision tree the object is accepted or rejected.

The information compiled in Table 1 is presented in detail, as well as the photometric data on which we base our classification and the links to the corresponding references, in the webpage: *"Dwarf Planets" Headquarters: http://www.astronomia.edu.uy/dwarfplanet.*

4. Characteristics of icy "dwarf planets" (plutoids)

The individual characteristics of the very large TNOs has been revised by Brown (2008). He presented a detailed discussion of the physical properties of the first 5 object in Table 1. Each of these objects presents particular features that deserve a detailed analysis. For further reading on these individual cases, the reader should refer to Brown's chapter and to the large number of papers dealing with observational data of these objects that have appeared in the literature in the recent years.

Instead, we decided to analyze the collective properties of the several tens largest TNOs. In the following section we review the scant information available regarding the physical and dynamical properties of this set of objects.

4.1. *The physical characteristics*

In Fig. 2 we have already presented the most reliable estimates of the geometrical albedo (p_V) against the diameter (D in km) coming from the Spitzer observations collected by Stansberry *et al.* (2008). Two clear sets can be identified: *i*): very large TNOs with sizes over 1000 km and very high albedos ($p_V \sim 0.6 - 0.9$); *ii*): objects with low albedos ($p_V \lesssim 0.2$) and diameters less than 1000 km. There might be a couple of objects with sizes over 1000 km but low albedos (Sedna and 84522); but the estimates are very uncertain since they are based on observations with a $SNR < 5$. The difference between larger and smaller TNOs can be interpreted as the capacity of larger objects to retain an atmosphere with a seasonal evolution. The larger objects could have deposits of fresh and high albedo ices on their surfaces, while the smaller objects present a weathered and darker surface. A physical modeling of this process would be desirable.

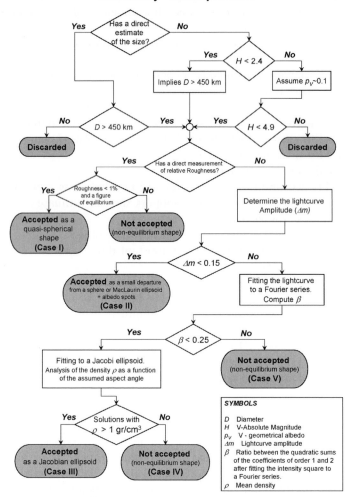

Figure 3. Decision tree to qualify a candidate as an icy "dwarf planet" (plutoid).

The results of a large observational campaign with photometric observations of several objects included in our list is being published by Thirouin *et al.* (2009). This data set constitutes the basis for a compilation of an extended database of light curve parameters that has appeared in the literature produced by Duffard *et al.* (2009). The authors obtain the mean rotational properties of the entire sample, determine the spin frequency distribution and search for correlations between different physical and orbital parameters (rotational period, peak-to-peak amplitude, semimajor axis, perihelion distance, aphelion distance, eccentricity, inclination). Among the conclusions presented by the authors we highlight the following ones: *i)* they found a correlation between the rotational period and the B-V color which might suggest that objects with shorter rotation periods may have suffered more collisions than objects with longer ones; *ii)* there is also a correlation between the amplitude and the absolute magnitude H which indicates that the smaller (and collisionally evolved) objects are more elongated than the bigger ones. Using their dataset, in Fig. 4a we show the later correlation between the amplitude and H. A vertical line at $H = 4.9$ is drawn to separate the "dwarf planet" candidates and the smaller objects.

The authors also conclude from the results of their model, that hydrostatic equilibrium is probably reached by almost all TNOs brighter than $H < 7$. In order to reassess this topic, we plot in Fig. 4b the single peak period vs the observed amplitude for the TNOs brighter than $H < 7$ in their dataset. The points can be divided in two groups: objects with low amplitudes and large periods, and objects with high-amplitudes and small periods. Three sets of objects are shown in the figure: *i)* full squares - objects with $H < 4.9$ not discarded as "dwarf planet" candidates; *ii)* empty squares - objects with $H < 4.9$ but discarded as "dwarf planet" candidates; and *iii)* empty diamonds - objects with $4.9 < H < 7$. We also draw a few lines that correspond to the relation between the period and the maximum amplitude of strengthless ellipsoidal figures of equilibrium with several densities (see Paper I for further details on how these lines are calculated). The lower curves correspond to the relation between the maximum lightcurve amplitude and half the period for Jacobi ellipsoids with densities $\rho = 0.5, 1, 2$ and 5 gcm^{-3}. The two upper horizontal lines correspond to a MacLaurin spheroid with density $\rho = 1$ and 2 gcm^{-3}, respectively.

Densities lower than 1 gcm^{-3} are required in order to be in hydrostatic equilibrium for most of the high amplitude objects ($\Delta m > 0.15$) with smaller sizes ($4.9 < H < 7$, empty diamonds in the plot). Even much lower densities are required in a few cases. Although we can not rule out that these smaller objects are in hydrostatic equilibrium, we point out that all the satellites of the outer planets larger than 100 km have densities higher than $\rho \gtrsim 1$ gcm^{-3}, with the exception of Hyperion†. In view of these evidences, we think that the conclusion of Duffard *et al.* (2009) that "hydrostatic equilibrium is probably reached by almost all TNOs brighter than $H < 7$ must deserve a more detailed analysis.

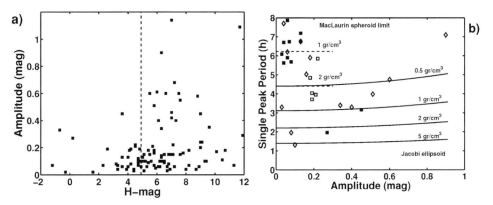

Figure 4. *a)* The lightcurve amplitude versus the absolute magnitude (H). The data is taken from Duffard *et al.* (2009). *b)* The amplitude versus the rotational period for the same data set. See the text for an explanation on the different symbols and the lines drawn in the plot.

4.2. *The dynamical characteristics*

In Fig. 5 a and b we plot the inclination (i) and eccentricity (e) versus the semimajor axis (a) for objects outside Uranus orbit. The objects are represented by a small black dot and a gray-shaded circle proportional to its diameter. The diameter scale is represented by an empty circle of 1000 km.

We note that most of the objects in Fig. 5 with noticeable diameters (typically larger than a few hundred km) are located in the so-called hot population of the transneptunian

---

† See e.g. the compilation of Physical Parameters of the Planetary Satellites at the Jet Propulsion Laboratory website and the references therein: http://ssd.jpl.nasa.gov/?sat_phys_par.

region, the region of high-i and high-e, e.g. $i \gtrsim 15$ deg and $e \gtrsim 0.1$. This result is reflected in the distribution of these parameters; while for the complete sample of objects with semimajor axis greater than Neptune's one, the mean inclination is $\bar{i}_{all} = 9.8$ deg, for the restricted sample of "dwarf planet" candidates is $\bar{i}_{dwarf} = 20.6$ deg. For the eccentricity the corresponding values for the two samples are $\bar{e}_{all} = 0.18$ and $\bar{e}_{dwarf} = 0.22$, respectively. The Kolmogorov-Smirnov test applied to both the distribution of i and e shows that the hypothesis that the two samples come from the same underlying one-dimensional probability is rejected at the 90% confidence level.

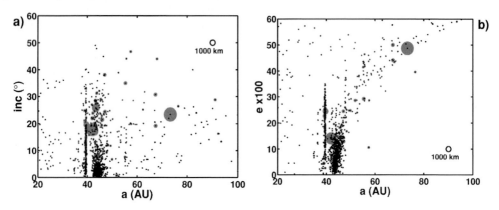

Figure 5. Plots of the dynamical parameters of TNOs. *a)* Inclination versus semimajor axis. *b)* Eccentricity versus semimajor axis.

The size segregation between large objects in the hot population and almost lack of large ones in the cold population has not been successfully explained by any of the prevailing cosmogonical models, like the Nice model (see e.g. Gomes 2003, 2009). It is an open problem yet to be solved.

4.3. *The size distribution*

Considering the accretional and collisional processes experienced by the transneptunian objects, it is expected that the cumulative size distribution should follow a power-law of an exponent close to 3. As mentioned above a proxy of the size is the absolute magnitude, but the relation between these two parameters depends on the geometrical albedo. For a fixed value of the albedo, a power-law cumulative size distribution of exponent α should correspond to a cumulative absolute magnitude distribution with a slope $s = \alpha/5$.

The size distribution of the TNOs has been revised by Petit *et al.* (2008). Most of the estimates of the size distribution relies on the computation of the luminosity function (LF), i.e. the cumulative number of TNOs brighter than a given apparent magnitude. Converting the LF into a size distribution involves the modeling of the dynamical and physical surface properties of the TNOs. Petit *et al.* (2008) compiled the results of several papers that address this question. A wide range of LF slopes has appeared in the literature from 0.3 to 0.9; which it can be transformed into an exponent of the cumulative size distribution in the range $\alpha \sim 3 - 4$ for bodies larger than a few tens of kilometers (Note that we use the cumulative exponent while Petit *et al.* (2008) use the differential one).

The cumulative absolute magnitude distribution is presented in Fig. 6a. An overabundance of bright objects over the dashed linear fit is observed, as it has already been noted by e.g. Brown (2008). A similar overabundance of large objects should appear in the size distribution if it is computed from the previous magnitude distribution and assuming a fixed albedo, as it is usual. Nevertheless, nowadays there is a large sample of TNOs with

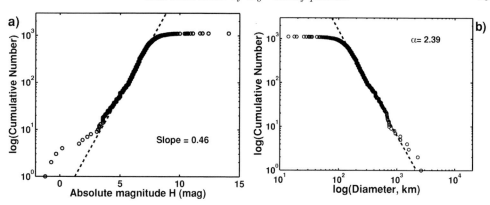

Figure 6. *a)* Cumulative distribution of absolute magnitude with the Number in logarithmic scale. *b)* Cumulative size distribution of in log-log scale.

reliable size estimates (Stansberry *et al.* 2008) that can be taken into account. Combining the diameters listed in Stansberry *et al.* (2008) dataset with the estimates derived from the absolute magnitude and a common albedo $p_V = 0.1$, we obtain the size distribution of the observed sample of TNOs presented in Fig. 6b. The overabundance of large TNOs has disappeared due to the correlation between sizes and albedo noted in Fig. 2. A good fit to a power-law is obtained for $D > 150$ km with an exponent $\alpha = 2.39$. Our dataset has not been corrected for observational biases, and this could partially explain the differences with previous estimates of the cumulative exponent. A new analysis with a proper correction of the observational biases is left for a future work; nevertheless we have shown that a better treatment of the observational data, which includes different sources of information, could improve the quality of the adjustment.

4.4. *How many plutoids are still missing?*

After the discovery of the third member of the transneptunian region in 1992 (1992QW1), the discovery rate had suffered a continuous increase up to the early 2000's (Fig. 7a). In the second half of this decade the discovery rate has decreased considerably, due to the fact that the number of wide-area surveys of TNOs has been drastically reduced (see Petit *et al.* (2008) for a list of the most relevant surveys). Until July 2009 there were discovered 904 TNOs brighter than $H > 8.1$ (larger than 100 km for $p_V = 0.1$). But Petit *et al.* (2008) estimated that there should be $\sim 10^4$ larger than this size; therefore, we are far to reach completeness.

A similar conclusion can be drawn from Fig. 7b where we plot the absolute magnitude H versus the discovery year. A clustering of the discoveries in the years 1999 to 2006 is observed. In Fig. 7b we also draw a horizontal dashed-line at $H = 4.9$, the limiting magnitude we have used for "dwarf planet" candidates. Almost all the "dwarf planet" candidates were discovered in the period 1999-2006; in particular 8 out of 10 of the absolute brightest ones were discovered by the survey lead by Brown (2008) which used the 48-inch Palomar Schmidt telescope (one object among this 8 was independently discovered by Aceituno *et al.* 2005).

If we assume that the size distribution with an exponent in the range $\alpha \sim 3-4$ extends down to objects of several hundred km, there should be between a few tens to a less than a couple of hundreds "dwarf planets" with $H < 4.9$ and sizes larger than 450 km. Therefore, we may be half the way to reach completeness for the large sample of TNOs.

The Palomar survey for large TNOs covered 20.000 deg^2 north of -30 deg declination (Brown 2008). Unfortunately there is no assessment of the efficiency of this survey.

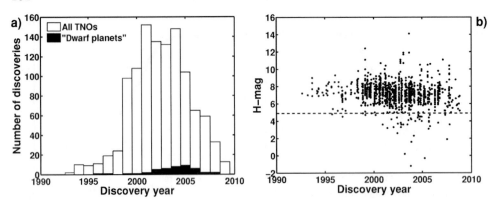

Figure 7. *a)* The number of discoveries per year as a function of the discovery year. *b)* The absolute magnitude (H) versus the discovery year. An horizontal dashed-line at $H = 4.9$ is drawn.

Assuming a 100% for the very bright objects $H < 1$ (the limit for the 4 IAU's "official" plutoids), there should be 3-4 similar objects yet to be discovered; one of them should be close to the galactic plane, since this region was not covered by the Palomar survey; and ~ 3 should be in the region south of $-30\,$deg declination.

5. Conclusions

We have reviewed the scientific grounds of the "Definition of a Planet in the Solar System" adopted by the XXVI General Assembly of the IAU. The two criteria used in the definition have been discussed: *i)* the dynamical criterion that separates planet and "dwarf planets"; and *ii)* the geophysical criterion that separates "dwarf planets" and the "small Solar System bodies". The classification scheme approved by the IAU reflects important dynamical and geophysical differences among the three set of objects in the Solar System.

We update the list of "dwarf planet" candidates that was initially presented by Tancredi & Favre (2008). The set of criteria to decide whether a candidate has a figure in hydrostatic equilibrium, and therefore it can be considered as a "dwarf planet", is presented for clarity as a decision tree.

After applying this decision tree to the list of candidates, we find that there are 15 very probable icy "dwarf planets" (plutoids), plus possibly 9 more, but we are lacking of reliable estimate of their sizes (they are listed in Table 1 with a *Yes?*). Three objects with preliminary estimated diameter larger than 450 km were discarded as "dwarf planets"; and one case where the observational evidence is conflicting. There are 18 objects with sizes possibly over the critical limit but they require further observations of the lightcurve and/or the size to consider them as "dwarf planets".

Finally, the most relevant physical and dynamical characteristics of the set of icy "dwarf planets" have been revised. We highlight the following conclusions:

(*a*) There is a remarkable difference between very large icy "dwarf planets" with high-albedos and smaller ones with low albedos. The difference can be interpreted as the capacity of larger objects to retain an atmosphere with a seasonal evolution, but a modeling of this process is encouraged.

(*b*) Objects with sizes estimates clearly over the limit of 450 km which present large amplitude lightcurves, have an elongated shape that is compatible with a Jacobi ellipsoids with densities $\rho > 1\,$gcm^{-3} But, in the case of objects in the size range 100 to 450 km with large amplitude lightcurves, densities much lower than 1 gcm^{-3} are required in order to be compatible with a Jacobi ellipsoid.

(*c*) There is a size segregation between large objects in the hot population and almost lack of large ones in the cold population that has not been successfully explained by any of the prevailing cosmogonical models.

(*d*) An overabundance of bright objects is observed in the cumulative absolute magnitude distribution; but this overabundance does not appear in the cumulative size distribution if the reliable sizes estimates based on IR data are included.

(*e*) There might be several tens up to more than a hundred objects larger than 450 km yet to be discovered. Among the very large ones ($H < 1$, the limit for the 4 IAU's "official" plutoids), there should be 3-4 similar objects yet missing.

To the end, we would like to raise the question: Should the IAU continue naming "dwarf planets"? In order to proceed cautiously, we suggest that the following objects could be included in the list of "official" "dwarf planets": (90377) Sedna, (90482) Orcus and (50000) Quaoar. These objects are clearly over the size limit of 450 km and the photometric observational evidence are in concordance with a figure in hydrostatic equilibrium.

References

Aceituno, J., Santos-Sanz, P., & Ortiz, J. L. 2005, *M.P.E.C.*, 2005-O36

Barucci, M. Capria, M., Harris, A., & Fulchignoni, M. 1989, *Icarus*, 78, 311

Brown, M. 2008, in: M. A. Barucci, H. Boehnhardt, D. P. Cruikshank & A. Morbidelli (eds.), *The Solar System Beyond Neptune* (University of Arizona Press, Tucson), p. 335

Chandrasekhar, R. 1987, *Ellipsoidal Figures of Equillibrium*, Dover Publications, New York

Duffard, R., Ortiz, J. L., Thirouin, A., Santos-Sanz, P., & Morales, N. 2009, *A&A*, 505, 1283

Gomes, R. S. 2003, *Icarus*, 161, 404

Gomes, R. S. 2003, *Celest. Mech. Dyn. Astr.*, 104, 39

Magnusson, P. 1991, *A&A*, 243, 512

Petit, J.-M., Kavelaars, J. J., Gladman, B., & Loredo, T. 2008, in: M. A. Barucci, H. Boehnhardt, D. P. Cruikshank & A. Morbidelli (eds.), *The Solar System Beyond Neptune* (University of Arizona Press, Tucson), p. 71

Petrovic, J. 2003, *J. Materials Science*, 38, 1

Soter, S. 2006, *AJ*, 132, 2513

Stansberry, J. A., Grundy, W. G., Brown, M., Cruikshank, D. P., Spencer, J., Trilling, D., & Margot, J. L. 2008, in: M. A. Barucci, H. Boehnhardt, D. P. Cruikshank & A. Morbidelli (eds.), *The Solar System Beyond Neptune* (University of Arizona Press, Tucson), p. 161

Stern, S. A. & Levison, H. F. 2002, *Highlights Astron.*, 12, 205

Tancredi, G. & Favre, S. 2008 (*Paper I*) *Icarus*, 195, 851

Thirouin, A., Ortiz, J. L., Duffard, R., Santos-Sanz, P., Aceituno, F. J., & Morales, N. 2009, *A&A*, submitted

Icy Bodies of the Solar Ststem
Proceedings IAU Symposium No. 263, 2009
J.A. Fernández, D. Lazzaro, D. Prialnik & R. Schultz, eds.

© International Astronomical Union 2010
doi:10.1017/S1743921310001729

Surface properties of icy transneptunian objects from the second ESO large program

Francesca E. DeMeo[1], Maria Antonietta Barucci[1], Alvaro Alvarez-Candal[2], Catherine de Bergh[1], Sonia Fornasier[1,3], Frédéric Merlin[4], Davide Perna[1,5,6], and Irina Belskaya[1,7]

[1]LESIA, Observatoire de Paris, 5, Place Jules Janssen, 92195 Meudon Cedex, France
email: francesca.demeo@obspm.fr
[2]European Southern Observatory, Alonso de Cordova 3107, Santiago, Chile
[3]Université de Paris 7 *Denis Diderot*, Paris, France
[4]Department of Astronomy, University of Maryland, College Park, MD 20742, USA
[5]INAF- Osservatorio Astronomico di Roma, Via Frascati 33, I-00040
Monte Porzio Catone, Italy
[6]Università di Roma *Tor Vergata,* Via della Ricerca Scientifica 1, I-00133 Roma, Italy
[7]Institute of Astronomy, Kharkiv National University, 35 Sumska str., 61022 Kharkiv, Ukraine

Abstract. An analysis is well underway for the data from the second Large Program (PI M. A. Barucci) dedicated to investigating the surface properties of Centaurs and Transneptunian objects through spectroscopic, photometric color, lightcurve, and polarimetric observations using the European Southern Observatory (ESO) Very Large Telescope (VLT) and New Technology Telescope (NTT). 45 objects were observed between 2006 and 2008, allowing a broad characterization of at least the largest and brightest objects among this population. In this report, we summarize all our findings, but focus on the analysis of the presence of ices such as methane, ethane, nitrogen, ammonia hydrate, methanol, and particularly H_2O which is so abundant throughout the outer solar system.

Keywords. Kuiper Belt, techniques: spectroscopic, techniques: photometric, techniques: polarimetric

1. Introduction

The Transneptunian and Centaur populations display a wide variety of characteristics. While all are considered small bodies, they range from tens to thousands of kilometers in diameter with densities from less than one for many small bodies to greater than 2 g/cm^3. Their spectra, revealing information about surface composition, have visible slopes ranging from nearly neutral to the reddest in the solar system. Near-infrared spectra of some objects have diagnostic absorption bands that can often be clearly linked to ices. TNO spectra are typically divided into several compositional groups. Featureless spectra have a wide range of slopes but no distinct absorption bands, thus leaving the surface composition largely unknown. H_2O ice rich spectra are abundant in the outer solar system (Barkume *et al.* 2008; Guilbert *et al.* 2009a). There are large differences in the strengths of the H_2O bands among bodies, where the strongest are seen on (136108) Haumea and its collisional family (Brown *et al.* 2007). All high signal-to-noise spectra of H_2O-rich spectra display the 1.65 - μm feature suggesting the crystalline phase, although the feature is not detected on smaller objects with admittedly lower quality spectra. Other ices are found only rarely. Methanol has been detected on two TNOs ((5145) Pholus and 2002 VE_{95}), NH_3 on Orcus and Pluto's satellite Charon, and methane on Pluto, Eris, Quoaoar, and Sedna. This diverse population has been under scrutiny since the first TNO discovery in 1992 since Pluto to understand its formation, evolution, and present state as well as what it can tell us about the solar system as a whole.

2. Overview of Large Program Results

A second Large Program (PI M. A. Barucci) dedicated to investigating the surface properties of Centaurs and Transneptunian objects was carried out between 2006 and 2008 using the European Southern Observatory (ESO) Very Large Telescope (VLT) and New Technology Telescope (NTT). Observational methods included spectroscopy, photometry (colors and lightcurves), and polarimetry. This variety of techniques was chosen to obtain complementary information about each body that any single method could not provide. Here we summarize the current state of the analysis of the 45 objects that were observed, and focus specifically on the presence of ices detected by features in near-infrared spectra.

2.1. *Polarimetric Measurements*

A behavior of linear polarization degree versus phase angle strongly depends on top surface layer properties, such as albedo, complex refractive index, particle size, packing density, and microscopic optical heterogeneity. 8 TNOs and Centaurs were observed: (136199) Eris, (136108) Haumea, (20000) Varuna, (38628) Huya, 26375 (1999 DE_9), (2060) Chiron, (5145) Pholus and (10199) Chariklo. Negative polarization was measured for all objects observed, varying from -0.3% to -2%. For TNOs two different behaviors of polarization phase dependencies were discovered (Bagnulo *et al.* 2008). Objects with a diameter smaller than 1000 km exhibit a negative polarization that rapidly increases (in absolute value) with the phase angle and reaches about 1% with a phase angle as small as 1 degree. The largest TNOs exhibit a small fraction of negative linear polarization which does not noticeably change in the observed phase angle range. Different polarimetric behaviors of these two groups, which indicate considerable differences in their surface properties, are probably related to the retention of volatiles (Bagnulo *et al.* 2008). The modeling of polarimetric behavior of the largest objects suggests that their topmost surface layer consists of large (compared to wavelength) inhomogeneous particles (Belskaya *et al.* 2008). Smaller size TNOs characterized by a pronounced branch of negative polarization revealed similar polarization behavior regardless of the fact that they have different surface albedos and belong to different dynamical groups. On the other hand, all three Centaurs observed so far show different polarization phase angle behaviors (Belskaya *et al.* 2009). It implies greater diversity in surface characteristics of Centaurs compared to that of TNOs.

2.2. *Lightcurves*

Measuring the lightcurves of TNOs enables the calculation of the rotational period when there is a significant periodic amplitude. Lower limits to the axis ratio a/b of an ellipsoid that could represent the body can also be found by assuming all magnitude changes are due to an elongated shape rather than albedo variations. One can also constrain the lower limits on the density based on the synodic periods and lightcurve amplitudes, which can provide important constraints on the internal structure of TNOs. Rotational periods also help reveal the collisional history of TNOs since the smaller bodies are expected to be collisional products of originally larger bodies.

There were about 40 TNOs and Centaurs that had determined rotational periods before this Large Program. 7 new objects were observed by Dotto *et al.* (2008) and Perna *et al.* (2009a). Dotto *et al.* (2008) measured 3 new objects ((65489) Ceto, (90568) 2004 GV_9, and (95626) 2002 GZ$_{32}$), and find quite short rotational periods and densities low enough to put these bodies within the limits for rotationally stable ellipsoids. Perna *et al.* (2009a) calculate 4 new rotational periods ((144897) 2004 UX$_{10}$, (145451) 2005 RM$_{43}$, (145453) 2005 RR$_{43}$, and 2003 UZ$_{413}$). 2003 UZ$_{413}$ is found to be a quite peculiar rapidly rotating object with a high estimated density. The results from these works have put into question the size-density trend found by Sheppard *et al.* (2008). While brighter objects are more dense, it now appears that fainter objects have a wider range of densities than previously thought (Perna *et al.* 2009a), which indicates the importance of improving TNO density statistics and albedo measurements in order to assess the existence of any size-density relationship.

2.3. *Photometric Colors*

Measuring photometric colors is the best methods of characterizing the surfaces of a large sample of TNOs. While spectroscopy provides more resolution, photometric measurements are often the only possibility for dimmer (smaller, darker, and/or farther) objects. Colors of TNOs and Centaurs in the visible and near-infrared were reported by DeMeo *et al.* (2009) and Perna *et al.* (2009b) for 45 objects, 19 of which had no previous photometric color measurements. TNOs and Centaurs are divided into four taxonomic groups, BB, BR, IR, and RR that span from neutral to very red.

Our measurements of three objects could not be classified in this system (26375, 145452, and 2007 UK$_{126}$), because their visible colors were red, matching best the IR and RR classes, while the near-infrared colors were bluer, consistent with the BB and BR classes. This suggests greater color diversity than what was found in the original sample used to create the taxonomy. In analysis of the distribution of taxonomic classes among dynamical classes and orbital elements, it was found that 8 of the 13 Centaurs in the sample were in the neutral to moderately red BB and BR classes while the other 5 were in the RR class, confirming the bimodality of the population (Peixinho *et al.* 2003). All IR class objects in our sample are classical objects, which is consistent with Fulchignoni *et al.* (2008) where they found the IR class confined to the resonant and classical population.

2.4. *Spectral Measurements - Featureless and H$_2$O-rich bodies*

Visible spectra for 43 TNOs and Centaurs were measured in Alvarez-Candal *et al.* (2008) and Fornasier *et al.* (2009), and 22 of these had no previously published spectra. Both authors find most spectra to be featureless. Alvarez-Candal *et al.* (2008) find that (208996) 2003 AZ$_{84}$, (10199) Chariklo, and (42355) Typhon have a broad, shallow feature centered at 0.65 μm for the first two and 0.6 μm for the latter. A similar band is seen on dark asteroids, and has been thought to be be due to the presence of aqueously altered minerals (Vilas *et al.* 1994), however, it is unknown how temperatures in the outer solar system could be high enough for aqueous alteration to occur.

Observations of (60558) Echeclus by Alvarez-Candal *et al.* (2008) had a visible slope 50% less than previous measurements. They also find that many spectra tend to flatten out past 0.75 μm relative to the visible slope. Observations of (42355) Typhon by Alvarez-Candal *et al.* (2009) confirm the previously identified feature, however, the spectrum of (208996) 2003 AZ$_{84}$ did not have the weak 0.65 - μm feature seen originally by Alvarez-Candal *et al.* (2008), suggesting surface heterogeneity. The visible slopes of 73 objects from these works and the literature were collected by Fornasier *et al.* (2009) and the mean was found to be 17.9 %/10^3 Å. The classical TNOs had a deficit of very red slopes compared with other populations with no slopes exceeding 35%/10^3 Å, although this is likely an observational bias because the redder objects tend to be smaller. They also find that the TNO/Centaur population has a wider distribution of slopes than the Trojan asteroids that may have originated in the Kuiper Belt.

21 near-infrared spectra were presented in Guilbert *et al.* (2009a). Of this sample, 4 are featureless, 6 show clear signatures of H_2O ice, 7 show more uncertain signatures of H_2O ice, 1 (Eris) has methane, and the signal-to-noise ratio of the last three is too low to analyze. The objects with rare spectral features are further discussed in the next section. Plutino 47171 was modeled by Protopapa *et al.* (2009) using Titan tholin, Triton tholin diluted in H_2O and either amorphous carbon or serpentine. Their spectrum differs slightly from two previously published spectra, suggesting surface heterogeneity. Alvarez-Candal *et al.* (2009) found that the spectrum of the scattered disk object (42355) Typhon could overall be well modeled with Titan and Triton tholins, H_2O ice, amorphous carbon, and serpentine. What could not be modeled, however, is the broad, shallow band around 0.6 μm. Merlin *et al.* (2009b) model 6 TNOs and find that the surfaces have H_2O ice and a dark material that is likely an irradiated product. They find objects could have both crystalline and amorphous H_2O ice on the surface, which is consistent with the balance between the states at low temperatures seen in the laboratory results of Zheng *et al.* (2008). However, the laboratory samples only undergo partial irradiation (e- only) and so other higher energetic irradiation sources must still be considered.

10199 Chariklo is a Centaur with a diameter of around 260 kilometers and an albedo of approximately 5.7% (Stansberry *et al.* 2008). Chariklo has been observed six times since 1997 and, interestingly, the first spectra display significant H_2O ice features, while the most recent from Guilbert *et al.* (2009a,b) are nearly featureless with insignificant amounts of H_2O. Since the rotational period of Chariklo is not known, it is unclear whether these differences are due to surface heterogeneity or if it is a temporal change (Guilbert *et al.* 2009b). Chariklo orbits the sun between 13 and 18.5 AU, so the spectral change is not likely due to its approach toward the sun because the distance change is not so great. Additionally, Guilbert *et al.* (2009b) report that no cometary signatures were found from recent images of Chariklo. In models of the spectra they show that Chariklo could be comprised of an irradiated coating and an underlying fresher layer that has been partially resurfaced by impacts.

2.5. *Spectral Measurements - Large Icy Bodies*

Three of the bodies we observed, which were also among the largest, have more unique spectral features. (90482) Orcus, (50000) Quaoar, and (136199) Eris have other ices besides H_2O on their surface. The results of the analysis of each body is presented here.

Barucci *et al.* (2008) observed the TNO (90482) Orcus in January of 2007 and present the best quality data to-date of this H_2O-rich object. Orcus shows deep H_2O features at 1.5 and 2 μm as well the 1.65 feature indicating crystallinity. They also detected a feature with a depth of 10% at 2.216 μm that resembles the ammonia hydrate feature seen on Charon. Barucci *et al.* (2008) model the spectrum using amorphous and crystalline H_2O ice, ammonia diluted in crystalline water ice and a blue component artificially created to reproduce the negative overall slope of Orcus' spectrum. It is difficult to firmly identify the ~2.2 - μm feature as ammonia hydrate based on one weak band. Further studies of objects with this band, however, are important because NH_3 is expected to be quickly destroyed by irradiation (Strazzulla and Palumbo 1998; Cooper *et al.* 2003). Proposed mechanisms to resupply NH_3 to the surface include cryovolcanism, impact gardening, and solid-state greenhouse or convection, with cryovolcanism being the favored mechanism (Cook *et al.* 2007). These phenomena could also explain the abundance of crystalline H_2O ice among TNOs, whose presence is intriguing because it should be amorphized by space weathering over the age of the solar system.

(50000) Quaoar has a diameter of ~840 km and is a relatively bright object among TNOs with an albedo of 19% (Stansberry *et al.* 2008). The visible spectrum is very red, with a slope of about 28% per 100 nm (Alvarez-Candal *et al.* 2008). Quaoar's near-infrared spectrum is unique because, although it displays clear H_2O features at 1.5, 1.65, and 2 μm and another feature near 2.2 μm as seen on Charon and Orcus, there are other weak features present that suggest the 2.2 - μm feature is due to methane rather than NH_3. Schaller and Brown (2007a) first suggested the presence of methane on Quaoar's surface while simultaneously detecting weak features that match those of ethane in the spectrum. Quaoar is a transition object that has not undergone complete volatile loss (Schaller and Brown 2007b). Dalle Ore *et al.* (2009) model a spectrum from the visible to near-infrared including the additional constraints of Spitzer data at 3.6 and 4.5 μm and find a best model consisting of an intimate mixture of crystalline and amorphous H_2O, CH_4, N_2, and C_2H_6 ices with Triton and Titan tholins. They find that amorphous H_2O ice increases the albedo at the wavelengths of the second Spitzer band, improving the fit, but the model still underestimates the reflectance, suggesting that either the quality of the optical constants used is not sufficient, or that there is another component on the surface missing in the model.

Merlin *et al.* (2009a) present two spectra of the largest TNO (136199) Eris, whose diameter is ~2400 km (Brown *et al.* 2006). Eris' spectrum in the near-infrared is dominated by deep and broad methane features suggesting large grain sizes. A detailed analysis of the spectra by Merlin *et al.* (2009a) reveals that there are small wavelength shifts in the position of methane bands compared to those obtained from laboratory measurements of pure methane ice. They propose several possible reasons for this shift: geographic versus intimate mixing of material, hydrated and diluted materials, temperature and physical state, and irradiation effects. Dilution of methane in N_2 is the most likely cause of the shift, even though N_2 cannot be directly recognized since it is probably in its alpha state (below 35.6 K) which has a very narrow band in the near-infrared that is not detectable at the spectral resolution of the data. The amount of nitrogen on Eris is expected to be much smaller than on Pluto and Triton because of the stronger 1.689 - μm pure methane feature and the less significant wavelength shifts. The deepest bands in in the visible wavelengths of Eris' spectrum are more shifted than the weakest suggesting a stratification of diluted methane ice.

3. Conclusions

The second ESO Large Program devoted to characterizing TNOs has contributed immensely to our understanding of these bodies by using many observational techniques. Advancement of the discovery and analysis of ices on TNO surfaces has been achieved particularly through near-infrared spectral measurements where diagnostic ice signatures are present. The presence or absence of these features has been determined for the 45 objects in this study, although more detailed modeling and analysis of about 20 objects in the sample is still underway. We await the final results from this program, as well as many more interesting discoveries and breakthroughs through the coming years and decades aided by discovery surveys and improvements in technology.

References

Alvarez-Candal, A., Barucci, M. A., Merlin, F., de Bergh, C., Fornasier, S., Guilbert, A., & Protopapa, S. 2009, *A&A*, submitted
Alvarez-Candal, A., Fornasier, S., Barucci, M. A., de Bergh, C., & Merlin, F. 2008, *A&A*, 487, 741

Bagnulo, S., Belskaya, I., Muinonen, K., Tozzi, G. P., Barucci, M. A., Kolokolova, L., & Fornasier, S. 2008, *A&A*, 491, L33

Barkume, K. M., Brown, M. E., & Schaller, E. L. 2008, *AJ*, 135, 55

Barucci, M. A., Merlin, F., Guilbert, A., de Bergh, C., Alvarez-Candal, A., Hainaut, O., Doressoundiram, A., Dumas, C., Owen, T., & Coradini, A. 2008, *A&A*, 479, L13

Belskaya, I., Bagnulo, S., Muinonen, K., Barucci, M. A., Tozzi, G. P., Fornasier, S., & Kolokolova, L. 2008, *A&A*, 479, 265

Belskaya, I., Bagnulo, S., Barucci, M. A., Muinonen, K., Tozzi, G. P., Fornasier, S., & Kolokolova, L. 2009, *Icarus*, submitted

Brown, M. E., Barkume, K. M., Ragozzine, D., Schaller, E. L. 2007, *Nature*, 446, 294

Brown, M. E., Schaller, E. L., Roe, H. G., Rabinowitz, D. L., & Trujillo, C. A. 2006, *ApJ*, 643, L61

Cook, J. C., Desch, S. J., Roush, T. L., Trujillo, C. A., & Geballe, T. R. 2007, *ApJ*, 663, 1406

Cooper, J. F., Christian, E. R., Richardson, J. D., & Wang, C. 2003, *EM&P*, 92, 261

Dalle Ore, C. M., Barucci, M. A., Emery, J. P., Cruikshank, D. P., Dalle Ore, L. V., Merlin, F., Alvarez-Candal, A., de Bergh, C., Trilling, D. E., Perna, D., Fornasier, S., Mastrapa, R. M. E., & Dotto, E. 2009, *A&A*, 501, 349

DeMeo, F. E., Fornasier, S., Barucci, M. A., Perna, D., Protopapa, S., Alvarez-Candal, A., Delsanti, A., Doressoundiram, A., Merlin, F., & de Bergh, C. 2009, *A&A*, 493, 283

Dotto, E., Perna, D., Barucci, M. A., Rossi, A., de Bergh, C., Doressoundiram, A., & Fornasier, S. 2008, *A&A*, 490, 829

Fornasier, S., Barucci, M. A., de Bergh, C., Alvarez-Candal, A., DeMeo, F. E., Merlin, F., Perna, D., Guilbert, A., Delsanti, A., Dotto, E., & Doressoundiram, A. 2009, *A&A*, accepted

Fulchignoni, M., Belskaya, I., Barucci, M. A., de Sanctis, M. C., & Doressoundiram, A. 2008, in: M. A. Barucci, H. Boehnhardt, D. Cruikshank & A. Morbidelli (eds.), *The Solar System Beyond Neptune*, (Tucson: Univ. of Arizona Press), p. 181

Guilbert, A., Alvarez-Candal, A., Merlin, F., Barucci, M. A., Dumas, C., de Bergh, C., & Delsanti, A. 2009a, *Icarus*, 201, 272

Guilbert, A., Barucci, M. A., Brunetto, R., Delsanti, A., Merlin, F., Alvarez-Candal, A., Fornasier, S., de Bergh, C., & Sarid, G. 2009b, *A&A*, 501, 777

Merlin, F., Alvarez-Candal, A., Delsanti, A., Fornasier, S., Barucci, M. A., DeMeo, F. E., de Bergh, C., Doressoundiram, A., Quirico, E., & Schmitt, B. 2009a, *AJ*,137, 315

Merlin, F., Barucci, M. A., de Bergh, C., Fornasier, S., Doressoundiram, A., Perna, D., & Protopapa, S. 2009b, *Icarus*, submitted

Peixinho, N., Doressoundiram, A., Delsanti, A., Boehnhardt, H., Barucci, M. A., & Belskaya, I. 2003, *A&A*, 410, L29

Perna, D., Dotto, E., Barucci, M. A., Rossi, A., Fornasier, S., & de Bergh, C. 2009a, *A&A*, accepted

Perna, D., Barucci, M. A., Fornasier, S., Demeo, F. E., Alvarez-Candal, A., Merlin, F., Dotto, E., Doressoundiram, A., & de Bergh, C. 2009b, *A&A*, submitted

Protopapa, S., Alvarez-Candal, A., Barucci, M. A., Tozzi, G. P., Fornasier, S., Delsanti, A., & Merlin, F. 2009, *A&A*, 501, 375

Schaller, E. L. & Brown, M. E. 2007a, *ApJ*, 659, L61

Schaller, E. L. & Brown, M. E. 2007b, *ApJ*, 670, L49

Sheppard, S. S., Lacerda, P., & Ortiz, J. L. 2008, in: M. A. Barucci, H. Boehnhardt, D. Cruikshank & A. Morbidelli (eds.), *The Solar System Beyond Neptune*, (Tucson: Univ. of Arizona Press), p. 129

Stansberry, J., Grundy, W., Brown, M., Cruikshank, D., Stansberry, J., Grundy, W., Brown, M., Cruikshank, D., Spencer, J., Trilling, D., & Margot, J. 2008, in: M. A. Barucci, H. Boehnhardt, D. Cruikshank & A. Morbidelli (eds.), *The Solar System Beyond Neptune*, (Tucson: Univ. of Arizona Press), p. 161

Strazzulla, G. & Palumbo, M. E. 1998, *P&SS*, 46, 1339

Vilas, F., Jarvis, K. S., & Gaffey, M. J. 1994, *Icarus*, 109, 274

Zheng, W., Jewitt, D., & Kaiser, R. I. 2008, *arXiv:0801.2805*

Icy Bodies of the Solar System
Proceedings IAU Symposium No. 263, 2009
J.A. Fernández, D. Lazzaro, D. Prialnik & R. Schulz, eds.

© International Astronomical Union 2010
doi:10.1017/S1743921310001730

The Dark Red Spot on KBO Haumea

Pedro Lacerda[1,2]

[1] Queen's University, Belfast BT7 1NN, United Kingdom.
email: `p.lacerda@qub.ac.uk`
[2] Newton Fellow

Abstract. Kuiper belt object 136108 Haumea is one of the most fascinating bodies in our solar system. Approximately $2000 \times 1600 \times 1000$ km in size, it is one of the largest Kuiper belt objects (KBOs) and an unusually elongated one for its size. The shape of Haumea is the result of rotational deformation due to its extremely short 3.9-hour rotation period. Unlike other 1000 km-scale KBOs which are coated in methane ice the surface of Haumea is covered in almost pure H_2O-ice. The bulk density of Haumea, estimated around 2.6 g cm^{-3}, suggests a more rocky interior composition, different from the H_2O-ice surface. Recently, Haumea has become the second KBO after Pluto to show observable signs of surface features. A region darker and redder than the average surface of Haumea has been identified, the composition and origin of which remain unknown. I discuss this recent finding and what it may tell us about Haumea.

Keywords. Kuiper Belt, techniques: photometric, infrared: solar system

---

1. Introduction

The Kuiper belt is currently the observational frontier of our solar system. Presumably the best kept remnants of the icy planetesimals that formed the outer planets, Kuiper belt objects (KBOs) have been the subjects of intense study in the past $\sim$15 years. One intriguing KBO is 136108 Haumea (formerly 2003 EL$_{61}$). First famous for its super-fast rotation and elongated shape, Haumea went on to surprise us with a host of interesting properties. Haumea's spin frequency of one rotation every $\sim$ 3.9 hr is unparalleled for an object this large (Sheppard *et al.* 2008). Its shape is rotationally deformed into a $2000 \times 1600 \times 1000$ km triaxial ellipsoid (Rabinowitz *et al.* 2006) to balance gravitational and centripetal accelerations. To attain such a fast rotation, Haumea might have suffered a giant impact at the time when the Kuiper belt was massive enough to render such events likely. Infrared spectroscopy has revealed a surface covered in almost pure H_2O ice (Trujillo *et al.* 2007) which gives Haumea an optically blue colour (Tegler *et al.* 2007). The surfaces of the remaining Pluto-sized KBOs (Eris, Pluto and Makemake) are covered in CH$_4$ ice instead, granting them the tag 'Methanoids'. Two satellites were discovered in orbit around Haumea (Brown *et al.* 2006), the largest of which is also coated in even purer H_2O ice (Barkume *et al.* 2006). The two satellites have nearly coplanar orbits with fast-evolving, complex dynamics due mainly to tidal effects from the primary (Ragozzine & Brown 2009). Haumea's bulk density, derived assuming it is near hydrostatic equilibrium, is $\rho \sim 2.6$ g cm^{-3} (Lacerda & Jewitt 2007). The surface material has density $\rho \sim 1$ in the same units implying that the interior must be differentiated and Haumea must have more rock-rich core. A number of KBOs showing signs of H_2O ice in their surface spectra all lie close to Haumea in orbital space (Schaller & Brown 2008); this, plus the unusually fast spin, the differentiated inner structure and the two small satellites also covered in H_2O ice, all have been taken as evidence that Haumea is the largest remnant of a massive collision that occured > 1 Gyr ago (Ragozzine & Brown 2007, Brown *et al.* 2007). However, several potential members of the collisional family have been eliminated based on infrared photometry (Snodgrass *et al.*, poster at this meeting).

2. The Dark Red Spot (DRS) on Haumea

We observed Haumea in mid-2007 using the University of Hawaii 2.2m telescope with the goal of measuring its lightcurve in two bands, B and R (Fig. 1a). Our high-quality photometry (Lacerda *et al.* 2008) shows two important features:

(*a*) The lightcurve is not symmetric as would be expected from a uniform ellipsoidal body. There is a clear asymmetry between the two sets of minima and maxima indicating the presence of a dark region on the surface (Fig. 1a). A model lightcurve generated by placing a dark spot on the equator of Haumea, visible at both minimum and maximum cross-section (Fig. 1b), successfully fits the data.

(*b*) Upon aligning the B and R lightcurve data we verify that the B points lie consistently below the R points precisely at the location of the dark spot. In other words, the dark spot is also redder than the average surface.

In the rest of the paper we use DRS to refer to the dark red spot. In our model (Fig. 1) the size and relative darkness of the DRS are degenerate: the spot may be as small as a few percent of the projected cross-section of Haumea and be about 20% as reflective as the rest of the surface, or it may be as large as to take a full hemisphere of Haumea being then only 5% less reflective than elsewhere. The same degeneracy applies to colour vs. spot size. However, assuming the DRS colour is within the range of values typically found in the solar system, $1.0 \lesssim B - R$ (mag) $\lesssim 2.5$, then when directly facing the observer the spot must take between 20% and 60% of the projected cross-section of Haumea, and have an albedo between 55% and 65%. This combination of colour and albedo is consistent with, e.g. Eris, Makemake and the bright regions on Pluto and on Saturn's satellite Iapetus; it is inconsistent with Pluto's darker regions, with Pluto's satellite Charon, with Saturn's irregular satellite Phoebe and with Centaurs Chiron and Pholus.

3. The DRS in the infrared

Prompted by the fact that Haumea is covered in H_2O ice, we set out to investigate how the properties of the ice changed close the DRS region by monitoring the infrared

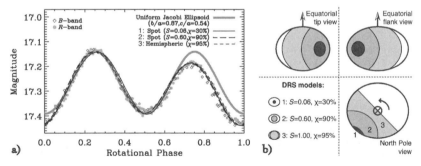

Figure 1. a) Lightcurve of Haumea in two filters. Both B and R data were taken over 3 nights to ensure repeatability. The effect of the dark red spot is apparent at rotational phases $0.7 \lesssim \phi \lesssim 1.0$: the maximum and minimum that bracket that region appear darker and the B-band flux is consistently lower than the R-band flux indicating the spot is redder than elsewhere. We measure a lightcurve period $P = 3.9155 \pm 0.0001$ hours and a photometric range $\Delta m = 0.29 \pm 0.02$ mag. The rotationally averaged colour is $B - R = 0.972 \pm 0.017$ mag. Best fit Jacobi ellipsoid models are overplotted: a thick solid grey line shows how the uniform surface model fails to fit the dark red spot region, and thinner lines show that a small ($S \ll 1$) and very dark ($\chi \ll 100\%$) spot or a large ($S \sim 1$) and not very dark ($\chi \sim 100\%$) spot fit the data equally well. **b)** Cartoon representation of the three spot models considered in a) showing the location of the spot on the surface of Haumea.

H_2O-ice absorption band at 1.5 μm (Fig. 2). We collected two sets of data: time-resolved, quasi-simultaneous broadband J (to probe the continuum) and medium band CH_4s (to probe the 1.5 μm) data using UKIRT, and a similar dataset at the Subaru telescope this time using broadband filters J and H. Both telescopes are located atop Mauna Kea.

Neither of the filters CH_4s or H fully probe the 1.5 μm band, and both are affected by a narrower band at 1.65 μm which, if present, indicates that the ice is mostly crystalline (Merlin et $al.$ 2007). Our UKIRT measurements indicate that the CH_4s/J flux ratio decreases by a few percent close to the DRS, while our Subaru measurement show that the H/J flux ratio increases also by a few percent. This apparent contradiction can be explained if the DRS is richer in crystalline H_2O-ice than the rest of the surface. That would change the shape of the 1.5 μm and 1.65 μm bands and cause exactly what is observed (see Fig. 2a). Fraser & Brown (2009) report HST NICMOS observations of Haumea using the broadband filters F110W and F160W centered respectively at 1.1 μm and 1.6 μm. They find that the F160W/F110W flux ratio decreases at the DRS, consistent with our UKIRT observations. Infrared spectra obtained at 4-m-class telescopes show no rotational variations, likely due to lack of sensitivity (Pinilla-Alonso et $al.$ 2009).

4. The DRS 4-band spectrum

Using the time-resolved BR data from the UH 2.2 m and JH data from Subaru we constructed 4-band spectra of Haumea as it rotates. We did this by interpolating each of the four lightcurves and measuring the relative differences between the bands. The result is shown in Fig. 2b where we plot 3 spectra away from the DRS ($\phi = 0.3, 0.4, 0.6$) and 3 spectra close to the DRS ($\phi = 0.7, 0.8, 0.9$). The Figure shows that the DRS material is a more efficient B absorber than the rest of Haumea, hence the redder $B-R$ colour, and that it is redder in visible-to-infrared colour. In §3 we saw that the DRS displays bluer or redder behaviour in the JH wavelength range depending on exactly which filter bandpasses are used.

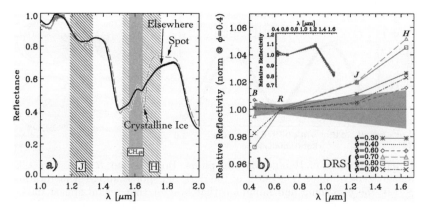

Figure 2. a) Near-infrared synthetic spectra of H_2O-ice. The orange dashed line shows a spectrum of crystalline ice (indicated by the 1.65 μm feature) while the solid black line corresponds to ice with a lower degree of crystallinity. **b)** Time-resolved 4-band spectrum of Haumea [adapted from Lacerda (2009)]. Each line is a spectrum at a given rotational phase. At rotational phases when the DRS faces the observer ($0.7 \lesssim \phi \lesssim 1.0$) the B band is depressed and the H band is enhanced. Spectra at each rotational phase are plotted relative to R band and all rotational phases have been normalised by $\phi = 0.4$. Inset shows spectra before normalisation at $\phi = 0.4$.

5. Interpretation

It is very unlikely that the dark spot is a topographical feature such as a mountain or valley as that would produce an achromatic change in brightness. Instead, the spot region exhibits slight but persistent visible and infrared colour properties that distinguish it from the rest of Haumea's surface. No atmosphere has been detected on Haumea rendering an explanation based on irregular condensation of gases on the surface unlikely. The fact that the DRS absorbs *B*-band light more efficiently could indicate the presence of hydrated minerals (Jewitt *et al.* 2007). Alternatively the redder tint could be due to the presence of irradiated organic materials which would also explain the blue behaviour observed in some infrared bandpasses (Fraser & Brown 2009). If confirmed, the higher degree of ice crystallinity at the DRS could signal a recent temperature rise.

6. Speculation

The DRS could be a region where material from Haumea's interior is trickling out. The high bulk density of Haumea indicates the presence of a more mineral-rich core. If warmer, deep-lying material would find its way to the surface it would appear darker and presumably redder than H_2O ice. The slight increase in temperature and the presence of H_2O are useful ingredients for mineral hydration and H_2O-ice crystallinization.

The DRS could also be the site of a recent impact of a $\sim 1 - 10$ km KBO onto Haumea. Small KBOs are dark and most are believed to be covered in red, irradiated organic mantles. The collision would locally raise the temperature, thereby accelerating the transition from amorphous-to-crystalline H_2O ice, and the impactor material would probably leave a visible trace on the surface.

The DRS could also be due to something completely different. Time-resolved spectroscopy of Haumea using $8 - 10$ m telescopes should help determine the composition of the DRS and help solve the mystery of its origin.

Acknowledgements

I am grateful to David Jewitt for comments on the manuscript and to the Royal Society for the support of a Newton Fellowship.

References

Barkume, K. M., Brown, M. E., & Schaller, E. L. 2006, *ApJ*, 640, L87
Brown, M. E., *et al.* 2006, *ApJ*, 639, L43
Brown, M. E., Barkume, K. M., Ragozzine, D., & Schaller, E. L. 2007, *Nature*, 446, 294
Fraser, W. C. & Brown, M. E. 2009, *ApJ*, 695, L1
Jewitt, D., Peixinho, N., & Hsieh, H. H. 2007, *AJ* 134, 2046
Jewitt, D. C. & Sheppard, S. S. 2002, *AJ*, 123, 2110
Lacerda, P. 2009, *AJ*, 137, 3404
Lacerda, P., Jewitt, D., & Peixinho, N. 2008, *AJ*, 135, 1749
Lacerda, P. & Jewitt, D. C. 2007, *AJ*, 133, 1393
Merlin, F., *et al.* 2007, *A&A*, 466, 1185
Pinilla-Alonso, N., *et al.* 2009, *A&A*, 496, 547
Rabinowitz, D. L., *et al.* 2006, *ApJ*, 639, 1238
Ragozzine, D. & Brown, M. E. 2007, *AJ*, 134, 2160

Ragozzine, D. & Brown, M. E. 2009, *AJ*, 137, 4766

Schaller, E. L. & Brown, M. E. 2008, *ApJ*, 684, L107

Sheppard, S. S., Lacerda, P., & Ortiz, J. L. 2008, in: M. A. Barucci, H. Boehnhardt, D. Cruishank & A. Morbidelli (eds.), *The Solar System Beyond Neptune*, (Tucson: University of Arizona Press), p. 129

Tegler, S. C., *et al.* 2007, *AJ*, 133, 526

Trujillo, C. A., *et al.* 2007, *ApJ*, 655, 1172

Icy Bodies of the Solar System
Proceedings IAU Symposium No. 263, 2009
J.A. Fernández, D. Lazzaro, D. Prialnik & R. Schulz, eds.
ⓒ International Astronomical Union 2010
doi:10.1017/S1743921310001742

Water alteration on (42355) Typhon?

Alvaro Alvarez-Candal[1] and Maria Antonietta Barucci[2]

[1]European Souther Observatory,
Alonso de Córdova 3107, Vitacura Casilla 19001, Santiago 19, Chile
email: `aalvarez@eso.org`

[2]LESIA/Observatoire de Paris
5, Place Jules Janssen, 92195, Meudon, France

Abstract. The visible spectra of (42355) Typhon showed evidence for aqueously altered materials. Therefore we seek to understand if such an event is possible.

We use data from the ESO / Very Large Telescope together with the Hapke Hapke radiative transfer model to interpret the surface composition of (42355) Typhon over the whole spectral range ($\sim 0.5 - 2.4\ \mu m$).

Our results points that (42355) Typhon could be a fragment from a larger parent body that suffered aqueous alteration.

Keywords. techniques: spectroscopic, Kuiper Belt

1. Introduction

Aiming at obtaining high quality data of the trans-Neptunian objects, TNOs, population, a large program for the observation of centaurs and TNOs was started using the facilities of the ESO / Very Large Telescope at Cerro Paranal in Chile (PI: M.A. Barucci). The program lasted from November 2006 until December 2008. In the framework of the large program we observed the scattered disk object, SDO, (42355) Typhon.

Its spectrum shows evidence of water ice in the near infrared (Guilbert *et al.* 2009a). Although no signature of ices are found in the visible, in a previous work we detected a subtle absorption feature possibly related to aqueously altered material (Alvarez-Candal *et al.* 2008, AC08 hereafter). (42355) Typhon is also the primary body of a binary system (Noll *et al.* 2006). Its diameter is about 175 km (Stansberry *et al.* 2008), however, this measurement could be overestimated due to its binary nature. Note that recently Benecchi et al. (2009) showed that the colors of both bodies in the system are similar.

2. Analysis

The data we analize here was obtained during two runs at the Very Large Telescope in January 2007 and April 2008, using FORS2 and ISAAC at the UT 1 and SINFONI at the UT 4. We use also data already published: VRIJHK photometry (DeMeo *et al.* 2009), visible spectroscopy by AC08, and H+K spectroscopy by Guilbert *et al.* (2009a). Therefore, in what follows, we concentrate on the results obtained from the modeling and our interpretation of the data.

Figure 1 shows the two composite spectra of (42355) Typhon, from 0.4 to 2.3 μm. To connect the entire spectral range we used the colors (V-J) and (V-H) from DeMeo *et al.* (2009). The spectral gradient, S', in the visible of both spectra are similar, 10.6 ± 0.6 and 11.7 ± 0.5 % $(0.1\ \mu m)^{-1}$, for the January 2007 and April 2008 observations, respectively.

As we suggested in AC08 that (42355) Typhon shows evidence of water alterarion, we processed both visible spectra adapting the technique described in Vilas *et al.*

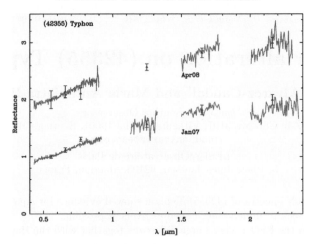

Figure 1. Composite spectrum of (42355) Typhon obtained with the data taken in the January 2007 and April 2008 runs. The spectra are normalized to unity at the V-filter, April 2008 spectrum is shifted by 1 for clarity. The triangles show the photometric colors used to set the reflectance scale. The region between 1.8 and 2.0 μm was removed due to the strong telluric absorptions.

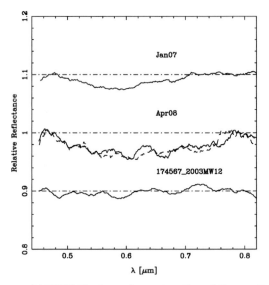

Figure 2. Visible spectra of (42355) Typhon after extraction of the continuum. The spectra are offset by a reflectance of 0.1 for comparison. The dashed line indicate the spectrum of (42355) Typhon as obtained dividing by a different solar analog observed the same night of the April 2008 run. To illustrate that the absorption we see is not due to our choice of calibration star, we apply the same analysis to spectra of (174567) 2003 MW$_{12}$, obtained the same night as our (43355) Typhon data, and find no evidence for an absorption.

(1993, 1994): We smoothed the spectra using a window of about 0.03 μm, then we estimated the slope of the continuum, tracing a line between 0.47 and 0.78 μm (e.g., Carvano *et al.* 2003). The spectra were then divided by this conventional continuum. These "normalized" spectra are shown in Fig. 2. There is evidence for the existence of an absorption feature in the April 2008 spectrum, even deeper than that observed in January 2007 although the shape and position are slightly different.

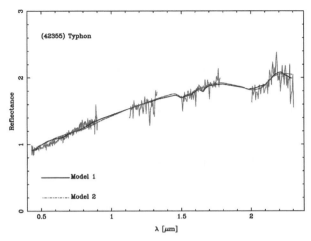

Figure 3. Composite spectra of (42355) Typhon taken in April 2008. The spectrum is normalized to unity at the V-filter. The best fitting models were obtained using intimate mixtures. The black line, Model 1, corresponds to a model including 12 % Serpentine (160 μm), 4 % of crystalline water ice (85 μm), 17, and 6 % of Triton and Titan tholins, respectively, and 61 % of amorphous carbon. This model corresponds to an albedo of 6.38 % at 0.55 μm. The dashed line, labeled Model 2, includes 6 % of crystalline water ice (45 μm), 28 % of Triton tholin (13 μm), 59 % of Titan tholin (1 mm), 6 % of amorphous water ice (10 μm), and 1 % of amorphous carbon. The albedo at 0.55 μm is 6.04 %.

In the near infrared we only see a feature in the H+K region which we associate with water ice. The apparent feature in the January 2007 spectrum at 2.25 μm is probably an artifact of the combination of decreasing efficiency of the detector and the faintness of the target.

The April 2008 data was modeled using a radiative transfer model, based on the Hapke theory (Hapke, 1993). The free parameters of the model are the compound grain size and abundance. The best fit model was obtained by minimizing χ^2 using a Marqvardt-Levenberg algorithm.

We considered: Triton and Titan tholins, to match the red continuum in the visible (Khare *et al.* 1984, 1993), water ice in crystalline and amorphous states (crystalline water ice constants at 40 K from Quirico and Schmitt 1997, amorphous ice at 38 K from Grundy and Schmitt 1998), amorphous carbon to match the low albedo (Zubko *et al.* 1996). The only hydrated mineral for which optical constants are available is Serpentine (R. Clark private communication). Both models describe, reasonably well, the whole spectral range, although none of them is able to reproduce the feature detected in the visible.

3. Discussion

The feature observed at about 0.6 μm resembles those found in some low albedo asteroids (Vilas *et al.*, 1994, Rivkin *et al.*, 2002), which are often associated to phyllosilicates, formed by the interaction of liquid water and silicates. It could indicate the presence of material that suffered aqueous alteration. Phyllosilicates could also have bands in the near infrared. The only ones covered by our observations are those between 2.2 and 2.4 μm. Unfortunately, our spectra do not have enough signal-to-noise ratio as to confidently confirm a detection of such features.

The decay of radiogenic material could produce enough heat as to melt the water ice in the interior of icy bodies (see McKinnon *et al.*, 2008). This is supported by the detection of hydrated silicates on comet nuclei (Lisse *et al.*, 2006). Nevertheless, (42355) Typhon is too small to have suffered this kind of process.

As (42355) Typhon is too small to have suffered alteration, it is still possible that it is a fragment of a larger parent body that did suffer it. SDOs are dynamically related to centaurs and JFCs, so further evidences could be find in these populations. Interestingly, Guilbert *et al.* (2009b) showed that the centaur (10199) Chariklo has a feature in the visible, probably indicating water alteration, similar to that of (42355) Typhon. (10199) Chariklo is twice as big as (42355) Typhon, therefore it is possible that it had enough internal heat to melt water ice, so as there exist a Chariklo, there could have existed a larger body that originated Typhon after a catastrophic collision.

References

Alvarez-Candal, A., Fornasier, S., Barucci, M. A., de Bergh, C., & Merlin, F. 2008, *A&A*, 487, 741

Benecchi, S. D., Noll, K. S., Grundy, W. M., Buie, W. M., Stephens, D. C., & Levison, H. F. 2009, *Icarus*, 200, 292

Carvano, J. M., Mothé-Diniz, T., & Lazzaro, D. 2003, *Icarus*, 161, 356

DeMeo, F., Fornasier, S., Barucci, M. A., Perna, D., Protopapa, S., Alvarez-Candal, A., Delsanti, A., Doressoundiram, A., Merlin, F., & de Bergh, C. 2009, *A&A*, 493, 283

Grundy, W. M. & Schmitt, B. 1998, *Journal Geophys. Res.*, 103, 25809

Guilbert, A., Alvarez-Candal, A., Merlin, F., Barucci, M. A., Dumas, C., de Bergh, C., & Delsanti, A. 2009a, *Icarus*, 201, 272

Guilbert, A., Barucci, M. A., Brunetto, R., Delsanti, A., Merlin, F., Alvarez-Candal, A., Fornasier, S., de Bergh, C., & Sarid, G. 2009b, *A&A*, 501, 777

Hapke, B. 1993, *Topics in Remote Sensing 3: Theory of reflectance and emittance spectroscopy*, (Cambridge: Cambridge Univ. Press)

Khare, B. N., Sagan, C., Arakawa, E. T., Suits, F., Callcott, T. A., & Williams, M. W. 1984, *Icarus*, 60, 127

Khare, B. N., Thompson, W. R., Cheng, L., Chyba, C., Sagan, C., Arakawa, E. T., Meisse, C., & Tuminello, P. S. 1993, *Icarus*, 103, 290

Lisse, C. M. & 16 colleagues 2006, *Nature*, 313, 635

McKinnon, W. B., Prialnik, D., Stern, S. A., & Coradini, A. 2008, in: M. A. Barucci, H. Boehnhardt, D., Cruikshank, & A. Morbidelli (eds.), *The Solar System Beyond Neptune*, (Tucson: Univ. of Arizona Press), p. 213

Noll, K. S., Grundy, W. M., Stephens, D. C., & Levison, H. F. 2006, *IAU Circulars*, 8689, 1

Quirico, E. & Schmitt, B. 1997, *Icarus*, 127, 354

Rivkin, A. S., Howell, E. S., Vilas, F., & Lebofsky, L. A. 2003, in: W. F. Bottke, A. Cellino, P. Paolicchi, & R. P. Binzel (eds.), *Asteroids III*, (Tucson: Univ. of Arizona Press, Tucson), p. 235

Stansberry, J., Grundy, W., Brown, M., Cruikshank, D., Spencer, J., Trilling, D., & Margot, J.-L. 2008, in: M. A. Barucci, H. Boehnhardt, D. Cruikshank, & A. Morbidelli, (eds.), *The Solar System Beyond Neptune*, (Tucson: Univ. of Arizona Press), p. 161

Vilas, F., Hatch, E. C., Larson, S. M., Sawyer, S. R., & Gaffey, M. J. 1993, *Icarus*, 102, 225

Vilas, F., Jarvis, K. S., & Gaffey, M. J. 1994, *Icarus*, 109, 274

Zubko, V. G., Mennella, V., Colangeli, L., & Bussoletti, E. 1996, *MNRAS*, 282, 1321

Icy Bodies of the Solar System
Proceedings IAU Symposium No. 263, 2009
J.A. Fernández, D. Lazzaro, D. Prialnik & R. Schulz, eds.

© International Astronomical Union 2010
doi:10.1017/S1743921310001754

Ground based observation of TNO targets for the Herschel Space Observatory

R. Duffard, J. L. Ortiz, A. Thirouin, P. Santos-Sanz, and N. Morales

Instituto de Astrofísica de Andalucía - CSIC, C/ Camino Bajo de Huetor,
50. Granada, 18008, Spain
email: duffard@iaa.es

Abstract. We have observed a subset of TNOs that are going to be studied by means of Herschel Space Telescope (HSO). More than 50 objects have been studied astrometrically and 30 with time series photometry. The main conclusion regarding the astrometry is that all the observed HSO targets have ephemerides uncertainties smaller than 5 arcsec, needed for the correct pointing of the space telescope. Concerning the time series analysis of the targets, most of the objects present low amplitude variability. This is an on-going program and more results are expected.

Keywords. Trans-neptunian Objects, photometry, astrometry

1. Introduction

Transneptunian Objects (TNOs) are believed to represent one of the most primordial populations in the solar system. The TNO population comprises (i) the main Kuiper Belt beyond the orbit of Neptune (32 - 50AU), consisting of objects in resonant and non-resonant orbits, (ii) the halo outskirts of "scattered" and "detached" bodies beyond 50AU, and (iii) the Centaurs, which are closer to the Sun and in transition towards the inner solar system where some of them are eventually captured as short-period comets in the Jupiter family (Gladman *et al.* 2008). About 1200 TNOs have been detected so far and, as detailed hereafter, new studies have started to reveal a richness of orbital and physical properties. These TNOs represent only a few percent of the estimated 30 000 TNOs brighter than 24 mag in the visible (Petit *et al.* 2008).

Herschel Space Observatory (HSO) that was launched this year is set up to make a key contribution to the study of dusty debris disks around other stars. Moreover, the approved Herschel program "TNOs are cool" (Mueller *et al.* 2009) will obtain thermal data of a set of 140 TNOs. As part of the international group with an approved TNO observational proposal of 372.8 hours in this space Telescope we need ground-based support for some of the selected targets. For the albedo estimations to be made using the HSO measurements, accurate visible photometry is required. Moreover, color information of the TNOs will be used for statistical analysis to study, for instance, correlations between size, albedo, photometric and dynamical parameters, etc. (Santos-Sanz *et al.* 2009). More than half of the HSO TNO targets require first time photometric measurements or refinement of existing results.

For successful observations with HSO, accurate pointing of the TNOs is fundamental to make best use of the available field of view for jittering and avoidance of bad pixels as well as to stay away from brighter disturbing background sources. This in turn requires a good knowledge of the orbit of the targets to the level of 5 arcsec for the time when HSO will observe the respective TNO. An analysis of the uncertainties in the ephemerides of the program targets has revealed that about half of the 140 program objects have prediction errors of 10 arcsec and more (some even several arcmin) in the critical time intervals for the HSO observations, partially because the last astrometry of the objects was obtained years ago and partially because the measured orbital arc is not sampled properly. In order to overcome this compromising situation for the success of the HSO observations, we measured new accurate astrometric positions of more than 50 TNO targets with uncertain ephemerides. The TNO images also allow obtaining photometry of the TNOs for a refinement of the absolute brightness of the targets in reflected sunlight, a result that will be used together with the HSO measurements to determine size and albedo of the objects. It goes without saying that the new astrometry of the HSO targets will contribute to the orbit refinement of the TNOs per se, thus improving the still unsatisfying situation for statistical analysis of the TNO orbits and dynamics. In this work we present some results of the observational campaign we are involved using different ground-based telescopes.

2. Observations

Our group at the Instituto de Astrofísica de Andalucía - CSIC started a vast program on lightcurves and astrometry of Kuiper Belt Objects (KBOs) in 2001. Observations were carried out from the 1.5m telescope at Sierra Nevada Observatory (OSN - Granada, Spain), from the 2.2m telescope at Centro Astronòmico Hispànico Alemàn (CAHA) at Calar Alto Observatory (Almeria, Spain) and from the 2.5m Isaac Newton Telescope (INT), the 2.5m Liverpool Telescope (LVPL), the 10.2m Gran Telescopio de Canarias (GTC) and the 4.2m William Herschel Telescope (WHT) at El Roque de Los Muchachos Observatory (La Palma, Spain).

Lightcurve observations, aiming to determine the rotational period and amplitude are mainly carried out at the OSN, INT and CAHA 2.2m. Photometry to determine the TNOs colors were done at the WHT and LVPL and finally, spectroscopic observations in the visible range were done at the GTC.

3. Results

All the TNO images taken, not only on purpose astrometric ones, were processed using the USNO-B1 catalogue. The corresponding right ascension and declination coordinates were submitted to the Minor Planet Center and a plot showing the offsets with respect the JPL Horizon system ephemerides is presented for comparison in Fig. 1.

The final time series photometry of each target was inspected for periodicities by means of the Lomb technique (Lomb 1976) as implemented in Press *et al.* (1992), but we also verified the results by using several other time series analysis techniques (such as PDM), the Harris *et al.* (1989) method and the CLEAN technique (Foster 1995). Concerning the amplitudes of the short-term variability, we used Fourier fits to the data in order to determine peak to valley amplitudes (full amplitudes). Two examples are shown in Fig. 2 and the complete set is presented in Thirouin *et al.* (2009). An analysis of the all available lightcurve database in literature is presented in Duffard *et al.* (2009).

Spectroscopy at GTC in an on-going observational program and the results are currently being analyzed.

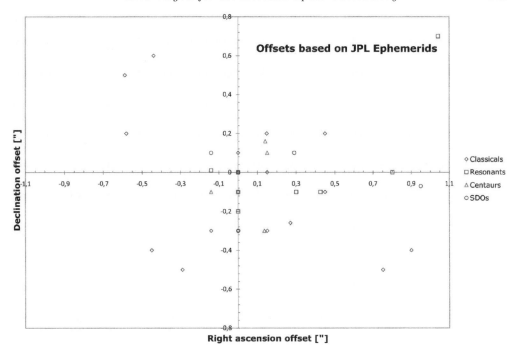

Figure 1. Offset in right ascension and declination, compared to the coordinates on JPL Horizon system ephemerides at the time of the observation.

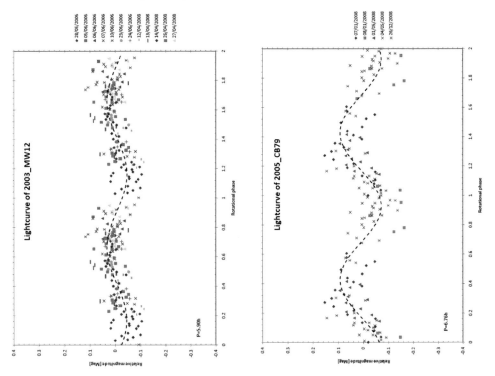

Figure 2. Two examples of lightcurves. A complete set of the newly determined lightcurves is presented in the work by Thirouin *et al.* (2009).

4. Conclusions

We have observed a subset of TNOs that are going to be studied by means of Herschel Space Telescope. Astrometry observations were aimed to improve the orbits of the selected targets whose positions must be known to high accuracy for Herschel pointing requirements. More than 50 objects have been studied astrometrically. The main conclusion regarding the astrometry is that all the observed HSO targets have ephemerides uncertainties smaller than 5 arcsec, needed for the correct pointing of the space telescope. There are no obvious systematic uncertainties.

Concerning the time series analysis of the targets, most of the objects present low amplitude variability. Low amplitude lightcurves are generally caused by albedo heterogeneity on the surfaces of the bodies (Duffard *et al.* 2008), although elongated objects seen at certain geometries can give rise to nearly flat lightcurves as well. The physical reason for the presence of many low amplitude rotators in the Kuiper Belt is investigated in more detail in a separate work (Duffard *et al.* 2009).

References

Duffard, R., Ortiz, J. L., Santos-Sanz, P., Mora, A., Gutierrez, P. J., Morales, N., & Guirado, D. 2008, *A&A*, 479, 877

Duffard, R., Ortiz, J. L., Thirouin, A., Santos-Sanz, P., & Morales, N. 2009, *A&A*, 505, 1283

Foster, G. 1995, *AJ*, 109, 1889

Gladman, B., Marsden, B. G., & Vanlaerhoven, C. 2008, in: M. A. Barucci, H. Boehnhardt, D. P. Cruikshank & A. Morbidelli (eds.), *The Solar System Beyond Neptune*, (Tucson: University of Arizona Press), p. 43

Harris, A. W., Young, J. W., & Bowell, E. 1989, *Icarus*, 77, 171

Lomb, N. R. 1976, *Ap&SS*, 39, 447

Muller, T. G., Lellouch, E., Bohnhardt, H., Stansberry, J., Barucci, A., Crovisier, J., Delsanti, A., Doressoundiram, A., Dotto, E., Duffard, R., & 23 coauthors 2009, *Earth, Moon, and Planets*, 105, 209

Petit, J.-M., Kavelaars, J. J., Gladman, B., & Loredo, T. 2008, in: M. A. Barucci, H. Boehnhardt, D. P. Cruikshank & A. Morbidelli (eds.), *The Solar System Beyond Neptune*, (Tucson: University of Arizona Press), p. 71

Press, W. H., Teukolsky, S. A., Vetterling, W. T., & Flannery, B. P. 1992, *Numerical recipes in FORTRAN. The art of scientific computing*, (Cambrigde: Cambridge University Press)

Santos-Sanz, P., Ortiz, J. L., Barrera, L., & Boehnhardt, H. 2009, *A&A*, 494, 693

Thirouin, A., Ortiz, J. L., Duffard, R., Santos-Sanz, P., Aceituno, F. J., & Morales, N. 2009, *A&A*, submitted

Part VI

Transition Objects

Icy Bodies of the Solar System
Proceedings IAU Symposium No. 263, 2009
J.A. Fernández, D. Lazzaro, D. Prialnik & R. Schulz, eds.

Dynamics, Origin, and Activation of Main Belt Comets

Nader Haghighipour

Institute for Astronomy and NASA Astrobiology Institute, University of Hawaii, 2680
Woodlawn Drive, Honolulu, HI 96822, USA
email: nader@ifa.hawaii.edu

Abstract. The discovery of Main Belt Comets (MBCs) has raised many questions regarding the origin and activation mechanism of these objects. Results of a study of the dynamics of these bodies suggest that MBCs were formed in-situ as the remnants of the break-up of large icy asteroids. Simulations show that similar to the asteroids in the main belt, MBCs with orbital eccentricities smaller than 0.2 and inclinations lower than 25° have stable orbits implying that many MBCs with initially larger eccentricities and inclinations might have been scattered to other regions of the asteroid belt. Among scattered MBCs, approximately 20% reach the region of terrestrial planets where they might have contributed to the accumulation of water on Earth. Simulations also show that collisions among MBCs and small objects could have played an important role in triggering the cometary activity of these bodies. Such collisions might have exposed sub-surface water ice which sublimated and created thin atmospheres and tails around MBCs. This paper discusses the results of numerical studies of the dynamics of MBCs and their implications for the origin of these objects. The results of a large numerical modeling of the collisions of m-sized bodies with km-sized asteroids in the outer part of the asteroid belt are also presented and the viability of the collision-triggering activation scenario is discussed.

Keywords. minor planets: asteroids, solar system: general, methods: n-body simulations

1. Introduction

The discovery of comet-like activities in four icy asteroids 7968 Elst-Pizzaro (133P/Elst-Pizzaro), 118401 (1999 RE_{70}, 176P/LINEAR), P/2005 U1 (Read), and P/2008 R1 (Garradd) has added a new item to the mysteries of the asteroid belt (Hsieh & Jewitt 2006; Jewitt, Yang & Haghighipour 2009). Known as Main Belt Comets (MBCs), these objects may be representatives of a new class of bodies that are dynamically asteroidal (i.e., their Tisserand parameters† are larger than 3), but have cometary appearance. As shown in Table 1, the orbits of these objects are in the outer half of the asteroid belt (Fig. 1) implying that they may contain sub-surface water ice. In fact the observation of the tail of 7968 Elst-Pizzaro by Hsieh, Jewitt & Fernández (2004) has indicated that the comet-like activity of this MBC is episodic (it is not the ejection of dust particles that were produced through an impact to this object) and is due to the dust particles that have been blown off the surface of this body by the drag force of the gas that was most likely produced by the sublimation of near-surface water ice.

† For a small object, such as an asteroid, that is subject to the gravitational attraction of a central star and the perturbation of a planetary body P, the quantity $a_P/a + 2[(1 - e^2) a/a_P]^{1/2} \cos i$ is defined as its Tisserand parameter T, where a is the semimajor axis of the object with respect to the star, e is its orbital eccentricity, i is its orbital inclination, and a_P is the semimajor axis of the planet. In general, $T < 3$ for comets (with respect to Jupiter), whereas those of asteroids are mostly $T > 3$.

Table 1. Orbital Elements of MBCs (Hsieh & Jewitt 2006)

MBC	a (AU)	e	i (deg.)	Tisserand	Diameter (km)
(133P)/7968 Elst-Pizzaro	3.156	0.165	1.39	3.184	5.0
118401 (176P/LINEAR)	3.196	0.192	0.24	3.166	4.4
P/2005 U1 (Read)	3.165	0.253	1.27	3.153	0.6
P/2008 R1 (Garradd)	2.726	0.342	15.9	3.216	1.4

The comet-like appearance of MBCs has raised questions regarding the origin of these objects. While the asteroidal orbits of these bodies, combined with the proximity of 7968 Elst-Pizzaro, 118401 (176P/LINEAR), and P/2005 U1 (Read) to the Themis and Beagle families of asteroids (Fig. 1), suggests that MBCs have formed in-situ as the remnants of collisionally broken larger objects, the cometary activities of these bodies may be taken to argue that MBCs are comets that were scattered inward from the outer regions of the solar system and were captured in their current orbits. Such a capture mechanism could not have occurred recently. Simulation of the dynamics of Kuiper belt object by Fernandez *et al.* (2002) have shown that, at the current dynamical state of the solar system, it would not be possible to scatter comets from regions outside the orbit of Neptune to the main asteroid belt. A primordial capture, on the other hand, may not be impossible. Recently Levison *et al.* (2009) have shown that within the context of the Nice model (Gomes *et al.* 2005; Morbidelli *et al.* 2005; Tsiganis *et al.* 2005) many trans-Neptunian objects could have been scattered inwards and captured in orbits in the asteroid belt as close in as 2.68 AU during the early state of the dynamics of the solar system. Whether these objects could be the source of MBCs is, however, uncertain. This paper evaluates this possibility, in particular in comparison with the in-situ formation model, by presenting the results of a numerical study if the dynamics of the currently known MBCs, and discussing their implications for the formation and origin of these objects.

As mentioned above, tails of MBCs are generated through the interactions of dust grains on the surface of these bodies with the gas produced by the sublimation of near-surface water ice (Hsieh, Jewitt & Fernández 2004). As shown by Schorghofer (2008), asteroids in the region between 2 AU and 3.3 AU can maintain sub-surface water ice for several billion years if their surfaces are covered by a layer of dust, even as thick as only a few meters. That implies that in order for an MBC to start its cometary activation, this dusty layer has to be removed. Hsieh & Jewitt (2006) have suggested that collisions between MBCs and objects as small as a few meter in size, can reveal the sub-surface water ice. Such collisions will result in the local exposure of ice which sublimates and creates a thin atmosphere and tail for an MBC. An activated MBC, on the other hand, may terminate its activity after sublimating all the ice at the location of its collision with a m-sized projectile. It may also start new activation if it is collided with a second m-sized object in a later time again. In other words, an MBC may be activated several times till it exhausts all its water ice, or is scattered to an unstable orbit and either leaves the asteroid belt or collides with another asteroid or a planet. It will therefore be useful to study the rate of the collisions of m-size bodies with km-size MBCs, in particular in the outer region of the asteroid belt. This paper presents the results of such simulations and discusses their implications for the activation of MBCs and the possibility of the detection of more of these objects.

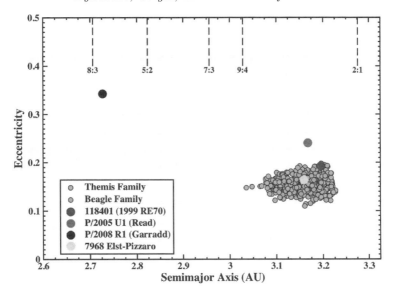

Figure 1. The four currently known MBCs and the Themis and Beagle families of asteroids. As shown here, 7968 Elst-Pizzaro and 118401 (Read) are within the Themis and Beagle families whereas P/2005 U1 is in their proximity. The MBC P/2008 R1 (Garradd) seems to be an object that was scattered out of its forming region. The locations of mean-motion resonances with Jupiter are also shown.

2. Orbital Integrations and Implications for the Origin of MBCs

To study the long-term stability of the four known MBCs, the orbits of these objects were integrated for 1 Gyr. Integrations included all the planets and Pluto, and treated MBCs as non-interacting objects. The effects of non-gravitational forces such as Yarkovsky, and the effect of the mass-loss of MBCs due to their cometary activities were not included. Since the activation of MBCs is episodic and intensifies during the perihelion passages of these objects, which is short compared to their orbital periods, the effect of the mass-loss may not alter the dynamics of these objects significantly. Integrations were carried out with Bulirsch-Stoer and with the Second-Order Mixed-Variable Symplectic (MVS) integrators in the N-body integration package MERCURY (Chambers 1999). The initial orbital elements of the MBCs and planets were obtained from documentation on solar system dynamics published by the Jet Propulsion Laboratory (http://ssd.jpl.nasa.gov/?bodies). The timestep of each integration was set to 9 days.

Figure 2 shows the results of the simulations. As shown here 7968 Elst-Pizzaro and 118401 (176P/LINEAR) maintain their orbits for 1 Gyr. However, P/2005 U1 (Read) and P/2008 R1 (Garradd) become unstable in approximately 20 Myr. Integrations were also carried out for different initial values of the semimajor axes and eccentricities of MBCs, changing these quantities in increments of $\Delta a = 0.0001$ AU and $\Delta e = 0.001$ within the ranges of their observational uncertainties. Similar results were obtained. 7968 Elst-Pizzaro and 118401 were stable whereas P/2005 U1 and P/2008 R1 became unstable in all simulations with a median lifetime of ~ 57 Myr. For more details on the results of the simulations, in particular on the analysis of the effects of mean-motion resonances on the dynamics of these MBCs, the reader is referred to Haghighipour (2009) and Jewitt, Yang & Haghighipour (2009).

As shown by Fig. 1, the orbit of P/2008 R1 (Garradd) is close to the influence zone of the 8:3 mean-motion resonance with Jupiter. Numerical simulations by Jewitt, Yang

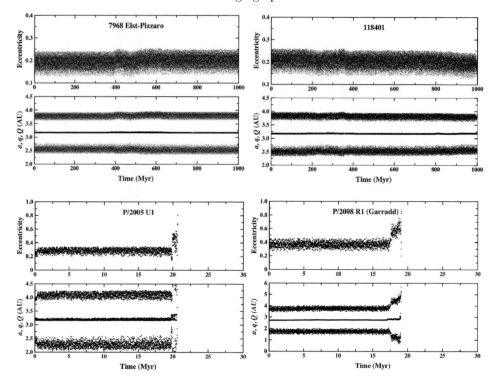

Figure 2. Graphs of the eccentricities, semimajor axes (a), perihelion (q), and aphelion (Q) distances of 7968 Elst-Pizzaro, 118401 (176P/LINEAR), P/2005 U1 (Read), and P/2008 R1 (Garradd). As shown here, 7968 Elst-Pizzaro and 118401 are stable for 1 Gyr whereas P/2005 U1 (Read) and P/2008 R1 (Garradd) become unstable in approximately 20 Myr.

& Haghighipour (2009) have shown that the region in the vicinity of P/2008 R1 is dynamically unstable implying that this MBC must have formed in another region of the asteroid belt and scattered to its current orbit. The orbital instability of P/2005 U1 (Read), on the other hand, may show a pathway to such scattering events. The proximity of P/2005 U1 to the Themis family and the location of the 1:2 mean-motion resonance with Jupiter suggest that this MBC was perhaps formed close to the influence zone of the 1:2 resonance. The original proximity of P/2005 U1 to this resonance has resulted in a gradual increase in its orbital eccentricity which will eventually make its orbit unstable. Such an instability might have also happened to the orbits of other MBCs and resulted in their scattering to other regions. To study this scenario, a large number of hypothetical MBCs were considered around the region where 7968 Elst-Pizzaro, 118401 (176P/LINEAR), and P/2005 U1 (Read) exist. The semimajor axes of these objects were varied between 3.14 AU and 3.24 AU, and their initial eccentricities were taken to be between 0 and 0.4. The initial orbital inclinations of these MBCs were chosen from a range of 0 to 40°.

The orbits of these hypothetical MBCs were integrated for 100 Myr. Figure 3 shows the results. In this figure, green circles correspond to MBCs with stable orbits whereas purple indicates instability. As shown here, 7968 Elst-Pizzaro and 118401 (176P/LINEAR) are in the stable region of the graph whereas P/2005 U1 (Read) is approaching the unstable area.

An interesting result depicted by Fig. 3 is the familiar role of secular resonances in establishing the boundaries of stable zones. Similar to the asteroids in the asteroid belt,

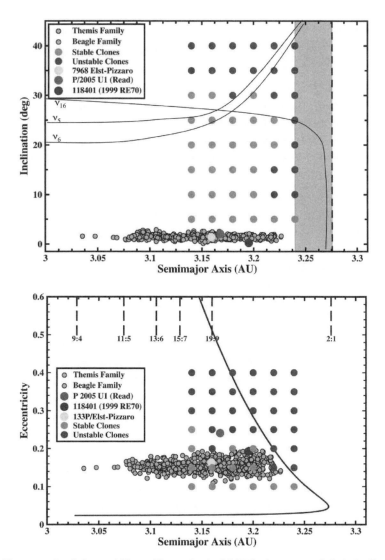

Figure 3. Top: graph of the stability of hypothetical MBCs in terms of their inclinations. The regions of secular resonances ν_5, ν_6, and ν_{16} corresponding to an eccentricity of 0.1 are also shown. Bottom: graph of the stability of hypothetical MBCs in terms of their eccentricities. The brown area in the top graph and solid line in the bottom graph show the region of the 2:1 MMR with Jupiter. Circles in green correspond to initial semimajor axes and eccentricities of stable MBC whereas those in purple show instability. Similar to the asteroid in the main belt, objects with inclinations larger than $\sim 25°$ and eccentricities larger than ~ 0.2 are unstable.

stability of an MBC depends on the values of its initial eccentricity and orbital inclination. Fig. 3 shows that for initial inclinations larger than $\sim 25°$, the orbit of an MBC becomes unstable due to the Kozai and the ν_5, ν_6 and ν_{16} secular resonances. For smaller values of inclination, the apastron distance of an MBC determines its stability. Those hypothetical MBCs close to or inside the region of the 2:1 MMR with Jupiter became unstable in a short time. An analysis of the orbits of the unstable objects indicates that approximately 80% of these bodies were scattered to large distances outside the solar system. This is a familiar result that has also been reported by O'Brien *et al.* (2007) and

Haghighipour & Scott (2008) in their simulations of the dynamical evolution of planetesimals in the outer asteroid belt. From the remaining 20% unstable MBCs, approximately
15% collided with Mars, Jupiter, or Saturn, and a small fraction ($\sim 5\%$) reached the
region of 1 AU implying that MBCs might have played a role in delivering water to the
Earth.

The stability analysis above has direct implications for the origin of MBCs and favors
the in-situ formation of these objects. In this scenario, MBCs are small asteroidal bodies
that were formed as a result of the collisional break-up of their larger precursor asteroids.
An alternative scenario based on the primordial capture of cometary bodies, although,
as shown by Levison et al. (2009), efficient in the inward scattering of D-type and P-type
asteroids and the delivery of these objects in particular to the region of Trojans, cannot
provide information on the inward scattering and distribution of C-type asteroids. That
is primarily due to the fact that C-type asteroids are mainly at small semimajor axes,
and the difference between their orbital distribution and that of D-type asteroids are
not known. Additionally, the colorless feature of MBCs, as indicated by Hsieh & Jewitt
(2006) and Hsieh, Jewitt & Fernández (2008) is not consistent with an origin model based
on the inward scattering of comets from the Kuiper belt region (the latter objects are
optically red).

The in-situ formation scenario is, however, consistent with MBCs orbital and spectral
properties. In this scenario, the break up of the precursor asteroids could have produced
many km-sized fragments, among which those with large inclinations and large eccentricities became unstable and were scattered to other regions. The remaining objects have
naturally asteroidal orbits (i.e. their Tisserand numbers are larger than 3), and similar
to their parent bodies, are C-type asteroids with no specific optical color. In regard to
7968 Elst-Pizzaro, 118401 (176P/LINEAR), and P/2005 U1 (Read), this scenario points
to the Themis family, and perhaps a smaller ~ 10 Myr sub-family (known as Beagle)
within these objects (Nesvorný et al 2008), as the origin of these MBCs. This scenario
also suggest that asteroid families, in particular those in the outer half of the asteroid
belt and with large parent bodies capable of differentiating and forming ice-rich mantles,
are the most probable places for detecting more MBCs. As indicated by the results of the
dynamical simulations, some members of such families may interact with giant planets
and reach orbits in other regions of the asteroid belt–a scenario that might explain the
existence of P/2008 R1 (Garradd) in its current orbit. All sky surveys such as those with
Pan STARRS 1 would be capable of detecting such individual MBCs, and are ideal for
carrying out targeted surveys for families of these objects.

3. Collision With Small Objects As The Activation-Triggering Mechanism

As mentioned in the introduction, it has been suggested that the tails of MBCs are
dust particles that have been carried away from the surfaces of these object by the gas
produced by the sublimation of water ice. This idea is based on the fact that the orbits
of the currently known MBCs are in the outer part of the asteroid belt where water ice
on the surface of asteroids can survive for billions of years when covered by a layer of
dust (Schorghofer 2008). A collision between an MBC and an object, even as small as a
m-sized boulder, can expose this ice. When such an MBC, with a locally exposed sub-
surface ice, approaches its perihelion, the ice sublimates and produces a weak atmosphere
which lifts and carries dust particles from the surface of the MBC, giving it a cometary
appearance.

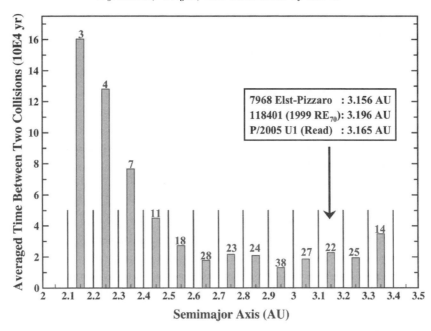

Figure 4. Graph of the averaged time between two successive collisions of m-sized objects with an MBC in the orbit of 7968 Elst-Pizzaro. The numbers on top of each bar indicate the percentage of the boulders of that region that collided with the MBC. As shown here, most of the collisions come from the vicinity of 7968 Elst-Pizzaro. The grand average of the time between two successive collision is approximately 40,000 years.

The number of m-sized boulders and the frequency of such collisions are not exactly known. However, it is possible to develop a simple computational model that can impose an upper limit to these collisions. In doing so, a heuristic model was developed based on the following assumptions.

(*a*) The asteroid belt was assumed to consist of only one asteroid, 7968 Elst-Pizzaro, and a disk of m-sized bodies. The surface density of the disk was set to have a $r^{-3/2}$ profile.

(*b*) The accumulative size distribution (N) of objects with diameter (D) was considered to be given by $N \propto D^n$, where n can have a value between -2 and -4. Following Dohnanyi (1969), it was assumed that $n = -2.5$.

(*c*) A total of 10^6 m-sized boulder were randomly distributed throughout the asteroid belt. The eccentricity of these objects were chosen from a range of 0 to 0.5, and their inclinations were taken to be between 0 and $25°$.

The orbits of the m-sized objects and that of the 7968 Elst-Pizzaro were integrated for 10 Myr. Similar to the previous simulations, integrations included all planets and Pluto. Results indicated that on average, one m-sized object collides with this MBC every 40,000 years. As shown in Fig. 4, a larger number of the colliding boulder come from the vicinity of 7968 Elst-Pizzaro. It is important to emphasize that this model is simplistic, and the results represent a high upper limit. In a more realistic model, the numbers of large bodies and the small boulders are much higher. As a result, many of the m-sized objects collide with their neighboring asteroids, or are ejected from the asteroid belt. It is expected that in such cases, the frequency of collisions between km-sized MBCs and m-sized boulders to decrease to approximately one every few thousand years.

4. Conclusions

• Current MBCs seem to have formed through the collision and break up of bigger asteroids. The results of the simulations of the dynamic of these objects point to the Themis family as the origin of 7968 Elst-Pizzaro, 118401 (176P/LINEAR), and P/2005 U1 (Read).

• Interaction with giant planet might have scattered MBCs from their original orbits to other locations in the asteroid belt. P/2008 R1 (Garradd) seems to be one of such scattered MBCs.

• More MBCs may exist in low inclinations and low eccentricities in the vicinity of asteroid families in the outer region of the asteroid belt.

• Collisions with small objects might have activated MBCs or eroded them.

• Many MBCs might have been active in the past and are either no longer active, or will become active if hit by a small body again.

• Many MBCs, with locally exposed sub-surface ice, may still be on their ways to their perihelion distances where they become active, or they may be awaiting collisions with smaller objects to get activated.

• All sky surveys such as Pan STARRS will be able to detect more MBC in near future.

Acknowledgements

I gratefully acknowledge fruitful discussions with H. Hsieh, D. Jewitt, H. Levison, K. Meech, D. Nesvorny, and N. Schorghofer. This work was partially supported by the NASA Astrobiology Institute under Cooperative Agreement NNA04CC08A at the Institute for Astronomy, NASA Astrobiology Central, the office of the Chancellor of the University of Hawaii, and a Theodore Dunham J. grant administered by Funds for Astrophysics Research, Inc. I am also grateful to Newton's Institute for Mathematical Science at the Cambridge University for their great hospitality during the preparation of this manuscript.

References

Chambers, J. E. 1999, *MNRAS*, 304, 793
Dohnanyi, J. S. 1969, *JGR*, 74, 2531
Fernandez, J. A., Gallardo, T., & Brunini, A. 2002, *Icarus*, 159, 358
Gomes, R., Levison, H. F., Tsiganis, K., & Morbidelli, A. 2005, *Nature*, 435, 466
Haghighipour, N. & Scott, E. R. D. 2008, *LPI Contribution* 1391, 1679
Haghighipour, N. 2009, to appear in *Meteor. Plant. Sci.* (arXiv:0910.5746)
Hsieh, H. H., Jewitt, D., & Fernández, Y. R. 2004, *AJ*, 127, 2997
Hsieh, H. H. & Jewitt, D. 2006, *Science*, 312, 561
Hsieh, H. H., Jewitt, D., & Fernández Y. R., 2008, *LPI Contribution*, 1405, 8200
Jewitt, D., Yang, B., & Haghighipour, N., 2009, *AJ*, 137, 4313
Levison, H. F., Bottke, W. F., Gounelle, M., Morbidelli, A., Nesvorný, D., & Tsiganis, K., 2009, *Nature*, 460, 364
Morbidelli, A., Levison, H. F., Tsiganis, K., & Gomes, R., 2005, *Nature*, 435, 462
Nesvorný, D. & Morbidelli, A., 1998, *AJ*, 116, 3029
Nesvorný, D., Bottke, W. F., Vokrouhlicky, D., Sykes, M., Lien, D. J., & Stansberry, J., 2008, *LPI Contribution*, 1405, 8265.
O'Brien, D. P., Morbidelli, A., & Bottke, W. F., 2007, *Icarus* 191, 434,
Schorghofer, N., 2008, *ApJ* 682, 697
Tsiganis, K., Gomes, R., Morbidelli, A., & Levison, H. F., 2005, *Nature*, 435.

Icy Bodies of the Solar System
Proceedings IAU Symposium No. 263, 2009
J.A. Fernández, D. Lazzaro, D. Prialnik & R. Schulz, eds.

© International Astronomical Union 2010
doi:10.1017/S1743921310001778

Are the main belt comets, comets?

Javier Licandro[1,2] and Humberto Campins[3]

[1]Instituto de Astrofísica de Canarias,
c/Vía Láctea s/n, 38200 La Laguna, Tenerife, Spain.
email: jlicandr@iac.es

[2]Departamento de Astrofísica, Universidad de La Laguna,
E-38205 La Laguna, Tenerife, Spain

[3]Physics Department, University of Central Florida,
Orlando, FL, 32816, USA.
email: campins@physics.ucf.edu

Abstract. We present the visible spectrum of asteroid-comet transition object 133P/Elst-Pizarro (7968), the first member of the new population of objects called Main Belt Comets (Hsieh & Jewitt 2006). The spectrum was obtained with the 4.2m William Herschel Telescope at the "Roque de los Muchachos" observatory. The orbital elements of 133P place it within the Themis collisional family, but the observed cometary activity during it last 3 perihelion passages also suggest a possible origin in the trans-Neptunian belt or the Oort Cloud, the known sources of comets. We found a clear similarity between our spectrum of 133P and those of other members of the Themis family such as 62 Erato, and a strong contrast with those of cometary nuclei, such as 162P/Siding-Spring. This spectral comparison leads us to conclude that 133P is unlikely to have a cometary origin. This conclusion is strengthened by spectral similarities with activated near-Earth asteroid 3200 Phaethon, and suggest that there are activated asteroids in the near-Earth asteroid and main belt populations with similar surface properties.

Keywords. asteroid, comet, spectroscopy

1. Introduction

Icy minor bodies are known to originate in the trans-neptunian Belt (TNB) and the Oort Cloud. Recently, a third class of objects has been discovered: the Main Belt Comets (MBCs) (Hsieh & Jewitt 2006). Currently, only 4 MBCs are known, two of them, 133P/Elst-Pizarro (7968) and 176P/LINEAR (118401), are within the Themis collisional family of asteroids, a third one, P/2005 U1 (Read) is almost within it, and P/2008 R1 (Garradd) is near the 8:3 mean motion resonance with Jupiter. All of them have Tisserand parameter $T_J > 3.15$, this suggests they have an unlikely cometary origin. 133P is the best characterized having been seen active at 3 perihelion passages, which supports the hypothesis that the activity is driven by sublimation of water ice (Hsieh & Jewitt 2006).

Understanding the asteroidal or cometary nature of these bodies is crucial. If they are formed in situ and, in particular if they are members of a collisional family and their activity is due to water ice sublimation, there should be water ice in many asteroids. If they are captured TNB or Oort cloud comets the mechanisms that drove them to their present orbits needs to be understood.

2. Observations

The spectrum of 133P in the visible spectral region was obtained on July 9, 2007, with the 4.2m William Herschel Telescope (WHT) at the "Roque de los Muchachos"

Observatorio (ORM, Canary Islands, Spain), using the double arm ISIS spectrograph. Spectra in the red and blue arm where obtained simultaneously, using the R300B grating in the blue arm, with a dispersion of 0.86Å/pixel, and the R158R gratting in the red arm, with a dispersion of 1.81Å/pixel. A 5" slit width was used, oriented at the parallactic angle to minimize the spectral effects of atmospheric dispersion. Two spectra of 1800s exposure time where obtained in both arms. The tracking was at the asteroid proper motion. Images were over-scan corrected, and flat-field corrected using lamp flats. The two-dimensional spectra were extracted, sky background subtracted, and collapsed to one dimension. The wavelength calibration was done using the Neon and Argon lamps.

To correct for telluric absorption and to obtain the relative reflectance, the G star Landolt (SA) 115-271, (Landolt 1992) was observed at different airmasses (similar to those of the object) before and after the 133P's observations, and used as a solar analogue star. The spectrum of 133P object was divided by those of the solar analogue star, and then normalized to unity around 0.55μm thus obtaining the normalized reflectance. The obtained spectrum, rebined to a dispersion of 100Å/pixel is shown in Fig. 1.

In order to compare with other Themis family objects, the spectra of Themis family asteroids 62 Erato, 379 Huenna and 383 Janina were observed on April 1, 2006 using the same instrument and configuration. Landolt (SA) 107-689 and the solar analogue star BS4486 were observed to obtain the relative reflectance of the asteroids. The three asteroids present very similar spectral characteristics. The relative reflectance of 62 Erato is shown in Fig. 1. Notice that it is very similar to the spectrum of 133P.

3. Discussion and conclusions

The spectrum of 133P (Fig. 1) present a slightly blue slope at wavelenghts larger than 5000Å, and a drop in the UV region. Notice that 133P & 62 Erato spectra are very similar, supporting that both are likely fragments of the collision that formed the Themis family. So, we conclude that 133P belongs to the Themis family as suggested also by its orbital elements. Notice also that the large majority of Themis family asteroids do not present any cometary like activity, suggesting that "activated asteroids" (or MBCs) in the family are scarce.

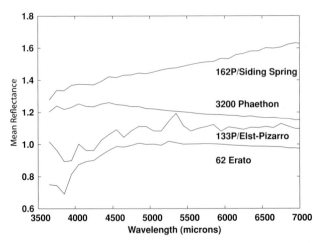

Figure 1. From top to bottom, visible spectrum of 162P/Siding Spring (Campins *et al.* 2006), 3200 Phaethon (Licandro *et al.* 2007), 133P/Elst-Pizarro and Themis family asteroid 62 Erato, shifted in reflectance axis for clarity.

Also in Fig. 1 it is evident that 133P's spectrum has similarities with that of the Near Earth Asteroid (NEA) 3200 Phaethon taken from Licandro *et al.* (2007). The shape of Phaethon's spectrum is similar, even though it is slightly bluer in the red and the UV drop is not as deep as in the 133P's spectrum. Phaethon is an asteroid-comet transition object that probably had past activity . Phaethon's surface contains hydrated silicates and this does not support its possible cometary nature (Licandro *et al.* 2007). The similarity between Phaethon's spectrum and that of 133P also suggests an asteroidal nature for the MBCs.

In Fig. 1 we also plot the spectrum of 162P/Siding-Spring from Campins *et al.* (2006), one of the best S/N spectrum of a comet nucleus, rebined to the same spectral resolution as the other objects, with the aim to explore the possible cometary origin of 133P. Notice that 162P's spectrum is very different to that of the Themis family asteroids and 3200 Phaethon. The large majority of comet nuclei with observed spectra, have an spectrum similar to that of 162P and compatible with that of P- or D-type asteroids (Jewitt 2002, Licandro *et al.* 2002, Campins *et al.* 2006, Snodgrass *et al.* 2008). P- and D-type spectra are featureless with a slightly red to red slope, and it is assumed they are composed of very primitive material. Finally, the asteroids in cometary orbits that more likely have a cometary origin (those with $T_J < 2.7$) are also P- and D-type (Licandro *et al.* 2008).

So, we conclude that 133P's spectrum and its dynamical properties show that this asteroid is a member of the Themis family of asteroids and is unlikely to have a cometary origin. If the activity is water-ice driven, these results suggest that there are some activated asteroids in the NEA and main belt population that were able to retain some water ice that sublimates under certain circumstances.

Exploring the volatile content of icy minor bodies is critical for understanding the physical conditions and the mechanisms of planetary formation, and also addresses the question of the origin of Earth's water. If the outer main belt has a large population of asteroids with ice, they could have contributed to the water on Earth. Additionally, this indicates the extent and origin of volatiles in asteroids that could be used as resources for space exploration.

Acknowledgements

JL gratefully acknowledges support from the spanish "Ministerio de Ciencia e Innovación" project AYA2008-06202-C03-02. HC acknowledges the support of NASA's Planetary Astronomy Program and of the National Science Foundation.

References

Campins, H., Ziffer, J., Licandro, J. Pinilla-Alonso, N., Fernandez, Y., de Leon, J., Mothé-Diniz, T., & Binzel, R. 2006, *AJ*, 132, 1346

Hsieh, H. & Jewitt, D. 2006, *Science*, 312, 561

Jewitt, D. 2002, *A..J*, 123, 1039

Landolt A. 1992, *AJ*, 104, 340

Licandro, J., Campins, H., Hergenrother, C., & Lara, L. M. 2002, *A&A*, 398, L45

Licandro, J., Campins, H., Mothé-Diniz, T., Pinilla-Alonso, N., & de León, J. 2007, *A&A*, 461, 751

Licandro, J., Alvarez-Candal, A., de León, J., Pinilla-Alonso, N., Lazzaro, D., & Campins, H., 2008, *A&A*, 481, 861

Snodgrass, C., Lowry, S. C., & Fitzsimmons, A. 2008, *MNRAS*, 385, 737

Icy Bodies of the Solar System
Proceedings IAU Symposium No. 263, 2009
J.A. Fernandez, D. Lazzaro, D. Prialnik & R. Schulz, eds.

© International Astronomical Union 2010
doi:10.1017/S174392131000178X

Material properties of transition objects 3200 Phaethon and 2003 EH₁

J. Borovička, P. Koten, P. Spurný, D. Čapek, L. Shrbený, and R. Štork

Astronomical Institute of the Academy of Sciences, CZ-25165 Ondřejov, Czech Republic
email: borovic@asu.cas.cz

Abstract. Asteroids 3200 Phaethon and 196256 (2003 EH$_1$) are connected with two major meteoroid streams, Geminids and Quadrantids, respectively. We have modeled the observed light curves and decelerations of Geminid and Quadrantid meteors and studied their spectra. In both cases, we have found typical bulk densities of about 2600 kg m^{-3}, much larger than in cometary meteoroids. Sodium was partially lost from Geminids and Quadrantids due to solar heating. The Quadrantid material was therefore not hidden deep inside the parent body 1500 years ago, when the perihelion was low enough for sodium loss to occur.

Keywords. meteors, meteoroids; minor planets, asteroids

1. Introduction

The close orbital similarity of asteroids 3200 Phaethon and 196256 (2003 EH$_1$) with the Geminid and Quadrantid meteoroid streams, respectively, leaves no doubt that there is a genetic relation between the asteroids and the streams (Whipple 1983; Jenniskens 2004). Since 2003 EH$_1$ is on a Jupiter-family-comet type orbit, it could be more easily considered a dormant comet. Phaethon, on the other hand, is on asteroidal orbit in the inner solar system and the origin of this body and of the Geminid stream has been a subject of considerable debate (e.g. Hsieh & Jewit 2005; Ryabova 2007; Licandro *et al.* 2007; Wiegert *et al.* 2008).

In this work, we studied trajectories, light curves, decelerations and spectra of Geminid and Quadrantid meteors. We estimated the likely mechanical properties (in particular porosity), of Geminid and Quadrantid meteoroids and compared them to meteoroids of clearly cometary origin. We also measured the content of volatile sodium, which is diagnostic of meteoroid thermal history.

2. Data analysis

We used the erosion model, which we recently developed and applied to Draconid meteors (Borovička *et al.* 2007). The model assumes that meteoroids are composed of grains, which are gradually released (eroded) during the atmospheric entry. The most important parameters of the model are the height at which the erosion started, the erosion and ablation coefficients, the size distribution of the grains, and the bulk density of the meteoroid. The grains were assumed to be spherical with the density of 3000 kg m^{-3}. In some cases, we needed two stage erosion to explain the data. In that cases, only certain fraction of meteoroid mass was involved in the initial erosion. The rest continued unaffected until (a part of it) was subject to the second stage erosion starting at a lower height.

In contrast to Draconids, Geminids and Quadrantids proved to be too faint before the start of the erosion to be detected. In consequence, the bulk density of the meteoroid

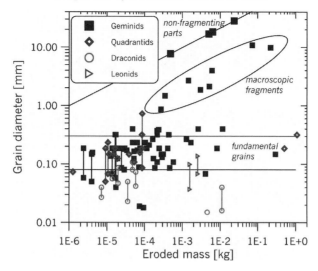

Figure 1. Sizes of grains or fragments in Geminid (full squares), Quadrantid (diamonds), Draconid (circles), and Leonid (triangles) meteoroids.

and the erosion coefficient could not be determined independently from our model. We formally fixed the meteoroid bulk density at 2000 kg m^{-3} and computed the formal erosion coefficient for this value of the bulk density. Since it is reasonable to assume that the energy necessary for the start of the erosion and for grain separation during the erosion is larger for less porous (i.e. more dense) meteoroids, we computed the likely densities of meteoroids (or their parts) from the following empirical formula (calibrated by Draconids):

$$\delta = \delta_{\mathrm{g}} / \left(1 + 3\,\bar{\eta}\,\frac{10^6}{E_{\mathrm{S}}} \right). \tag{2.1}$$

Here δ_{g} is the grain density, $\bar{\eta}$ is the formal erosion coefficient (units: s^2 km^{-2} = kg MJ^{-1}) for the bulk density fixed at 2000 kg m^{-3}, and E_{S} is the energy received per unit cross-section before the start of erosion (J m^{-2}).

We applied the erosion model to 37 Geminid meteors of magnitudes from +4 to −2 observed by image intensified video cameras in 2006. The corresponding meteoroid masses and sizes were 10^{-6} to 10^{-3} kg, and 1 – 10 mm. To extend the analysis to larger bodies, we also analyzed 7 Geminid fireballs (magnitudes −5 to −9, masses 0.01 to 1 kg, sizes 2 – 10 cm) photographed within the scope of the European Fireball Network. For Quadrantids, we used 10 video meteors (+1.5 to −2 mag, 10^{-5} to 10^{-4} kg) and one photographic fireball (−11 mag, 2 kg) observed in 2009. We further analyzed 2 Leonid fireballs observed in 1999, which showed significant deceleration. Draconid data were taken from Borovička *et al.* (2007)

For a majority of video meteors, we also obtained spectra with an additional video camera. The spectra were analyzed for the relative content of magnesium, sodium, and iron in a similar way as we did previously for other meteors (Borovička *et al.* 2005). Spectra were not available for the photographic fireballs.

3. Results

The resulting sizes of grains in various meteoroids are presented in Fig. 1. We tried to fit each meteor data with only one size of grains. If it was not possible, the upper and

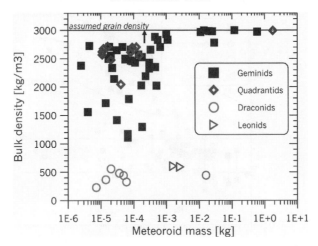

Figure 2. Inferred bulk densities for Geminid (squares), Quadrantid (diamonds), Draconid (circles), and Leonid (triangles) meteoroids. One iron-rich Geminid had much larger density and is not shown.

lower limit of grain sizes, connected with vertical bar, are shown. In cases of two stage erosion, both stages were plotted as independent events.

A majority of grain sizes of Geminids and Quadrantids lies in the range of $80 - 300$ μm. This can be considered as the typical size of fundamental grains. Typical grain sizes of Draconids and Leonids are only $20 - 100$ μm. In the second stage, some large Geminids did not fragment into grains but into macroscopic pieces of $\sim 1 - 10$ mm. In four cases, a cm-sized parts of the meteoroids did not fragment at all.

The bulk densities, computed according to Eq. (2.1), are plotted in Fig. 2. In cases of two stage erosion, both parts were combined together to compute the density of the whole meteoroid. Typical densities of Geminids smaller than 1 gram is about 2600 kg m^{-3} (porosity 15%), although porous with densities down to 1000 kg m^{-3} ($p = 60\%$) were observed as well. Geminids larger than 1 gram were found to be very compact with bulk densities approaching the grain density. Densities of Quadrantids are similar to Geminid densities, while densities of Draconids and Leonids are much lower, about 500 kg m^{-3}. Babadzhanov & Kokhirova (2009) obtained densities similar to ours for Leonids and Geminids and somewhat lower for Quadrantids.

Sodium was found to be depleted by almost an order of magnitude (relatively to magnesium and chondritic abundances) in Geminids smaller than 10^{-4} kg (Fig. 3). The depletion is lower for 10^{-3} kg Geminids. In Quadrantids, the trend is similar but Na depletion is generally lower than in Geminids.

The Geminid orbit is remarkable by its low perihelion (0.14 AU). The computation of Čapek & Borovička (2009) showed that Na can be lost from Geminids by thermal desorption provided that meteoroids are composed from grains not larger than several hundreds of microns and that the pores between the grains are interconnected. They suggested that variation of Na content in Geminids may be due to varying grain sizes. However, the grain sizes proved to be relatively uniform (Fig. 1). We have found that the Na content in Geminids is correlated with the mean pore size. The smaller pores, the larger content of Na. In cases of two stage erosion, the pore size in the denser part of the meteoroid is important. We have spectra only for two meteors which did not fragment into grains and, expectably, they have high Na content.

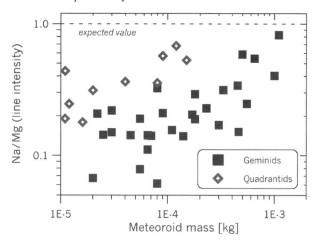

Figure 3. Measured intensity ratio of Na and Mg lines in Geminid (squares) and Quadrantid (diamonds) meteors as a function of meteoroid mass. The expected line intensity ratio for chondritic meteoroid composition is also shown.

The perihelion of Quadrantid orbit is large (0.98 AU) but it was quite small ∼ 1500 years ago (Porubčan & Kornoš 2005). However, the Quadrantid stream can be only 500 years old or even less (Jenniskens 2004; Wiegert & Brown 2005). The fact that there is Na depletion, can be possibly explained by thermal desorption at the surface of the parent body during the low perihelion era.

4. Summary

We have found that Geminid and Quadrantid meteoroids have similar structure. They are composed from grains. The grains are larger than in cometary Draconids. In contrast to Draconids, the porosity is low, typically 10–20%, although it can reach up to 60%. Larger meteoroids (> 1 cm) have macro-porosity lower than 10% and, in case of Geminids, contain compact, non-granular parts. Partial loss of sodium due to solar heating in close vicinity to the Sun occurred in Geminids and, to a lesser extend, in Quadrantids, too. It follows that Quadrantid material was exposed to solar radiation 1500 years ago, i.e. it was not hidden deep inside the parent body at that time.

Acknowledgements

This work was supported by grants IAA 300030813, GA 205/09/1302, and MRTN-CT-2006-035519.

References

Babadzhanov, P. B. & Kokhirova, G. I. 2009, *A&A*, 495, 353
Borovička, J., Koten, P., Spurný, P., Boček, J., & Štork, R. 2005, *Icarus*, 174, 15
Borovička, J., Spurný, P., & Koten, P. 2007, *A&A*, 473, 661
Čapek, D. & Borovička, J. 2009, *Icarus*, 202, 361
Hsieh, H. H. & Jewit, D., 2005 *ApJ*, 624, 1093
Jenniskens, P. 2004 *AJ*, 127, 3018

Licandro, J., Campins, H., Mothé-Diniz, T., Pinilla-Alonso, N., & de León, J. 2007, *A&A*, 461, 751

Porubčan, V. & Kornoš, L. 2005, *Contrib. Astron. Obs. Skalnaté Pleso*, 35, 5

Ryabova, G. O. 2007, *MNRAS*, 375, 1371

Whipple, F. L. 1983, *IAU Circ.* no. 3881

Wiegert, P. & Brown, P. 2005, *Icarus*, 179, 139

Wiegert, P. A., Houde, M., & Peng, R. 2008, *Icarus*, 194, 843

Icy Bodies of the Solar System
Proceedings IAU Symposium No. 263, 2009
J.A. Fernández, D. Lazzaro, D. Prialnik & R. Schulz, eds.

© International Astronomical Union 2010
doi:10.1017/S1743921310001791

Modeling the effects of a faint dust coma on asteroid spectra

Jorge Márcio Carvano[1] and Silvia Lorenz-Martins[2]

[1]Observatório Nacional (COAA), rua Gal. José Cristino 77, São Cristóvão, CEP20921-400 Rio de Janeiro RJ, Brazil.
email: carvano@on.br

[2]Universidade Federal do Rio de Janeiro/Observatório do Valongo. Lad.Pedro Antônio, 43 - 20080-090 Rio de Janeiro, Brazil

Abstract. In this work we use a simple model to study the influence of a faint dust coma on asteroid spectra, in an effort to reproduce the unusual spectral behavior seen on the asteroid (5201) Ferraz-Mello and other objects.

Keywords. minor planets, asteroids

1. Introduction

Several lines of evidence suggest the possibility of a link between comets and asteroids: (a) the existence of icy objects, such as the trans- Neptunians and Centaurs, that are unlikely to develop a coma, since they never come close enough to the Sun; (b) the discovery that sublimation can stop on comets, either by the depletion of the volatile material or by the growth of a surface crust of refractory material; (c) the observation of objects in the main belt of asteroids that present temporal burst of activity, such as (7968) Elst-Pizarro, P/2005 U1 (Read) and (118401) 1999 RE70; (d) the association of meteor showers to asteroids, such as (3200) Phaeton and others; (e) the presence of asteroids with no visible coma in orbits with Tisserand parameter smaller than 3.

Recently, Carvano *et al.* (2008) observed the g, r, i, z colors of the asteroid (5201) Ferraz-Mello, whose dynamical behaviour strongly suggested it to be a captured Jupiter family comet. The colors were transformed to reflectance and comparison with a sample of similarly derived spectra of asteroids, Centaurs, TNOs and cometary nuclei revealed that reflectance spectrum of (5201) Ferraz-Mello is unusual (Fig. 1a), with a steep spectral gradient that is comparable to TNOs and Centaurs, but with an increase in the reflectance in the g band that is not common in those populations. A similar behaviour is however seen in cometary nuclei that were observed in the presence of a faint dust coma (Fig. 1b). This suggests that the presence of a (unseen) faint dust coma might be the culprit of that peculiar increase in the reflectance in the g band.

In this work we use radiative transfer models in order test that hypothesis. We assume a spherical asteroid surrounded by a spherically symmetric dust coma. The reflectance of the asteroid is modelled using Hapke theory (Hapke (1993)), while the scattering due to the dust is modelled via a Monte-Carlo method.

2. Model

In a first attempt to model the effects of a faint dust coma to the reflectance spectrum of an asteroid we assume that the observed spectrum is the sum of two components: the Sun light reflected at the surface of the asteroid, attenuated as it traverses the coma in

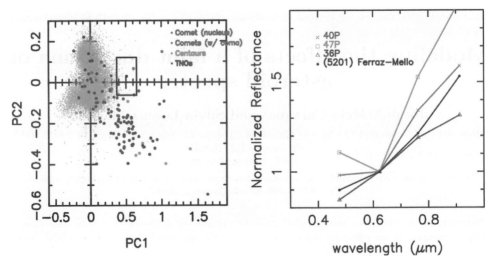

Figure 1. SDSS spectrum of (5201) Ferraz-Mello, compared with spectra of asteroids, Centaurs, TNOs and cometary Nuclei. **(a)** Principal component plot for (5201) Ferraz-Mello - black dot, uncertainties denoted by a box, asteroids - gray dots, Centaurs, TNOs, and cometary nuclei. Asteroid colors were obtained from the SDSS MOC3 (Gunn (1999)), TNOs and Centaurs from Neese (2006) and cometary nuclei from Snodgrass *et al.* (2007). Figure shows that (5201) Ferraz-Mello plots outside the main trend of small Solar System objects. **(b)** discrete spectra of (5201) Ferraz-Melo together with spectra of comet nuclei obtained in the presence of a faint dust coma.

its way in from the Sun and out towards the observer; and the Sun light that is scattered by the coma into the line-of sight of the observer.

To model the first component we consider a spherical asteroid and assume that the reflectance at its surface is described by Hapke's model (Hapke (1993)), disregarding the opposition effect and the macroscopic roughness of the surface. Therefore, the reflectance of each surface element of the asteroid is described by:

$$R_\lambda = \frac{w_\lambda}{4\pi} \frac{\mu_0}{\mu_0 + \mu_e} \left[p(\theta) + H(\mu_0)H(\mu_e) - 1 \right] \qquad (2.1)$$

where $w(\lambda)$ is the single scatter albedo at a given wavelength, p is the volumetric phase function (assumed to be a one-lobed Heyney-Greenstein function), mu_0 and mu_e are, respectively, the cosines of the incidence and emission angles and H is the Chandrasekar function and θ is the solar phase angle. The single scatter albedo is calculated from optical constants for a given grain sized_A, using a geometric optics approximation. To compute the fraction of incident Sun light reflected at the surface of the asteroid we then consider a sphere made up of triangular facets and sum up the contribution of the reflectance of every facet that is visible and illuminated.

To model the contribution of the coma we consider a spherical cloud composed of spherical homogeneous particles whose volumetric density fall with the inverse of the square of the distance to the center of the asteroid and use a Monte Carlo method to calculate the amount o radiation scattered towards the observer, assuming single scattering. The code essentially shoots a energy parcel at a random position on the cloud, picks a random optical depth for the interaction and then calculates the amount of energy that is removed by extinction and the amount that is scattered in the direction of the observer. For a individual parcel the fraction of energy that is sent towards the observer is

$$S_{i\lambda} = Q_{S_\lambda} \pi d_c^2 \frac{N_0}{r_i^2} p_\lambda(\theta) e^{-2\tau_{i\lambda}} \tag{2.2}$$

where N_0 is the number of particles, d_C is the diameter of the particles in the coma, Q_{S_λ} and $p_\lambda(\theta)$ are, respectively, the scattering efficiency and phase function of the grains, calculated using Mie theory from optical constants for a given grain size.

The total reflectance from the asteroid and the coma is then calculated as

$$E_\lambda = R_\lambda e^{-2\tau_\lambda} + M^{-1} \left(\frac{r_C}{r_A}\right)^2 \sum_i^M S_{i\lambda} \tag{2.3}$$

where r_C and r_A are respectively the radius of the coma and of the asteroid and M is the number of parcels shot at the cloud.

Finally, in order to simulate a material with spectral behaviour akin to outer Main Belt asteroids we follow Grundy (2009) to derive optical constants for a mixture of water ice and tholins using a Maxwell-Garnet model; the optical constants for the end-members were taken from Khare *et al.* (1984) and G. Hansen (personal communication). The same optical constants were used for the grains in the coma and on the asteroid surface.

3. Preliminary results and perspectives

The resulting model has 5 free parameters: N_0, d_C, d_A, the coefficient ψ of the Heyney-Greenstein function used in 2.1 and η, the volumetric fraction of tholins in the grains used in the Maxwell-Garnet model. At this stage we consider only the case where the asteroid is observed at zero phase angle – the lack of opposition coefficients in 2.1 is justified since we are interested here in the spectral behaviour of the reflected light, and the opposition effect is independent of wavelength. Likewise, since a gray phase function for the surface is assumed, we use an arbitrary value of $\psi = 0.1$ for the Heyney-Greenstein coefficient.

After some experimentation, we adopted $\eta = 0.05$. From the remaining parameters, N_0 and d_C have the most influence of the resulting spectra. Numerical tests show that a spectral behaviour similar to the objects discussed in section 1 can be reproduced using sub-micron grains in the coma. Figures 2a-c show the effect of grains with $d_C = 0.2\ \mu m$ for increasing values of N_0, and the resulting SDSS spectra for the highest N_0 is shown in Figure 2d.

It is clear that the simple model used here is able to produce an increase in the reflectance in the shorter wavelengths, as expected, but the resulting spectra tend to be bluer than the asteroid spectrum without the coma.

The results presented here suggest that the presence of a faint coma composed of sub-micron particles can produce the unusual reflectance spectra that is observerd in (5201) Ferraz-Mello and other objects. To fully understand the problem however it is necessary a more accurate model, that includes multiple scattering in the coma, a size distribution for the grains and possibly a two component coma. It is also desirable to study the effects of the solar phase angle of the observations on the reflectance spectra of the objects.

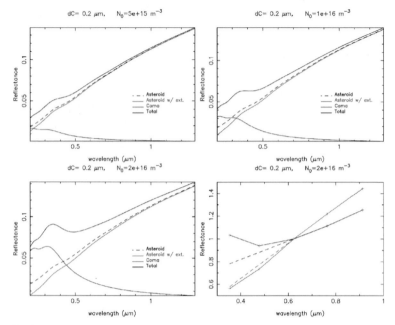

Figure 2. (a-c) Resulting spectra for increasing dust density in the coma, with $d_C = 0.2\ \mu m$, $d_A = 10\ \mu m$. (d) Convolution of the model spectra with the SDSS bandpasses without a coma and with the same model as in (c); the dotted lines are extensions towards the u filter of the spectral inclination between the r and i filters.

References

Carvano, J. M., Ferraz-Mello, S., & Lazzaro, D. 2008, *A&A*, 489, 811

Grundy, W. M. 2009, *Icarus*, 199, 560

Gunn, J. E. 1999, *Bulletin of the American Astronomical Society*, 31, 1418

Hapke, B. 1993, *Theory of reflectance and emittance spectroscopy*, (Cambridge: Cambridge University Press)

Khare, B. N., Sagan, C., Arakawa, E. T., Suits, F., Callcott, T. A., & Williams, M. W. 1984, *Icarus*, 60, 127

Neese, C. 2006, *NASA Planetary Data System, EAR-A-COMPIL-3-TNO-CEN-COLOR-V3.0*, 57

Snodgrass, C., Lowry, S. C., & Fitzsimmons, A. 2007, *ArXiv e-prints*, 712

Icy Bodies of the Solar System
Proceedings IAU Symposium No. 263, 2009
J.A. Fernàndez, D. Lazzaro, D. Prialnik & R. Schulz, eds.

© International Astronomical Union 2010
doi:10.1017/S1743921310001808

The unusually frail asteroid 2008 TC3

Peter Jenniskens[1], Muawia H. Shaddad[2] and The Almahata Sitta Consortium

[1]SETI Institute, 515 N. Whisman Road Mountain View, CA 94043, USA
email: petrus.m.jenniskens@nasa.gov

[2]Physics Department, Faculty of Science, University of Khartoum, P.O.Box 321, Khartoum 11115, Sudan
email: shaddadmhsh@yahoo.com

Abstract. The first asteroid to be discovered in space and subsequently observed to impact Earth, asteroid 2008 TC3, exploded at a high 37 km altitude and stopped ablating at 32 km. This would classify the fireball as of Ceplecha's PE-criterion IIIb/a, meaning "cometary" in nature. In this case, the structural weakness may have come from pores found in some of the recovered meteorites, called *"Almahata Sitta"* (= Station 6 in Arabic). The explosion turned most of the asteroid mass to dust and vapor, only a tiny fraction shattered into macroscopic meteorites, the heaviest of which was 283 gram. Other similarly frail asteroids may be related to main belt comets.

Keywords. Meteoroids, comets, asteroids

1. Introduction

On October 6, 2008, a small 3-4 meter sized asteroid was discovered by the Catalina Sky Survey program at Mount Lemmon (Kowalski *et al.* 2008). The asteroid, designated 2008 TC3, turned out to be on a collision course with Earth. Some 570 astrometric positions were determined, from which the impact trajectory was calculated (Chesley 2008). One 0.55–1.0 μm reflection spectrum was measured (flat in the visual with a weak 0.9 μm pyroxene band), which suggested the asteroid was of "C", "B", or "F" taxonomic class (Jenniskens *et al.* 2009, for a popular account see Kwok 2009).

The impact occurred 20 hours later over the Nubian Desert of northern Sudan. It was seen by KLM pilot Ron de Poorter, as well as by thousands of Sudanese along the river Nile awake for Morning Prayer. US government satellites detected the fireball first at 65 km, penetrating down to 37 km where the object exploded in three bright flares. From METEOSAT 8 images, it was deduced that fragmentation may have started with a small flare around 53 km (Borovicka & Charvat 2009). Two bright flares occurred with peak brightness of -18.8 magn. at 45 km and -19.7 magn. at 37.5 km. After this, the fireball penetrated for another second to end in a final weaker flare at $\sim$32.7 km. Two dust clouds of silicate smoke were deposited around 44 and 36 km, with a total mass of about 3,100$\pm$600 kg. More mass was likely lost in the form of larger and colder grains than seen by METEOSAT 8. From the fireball brightness, the initial mass of 2008 TC3 was estimated between 35,000–65,000 kg (Borovicka & Charvat 2009, Jenniskens *et al.* 2009).

The fireball PE-criterion, which uses the fireball's observed end height of about 32 km, velocity (12.4 km/s), ass and entry angle (20°) as a proxy for estimating its physical structure, would make this a IIIb/a-type, normally associated with cometary debris (Ceplecha *et al.* 1998). Of course, cometary debris is known to disrupt at even lower pressures

Figure 1. Search strategy for finding fragments of 2008 TC3 in the Nubian Desert of northern Sudan: students of the University of Khartoum comb the desert gravel.

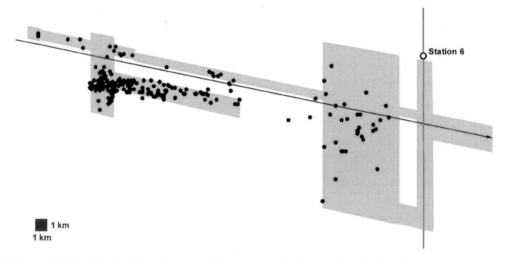

Figure 2. The distribution of recovered fragments relative to the calculated impact trajectory of asteroid 2008 TC3 (arrow pointing right) and the location of Station 6 (on the north-south railroad from Wadi Halfa to Abu Hamad). Gray areas were searched.

(< 0.1 MPa) than that at which 2008 TC3 disrupted (0.1–0.2 MPa), but ordinary chondrites are expected to break at around 5 Mpa (~ 25 km altitude). It is clear that this asteroid must have had a low cohesive strength or that it was exposed to unusually high thermal or mechanical stresses between 65 and 37 km.

For two months, the explosion of the asteroid in the atmosphere appeared to be the final word on the 2008 TC3 story. No meteorite had ever been recovered from a fireball ending this high in the atmosphere. In early December 2008, however, a search by students and staff of the University of Khartoum succeeded in recovering 15 meteorites along the calculated approach path. To find small fragments in a gravelly desert, a search strategy was adopted whereby the desert floor was combed by foot (Fig. 1). In subsequent searches in late December, and in February/March 2009, the total number of recovered meteorites rose to about 300, with a total mass of some 5 kg (Fig. 2). The largest recovered meteorite was 283 gr (#27, Fig. 3).

The meteorite strewnfield is unusual too, in that the meteorites are spread over a much larger area around the impact trajectory than commonly found. The strewnfield covered an area of at least 29 x 8 km (Fig. 2). We now understand that this was because of the

Figure 3. Examples of recovered meteorites (#1, 4, and 27) showing a wide array of textures and albedoes.

high explosion altitude, so that the asteroid had not yet significantly slowed down at the time of breakup. The range of recovered fragments is also larger than in most other falls, with many fragments 1 cm or smaller in diameter.

The meteorites are of a type called "anomalous polymict ureilite", a non-basaltic type of achondrite (Figure 3). A large range of textures and albedoes were found, with some meteorites more rich in pyroxene than others, some even showing pyroxene-rich layers (Fig. 3). The material is anomalous because of the abundant presence of pores and high $\delta^{17}O$ in some of the meteorites, and a large concentration of highly sintered organic matter. The organic matter is mostly graphitized, with small amounts of nano-diamonds, PAHs, and even amino acids. The pores are present in interconnected sheets and appear to form the outlines of partially sintered grains (Zolensky *et al.* 2010).

Most meteorites were dark, with an albedo of 0.046 ± 0.005 for the darkest components and values of 0.08–0.15 for lighter parts (Jenniskens *et al.* 2009). Combined with the measured absolute brightness of the asteroid, this would give a volume of $\leqslant 2.8 \pm 6 \text{ m}^3$, or a bulk density of $\geqslant 1.8 \pm 0.6 \text{ g/cm}^3$ (Scheirich *et al.* 2010).

The meteorite bulk density was measured by using fine sand to determine the volume displacement (Shaddad *et al.* 2010). Care was taken to shake the sand just enough so it settled into a rigid mass. From this, we have a significant range of densities for individual meteorites (Fig. 4). The mean value is 2.8 g/cm^3, with a significant variation in the range of 1.7 to 3.3 g/cm^3. These values are preliminary, while methods are being pursued to obtain better values.

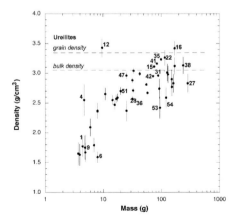

Figure 4. Measured mean densities of Almahata Sitta as a function of mass.

Other ureilites have a bulk density of 3.05 ± 0.22 g/cm^3 and a micro-porosity of 9%, assuming an average ureilite grain density of 3.35 g/cm^3 (Britt & Consolmagno 2003). Many larger masses of Almahata Sitta have similar bulk density. Indeed, many show the large olivine crystals and same $\delta^{17}O/\delta^{18}O$ ratios typical of known ureilites. On the other hand, some have significantly lower density, with porosities in the range of 20–50%. Similar high porosities are found in carbonaceous chondrites, but here the presence of large olivine crystals and partially sintered grains produces a particularly fragile material.

The shape of 2008 TC3, too, may have played a role. Inversion of the asteroid's light curve implies a shape like a loaf of bread with one large flat surface (Scheirich et al. 2010). Entry modeling shows that the asteroid re-oriented with this flat side forward between 65–37 km, which can cause high mechanical stresses (Bent et al. 2010).

Based on the astronomical spectrum and the reflection properties of the meteorites, the asteroid is now classified as of taxonomic class "F", or possibly "B" (Jenniskens et al. 2009). Many "B"-class asteroids, have hydrated minerals, not found in Almahata Sitta. One such "B"-class asteroid is 3200 Phaeton, parent body of the Geminid meteor shower, now linked to the Pallas family of "F" and "B" class asteroids (Jenniskens et al. 2010). Hence, it is possible that some asteroids, notably F- and B-class asteroids, are so frail that they resemble cometary matter in the event of a collision, possibly leading to main belt comets (Hsieh & Jewitt 2006).

References

Bent, F., et al. 2010, Meteoritics and Plan. Sci., (in preparation)

Borovicka, J. & Charvat, Z. 2009, A&A, 507, 1015

Britt, D. T. & Consolmagno, S. J. 2003, Meteorit. Planet. Sci., 38, 1161

Brown, P. G., ReVelle, D. O., Tagliaferri, E., & Hildebrand, A. R. 2002, Meteoritics Planet Sci., 37, 661

Ceplecha, Z. et al. 1998, Space Sci. Rev., 84, 327

Chesley, S. 2008, Minor Planet Electronic Circ., 2008-T50 (issued Oct. 6, 14:59 UT)

Hsieh, H. H. & Jewitt, D. 2006, Science, 312, 561

Jenniskens, P., Shaddad, M. H., Numan, D. et al. 2009, Nature, 458, 485

Jenniskens, P., Vaubaillon, J., Binzel, R. P., DeMeo, F. E., Nesvorný, D., Fitzsimmons, A., Hiroit, T., Marchis, F., Bishop, J. L., Zolensky, M. E., Herrin, J. S., & Shaddad, M. H. 2010, Meteoritics and Plan. Sci., submitted

Kowalski, R. A. et al. 2008, Minor Planet Electronic Circ., 2008-T50, Minor Planet Center, Smithsonian Astrophysical Observatory

Kwok, R. 2009, Nature 458, 401

Scheirich, P. et al. 2010, Meteoritics and Plan. Sci., (in preparation)

Shaddad, M. H., Jenniskens, P. et al. 2010, Meteoritics and Plan. Sci. (in preparation)

Zolensky, M. et al. 2010, Meteoritics and Plan. Sci., (in preparation)

Icy Bodies of the Solar System
Proceedings IAU Symposium No. 263, 2009
J.A. Fernández, D. Lazzaro, D. Prialnik & R. Schulz, eds.

© International Astronomical Union 2010
doi:10.1017/S174392131000181X

Searching for minor absorptions on D-type asteroids

Thais Mothé-Diniz

Universidade Federal do Rio de Janeiro, Observatório do Valongo
Rio de Janeiro, RJ, Brazil
email: `thais.mothe@astro.ufrj.br`

Abstract. Preferably located in the outer main belt, D-type asteroids experienced less heating and represent an important population for studies on the origin and evolution of the asteroid belt, as well as the relations between asteroidal and cometary bodies. Their surface mineralogy is currently related to a mixture of organics, anhydrous silicates, opaque material and ice. However, like other taxonomic classes, a large spectral diversity can be seen among D-type objects. We use the Visible spectra of 100 D-type objects available in the literature to search for minor absorptions in those objects. The presence of minor absorptions around 0.6 and 0.8 microns is reported for a large number of objects in the sample. The presence of such bands is not related to the heliocentric distance of the objects, since the absorptions can be seen in the whole main belt, up to the the Trojans region.

Keywords. Asteroids, D-types, absorptions

1. Introduction

D–type asteroids probably experienced less heating and represent important relics from early epochs of Solar System formation. Their study may help to better understand the origin and evolution of the asteroid belt, as well as the relations between asteroidal and cometary bodies. Inferring the composition of D–type asteroids is specially difficult since, apart from the recently discovered meteorites Tagish Lake and WIS91600, no other meteorite analogue has been found on Earth. From their spectral slope and apparent absence of features, the surface mineralogy of D–type asteroids is currently related to a mixture of organics, anhydrous silicates, opaque material and ice. These objects dominate the external region (from $\sim$ 3.2 AU) of the main asteroid belt, being abundant among the Cybeles (between 3.3 and 3.5 AU), the Hildas (around 4.0 AU, in the 3:2 resonance with Jupiter) and the Trojans (at 5.2 AU, around the Lagrangian points L_4 and L_5 of Jupiter). However, a non-negligible amount of these objects can also be found in the inner regions of the belt, from about 2.1 AU.

Previous spectroscopic studies of D–types have been preferably done in the ambit of the study of dynamical associations. The Hilda and Cybele asteroids were investigated by Dahlgren *et al.* (1997) and Lagerkvist *et al.* (2005). The Trojans by Jewitt & Luu(1990), Emery & Brown (2003), Fornasier *et al.* (2004), Dotto *et al.* (2005). Other studies of low-albedo asteroids were performed by Vilas & McFadden(1992), Vilas *et al.* (1993), Dumas *et al.* (1998). Besides that, some D-type asteroids have been associated to low-activity comets through dynamical, physical and spectroscopical constraints (Campins *et al.* 1987; Millis *et al.* 1988; Luu 1993; Harris *et al.* 2001; Abell *et al.* 2005; Campins *et al.* 2005).

Investigations by Jones *et al.* (1990) in the region around 3 μm and by Vilas *et al.* (1994) in the visible, suggested that these objects lack hydrated minerals in their surfaces.

Cruikshank *et al.* (2001), on the other hand, proposed that if the surfaces of the D–type asteroids are rich in opaque phases, they may contain significant amounts of hydrosilicates. The authors, however, did not show any detectable absorption band in the visible and near-infrared (NIR) in that work. Indeed, recently Kanno *et al.* (2003) reported the first detection of a 3 μm band in a D–type asteroid: (773) Irmintraud. They also showed that the shape of the band on (773) was very similar to that on the asteroid (1) Ceres.

Differences among D–type asteroids have been investigated by several authors, like Lagerkvist *et al.* (1993) who reported a clear correlation between the color index and the heliocentric distance, and Carvano *et al.* (2003) who noticed that inner main-belt D–types have more concave spectral shapes and higher albedos than those in the outer main-belt. In this work we study the spectra of 100 D–type asteroids, obtained from the public spectroscopic surveys SMASSII (Bus & Binzel 2002), S3OS2 (Lazzaro *et al.* 2004), and from the work of Fornasier *et al.* (2004), where it is possible to find the details of the respective observations. The data are composed of low resolution spectra on the 0.44/0.5 –0.92 μm range, all classified according to the Bus taxonomic system (Bus & Binzel 2002). Although belonging to the Centaurs group, the object (10199) Chariklo observed by the S3OS2 survey was included in the search for absorptions, since its spectral slope is similar to that of some trojan D-types. The relevant orbital and physical parameters for all objects in this study can be seen in Table 1.

Our first step was to search for correlations among physical and orbital parameters, including the Tisserand parameter T_J. Subsequently, a search for small absorption bands was performed, and we examined the possible correlations between the presence/absence of bands with all the parameters available. The results of these searches are presented in the sections that follow.

2. Search for absorptions and correlations on D-type asteroids

Search for minor features. In the search for minor absorption bands, we followed the approach of Vilas *et al.* (1993) to enhance minor spectral features present in the spectra. We have first smoothed each spectrum with a running-box average in order to reduce the spectral resolution , and then divided by a linear continuum across the spectral interval of 0.5–0.92 μm. With this procedure, we have detected subtle features near 0.60–0.65 μm and 0.80–0.9 μm, with depths down to about 2% of the background continuum, on many objects in our sample. These features are illustrated in Fig. 1. A confidence code was assigned according to the noise level of the spectra and repeatability of the feature between different spectra of the same asteroid. Whenever an absorption band deeper than our detection limit was found in one or more spectra of an object after the continuum removal, then it was considered a "positive" identification. Many spectra presented features that were not well defined, with depths smaller than the detection limits and that were absent in other spectra from different nights/observing times. In this case, we considered the feature as "doubtful". The detection limits are determined from the S/N of the spectra in the region of the bands, which are, in average, between 1–3%. The "positive" identifications are represented in Table 1 with a cross (x). Objects presenting a doubtful feature in either or both regions are: 267, 612, 666, 717, 729, 732, 891, 1361, 1689, 2105, 2246, 2263, 2569, 2872, 3063, 3248, 3709, 3793, 4063, 4068, 4103, 4489, 4617, 4744, 4833, 4835, 5264, 5461, 5818, 6545, 7516, 8513, 13463, and 30698. Objects showing no absorption at all are: 361, 520, 565, 911, 1094, 1118, 1226, 1293, 1328, 1535, 2086, 2454, 2677, 2891, 3015, 3141, 3333, 3990, 4035, 4902, 5258, 5362, 5914, 7635, 9430, 11351, 12917, 14465, 15502, 20738, and 24390. Object (818) Kapytenia presents

Table 1. Main physical parameters of the asteroids studied. The positive identification of the 0.6 and/or 0.8 μm bands is reported with a "x", while a "?" denotes a doubtful feature. Any dynamical association is also listed.

Number	Name	a	p_ν	D(km)	T_J	Dyn.Assoc.	Slope	Concavity	0.6?	0.8?
12	Victoria	2.34	0.176*	112.77	3.52		10.1	ccv	x	x
510	Mabella	2.61	0.07	57.40	3.36	-	7.0	ccv	x	-
547	Praxedis	2.77	0.06	69.60	3.23	-	6.3	ccv	x	-
579	Sidonia	3.01	0.175	85.56	3.22	Eos	7.9	ccv	x	x
721	Tabora	3.55	0.060	76.07	3.09	Cybele	8.3	flat	x	x
726	Joella	2.56	0.05	44.00	3.33	-	6.3	flat	x	x
775	Lumiere	3.01	0.108	33.59	3.23	Eos	8.8	ccv	-	x
798	Ruth	3.01	0.16	43.20	3.23	Eos	6.4	ccv	x	x
818	Kapteynia	3.17	0.165	49.45	3.14		8.6	ccv	?	x
1006	Lagrangea	3.15	0.07	29.60	3.08	-	6.7	flat	x	-
1172	Aneas	5.19	-	142.80	2.91	Trojan*	9.1	flat	x	x
1209	Pumma	3.17	-	-	3.18	-	6.9	ccv	x	-
1275	Cimbria	2.68	0.11	29.20	3.32	Eunomia	6.4	cvx	-	x
1284	Latvia	2.65	0.104	36.81	3.35		8.5	ccv	x	x
1321	Majuba	2.94	0.143	36.30	3.23		8.3	ccv	-	x
1328	Devota	3.51	0.04	57.20	3.10	-	14.5	flat	-	-
1400	Tirela	3.12	-	-	3.12	Tirela	8.2	ccv	x	x
1481	Tubingia	3.01	0.117	33.26	3.24		7.8	ccv	x	-
1542	Schalen	3.10	0.065	45.19	3.21		9.0	ccv	x	x
1574	Meyer	3.54	0.039	58.68	3.07	Cybele	10.2	flat	x	-
1609	Brenda	2.58	0.115	29.64	3.31		9.8	ccv	x	x
1647	Menelaus	5.24	-	-	2.99	Trojan*	6.7	flat	x	-
2235	Vittore	3.22	0.047	44.45	3.07		10.1	flat	-	x
2266	Tchaikovsky	3.39	0.038	46.94	3.08	Cybele	10.0	flat	x	x
2448	Sholokhov	2.79	0.13	30.20	3.25	-	8.3	flat	x	-
2498	Tsesevich	2.92	-	-	3.28	Koronis	7.3	ccv	x	x
2867	Steins	2.36	-	-	3.51	Inner	7.0	ccv	-	x
2959	Scholl	3.95	0.05	34.20	2.99	-	14.2	flat	x	-
3140	Stellafane	3.02	0.126	24.75	3.21	Eos	8.2	flat	x	x
3152	Jones	2.63	0.05	33.20	3.37	-	8.3	flat	x	x
3453	Dostoevsky	2.39	-	-	3.53		7.3	flat	x	x
3682	Welther	2.75	0.12	19.40	3.23	-	6.2	ccv	x	-
3906	Chao	2.93	0.04	47.00	3.12	-	5.2	flat	x	-
5648	1990VU1	5.17	-	-	2.82	-	12.8	flat	x	-
10199	Chariklo	15.79	-	-	3.48	Centaur	10.3	flat	x	x
15535	2000AT177	5.16	-	-	2.94	Trojan*	11.7	flat	x	x

one positive identification around 0.8 μm and one doubtful identification around 0.6 μm. The later is denoted on Table 1 as a "?".

Considering only the positive identifications, $\approx$ 12% of our sample presents only the 0.6–0.65 μm band, while 5% only the 0.8–0.9 μm band and $\approx$ 18% of the sample has both bands. If we take into account also the doubtful features , these percentages raise to 21%, 12% and 32% respectively. Inspection of the relations between the presence of features and orbital and physical parameters reveals that unambiguous positive identifications were found in the whole main-belt and Trojan population.

Testing for correlations. Correlations were searched among the parameters using the Spearman rank order correlation. The following orbital and physical parameters were used: semi-major axis a, absolute magnitude H, albedo ρ_ν, diameter D and slope S'. The spectra were normalized to unity at 0.6 μm, and a linear reflectance slope S' of each

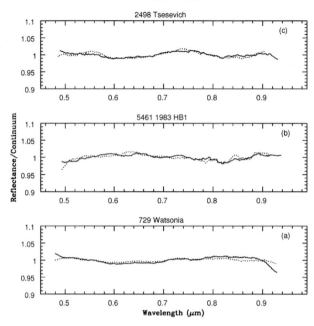

Figure 1. Reflectance spectra of three D–type asteroids smoothed with a running-box average. The spectra have also been divided by a 2-degree polynomial fit, in order to remove the continuum. Different line types denote spectra from different nights/observing runs. (a) spectrum with a 0.6 μm band, (b) with a 0.8 μm band and (c) with both the 0.6 and 0.8 μm bands.

spectrum was defined as the angular coefficient of a linear fit to the spectrum in the 0.50–92 μm range, given in $\%10^3 \mathring{A}$. The morphology of the spectra was also inspected, and each object was classified in one of the three groups: concave (ccv) ∩, convex (cvx) ∪ or flat /. The curvatures for the spectra presenting features are shown on Table 1 . Within a confidence level of 99% and using the entire sample, a correlation was found between the following pairs of parameters: ρ_ν and T_J, H and T_J. Anti-correlation is seen between ρ_ν and D, ρ_ν and the semi-major axis a, and between D and T_J. At a confidence level of 95% an anti-correlation between T_J and S' is also reported, as well as a correlation between a and S'. It is important to notice that Carvano *et al.* (2003) remark that the correlation between a and S' depends on the normalization wavelength, and therefore, cannot be considered significant. Only one "cvx" spectrum in our sample presents absorptions, and we report a predominance of "flat" spectra with these features.

3. Discussion

Absorption features centered near 0.6-0.65 μm and 0.8-0.9 μm have been detected for the first time in a large number of D-type asteroids. Previously, Vilas *et al.* (1993) reported both bands in a small number of asteroids: 165, 225, 334 ,368, and 877, which are all low-albedo and not steeply sloped. They attributed those bands respectively to $6A1 \rightarrow 4T2(G)$ and $6A1 \rightarrow 4T1(G)$ charge transfer transitions ($Fe3+ \rightarrow Fe$) in minerals such as the oxyhydroxide goethite, the Fe oxide hematite and the sulfate jarosite. They are products of the aqueous alteration of anhydrous silicates. Analogues for these features are seen in spectra of CM2 carbonaceous chondrites Cold Bokkeveld and Murray.

In this work the literature (eg. Burns 1993) was searched for alternatives for both features. The only mineral found with both absorptions seem to be the Andradite, which is a calcium-iron silicate, where they occur via crystal field transitions. This mineral is found on the CV3 Chondrite Allende. Recently Vilas *et al.* (2006) reported aqueous alteration (0.7 μm band) on some satellites of the Jovian planets suggesting, suggesting that those bodies did not form in their present locations, but could have formed in the "aqueous alteration zone", from 2.6 - 3.5 μm, (Vilas *et al.* 1994) in the main-belt. If the 0.6-0.65 μm and 0.8-0.9 μm bands are due to aqueous alteration, their presence on D-types (main-belt, Trojans, and Chariklo) suggests that this kind of thermal processing on asteroids was not limited only to the 2.6 - 3.5 μm zone, but happened in the whole main-belt, at least up to 5.2 AU and in objects of the Centaur group.

Acknowledgements

This work has been supported by the *Conselho Nacional de Desenvolvimento Científico e Tecnológico* – CNPq/Brasil and *Fundação de Amparo à Pesquisa do Estado do Rio de Janeiro* – FAPERJ.

References

Abell, P. A., Fernández, Y. R., Pravec, P. *et al.* 2005, *Icarus*, 179, 174

A'Hearn, M. F., Campins, H., Schleicher, D. G., & Millis, R. L. 1989, *ApJ*, 347, 1155

Burns, R. G. 1993, *Mineralogical Applications of Crystal Field Theory*, Cambridge Topics in Mineral Physics and Chemistry (Cambrigde: Cambridge University Press)

Bus, S. J. 1999, *Compositional structure in the asteroid belt: results of a spectroscopic survey*. PhD Dissertation. Massachusets Institute of Technology.

Bus, S. J. & Binzel, R. P. 2002, *Icarus*, 158, 106

Barucci, M. A., Doressoundiram, A., Fulchignoni, M. *et al.* 1998, *Icarus*, 132, 388

Campins, H., A'Hearn, M. F., & McFadden, L.-A. 1987, *ApJ*, 316, 847

Campins, H., Licandro, J., Ziffer, J. *et al.* 2005, *IAU Symposium 229, abstracts book*

Carvano, J. M., Mothé-Diniz, T., & Lazzaro, D. 2003, *Icarus*, 161, 356

Cruikshank, D. P., Dalle Ore, C. M., Roush, T. L. *et al.* 2001, *Icarus*, 153, 348

Dahlgren, M., Lagerkvist, C. I., Fitzsimmons, A., Williams, I. P., & Gordon, M. 1997, *A&A* 323, 606

Dotto, E., Fornasier, S., Barucci, M. A. *et al.* 2005, *Icarus*, submitted.

Dumas, C., Owen, T., & Barucci, M. A. 1998, *Icarus*, 133, 221

Emery, J. P. & Brown, R. H. 2003, *Icarus* 164, 104

Fitzsimmons, A., Dahlgren, M., Lagerkvist, C.-I., Magnusson, P., & Williams, I.P. 1994, *A&A*, 282, 634

Fornasier, S., Dotto, E., Marzari, F. *et al.* 2004, *Icarus*, 172, 221

Harris, A. W., Delbó, M., Binzel, R. P. *et al.* 2001, *Icarus*, 153, 332

Jones T. D., Lebofsky L. A., Lewis J. S., & Marley M. S. 1990, *Icarus* 88, 172

Jewitt D. C. & Luu J. X. 1990, *AJ*, 100, 933

Kanno, A., Hiroi, T., Nakamura, R. *et al.* 2003, *Geophys. Res. Letters*, 30, 2

Lagerkvist, C.-I.,Fitzsimmons, A., Magnusson, P., & Williams, I. P. 1993, *MNRAS*, 260, 679

Lagerkvist, C.-I., Moroz, L., Nathues, A. *et al.* 2005, *A&A*, 432, 349

Lazzaro, D., Angeli, C. A., Carvano, J. M. *et al.* 2004, *Icarus*, 172, 179

Lederer, S. M., & Vilas, F. 2003, *Earth Moon and Planets*, 92, 193

Luu, J. X. 1993, *Icarus*, 104, 138

Millis, R. L., A'Hearn, M. F., & Campins, H. 1988, *ApJ*, 324, 1194

Morbidelli, A., Levison, H. F., Tsiganis, K., & Gomes, R. 2005, *Nature*, 435, 462

Rivkin, A. S., Howell, E. S., Vilas, F., & Lebofsky, L. A. 2003, in: W. F. Bottke Jr., A. Cellino, P. Paolicchi & R. P. Binzel (eds.), *Asteroids III*, (Tucson: Univ. of Arizona), p. 235

Vilas, F. & McFadden, L. A. 1992, *Icarus*, 100, 85

Vilas, F., Hatch, E. C., Larson, S. M., Sawyer, S. R., & Gaffey, M. J. 1993, *Icarus*, 102, 225

Vilas F., Jarvis K. S., & Gaffey, M. J. 1994, *Icarus*, 109, 274

Vilas F., Lederer S. M., Gillb S. L, Jarvisc K. S. & Thomas-Osip, J.E. 2005, *Icarus*, 180, 453

Icy Bodies of the Solar System
Proceedings IAU Symposium No. 263, 2009
J.A. Fernández, D. Lazzaro, D. Prialnik & R. Schulz, eds.
© International Astronomical Union 2010
doi:10.1017/S1743921310001821

The Distribution of Main Belt Asteroids with Featureless Spectra from the Sloan Digital Sky Survey Photometry

Anderson O. Ribeiro and Fernando Roig

Observatório Nacional, Rio de Janeiro, 20921-400, RJ, Brazil
email: anderson@on.br
email: froig@on.br

Abstract. In this work, we propose to analyse the existence of possible correlations between the taxonomic classes of asteroids showing featureless spectra –i.e. a flat continuum with no absorption bands– and their orbital properties. We compute the mean spectral slope of 14 753 asteroids using the photometric data from the Sloan Digital Sky survey Moving Objects Catalog (SDSS-MOC4). Although the quality of these data is not comparable in resolution to the spectroscopic data, the amount of observations in the SDSS-MOC4 is more than 20 times larger that in the available spectral databases. This allows us to obtain a statistically significant result.

Keywords. minor planets, asteroids – surveys

1. Introduction

Several studies about the surface properties of Main Belt asteroids, based on spectroscopic observations, indicate that there exist correlations between the different taxonomic classes and their orbital distribution (Mothé-Diniz, Carvano & Lazzaro, 2003; Carvano, Mothé-Diniz & Lazzaro, 2003). However, these studies are limited by the low number of available spectroscopic data. Some 20 years ago, the known distribution of taxonomic classes was compatible with the idea that asteroids' mineralogy was correlated to the temperature gradient of the primordial nebula. Asteroids closer to the Sun showed evidences of more active heating processes, that were not detected in the more distant asteroids. But in recent years, the increasing number of observations reveal that the taxonomic classes appear uniformly distributed among the Main Belt, showing no apparent correlation to the temperature gradient of the nebula.

We propose here to analyse such correlations using the 4th release of the Moving Objects Catalog of the Sloan Digital Sky Survey (SDSS-MOC4). This catalog provides 5 band photometry of a sample of asteroids which is about 20 times larger than the spectroscopic sample.

Our analysis is limited to the so-called background asteroids, i.e. those that do not belong to any dynamical family of asteroids, and we focus on bodies with featureless spectra, i.e. those that show a flat spectrum with no evidences of absorption bands in their spectra.

2. Methodology

The SDSS-MOC4 provides calibrated magnitudes in the filters u,g,r,i,z, centred at 3540 Å, 4770 Å, 6230 Å, 7630 Å, and 9130 Å, respectively. From these magnitudes we compute reflectance colors and the corresponding reflectance fluxes, normalized to 1 at the r band. Observations with errors > 10% in any band were discarded.

To separate the observations corresponding to featureless spectra, we apply the Principal Components Analysis (PCA). Featureless observations can be easily identified in the space os the two first principal components of the reflectance fluxes distribution.

The featureless observations were further separated into those corresponding to the members of the major dynamical families in the Main Belt and those corresponding to the background population of asteroids.

Finally, in order to characterize each featureless observation, we compute the spectral slope by fitting a straight line to the reflectance fluxes *vs.* wavelength.

3. Results

We analyze behavior of the population of background asteroids in the three regions of the Main Belt: inner, middle, and outer. Our results indicate that there are very weak correlations between the mean spectral slope and the orbital eccentricities and inclinations, both for the inner and middle Belt populations. More significant correlations are observed in the outer Belt, especially in terms of orbital inclination (Fig. 1). The wavy behavior at high inclinations shown in Fig. 1, right panel, is particularly notorious.

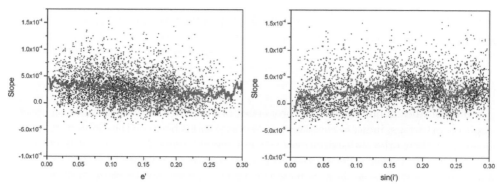

Figure 1. Correlations between mean spectral slope and proper orbital eccentricity (left), and sinus of proper orbital inclination (right), for background asteroids in the outer Main Belt with featureless spectra. The full line is the running average of the data.

A comparison with the distribution of spectral slopes of asteroids members of dynamical families in the outer Main Belt (Fig. 2), seems to indicate that the correlations observed for the background population are driven by the two major families in the region: the Themis family (high eccentricity, low inclination) and the Eos family (low eccentricity, high inclination). We propose that these two families contaminate the background population with a significant amount of fugitive asteroids, that left the presently detected dynamical families due to long term dynamical evolution, but remain in the neighborhood of their parent families.

4. Conclusion

We do not find any evidences of significant correlations between the spectral slopes and the proper orbital elements of background asteroids with featureless spectra in the inner and middle Belt.

In the outer Belt, the stronger correlations seem to be driven by the background contamination from the major asteroid families. We believe that this is not an artefact of the families identification process, but rather a real effect caused by the presence

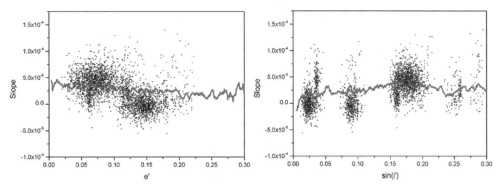

Figure 2. Correlations between mean spectral slope and proper orbital eccentricity (left), and sinus of proper orbital inclination (right), for asteroids with featureless spectra that belong to dynamical families in the outer Main Belt. The full line is the same shown in Fig. 1.

of former family members that dynamically evolved beyond the present edges of these families.

Our results tend to indicate that the actual background population of asteroids with featureless spectra are uniformly distributed in terms of eccentricities and inclinations all over the Main Belt. No correlation between spectral slopes and mean heliocentric distances has been observed either.

References

Carvano, J. M., Mothé-Diniz, T., & Lazzaro, D. 2003, *Icarus*, 161, 356

Mothé-Diniz, T., Carvano, J. M., & Lazzaro, D. 2003, *Icarus*, 162, 10

Icy Bodies of the Solar System
Proceedings IAU Symposium No. 263, 2009
J.A. Fernández, D. Lazzaro, D. Prialnik & R. Schulz, eds.

© International Astronomical Union 2010
doi:10.1017/S1743921310001833

Dynamical Maps of the Inner Asteroid Belt

Tatiana A. Michtchenko[1], Daniela Lazzaro[2], Jorge M. Carvano[2], and Sylvio Ferraz-Mello[1]

[1]Instituto de Astronomia, Geofísica e Ciências Atmosféricas, USP, Rua do Matão 1226, 05508-900 São Paulo, Brazil
email: tatiana@astro.iag.usp.br

[2]Observatório Nacional, R. Gal. José Cristino 77, 20921-400, Rio de Janeiro, Brazil

Abstract. We construct the dynamic portrait of the inner asteroidal belt using the information about the distribution of the test particles, which were initially placed on a perfectly rectangular grid of initial conditions, after 4.2 Myr of gravitational interactions with the Sun and five planets, from Mars to Neptune. Using the Spectral Analysis Method introduced by Michtchenko *et al.* (2002), we illustrate the asteroidal behaviour on the dynamical maps. We superpose over the maps the information on the proper elements and proper frequencies of the real objects, extracted from the database *AstDyS* (Milani & Knežević 1994; Knežević & Milani 2003). The comparison of the maps with the distribution of the real objects allows us to detect possible dynamical mechanisms acting in the domain under study: these mechanisms are related to mean-motion and secular resonances. Their long-lasting action, overlaying with the Yarkovsky effect, may explain many observed features of the distribution of the asteroids.

Keywords. minor planets; asteroids, dynamics

1. Introduction

It is nowadays accepted that many features of the asteroid distribution may be explained by long-lasting actions of dynamical mechanisms, present in the inner main belt. With purpose to identify these mechanisms, we elaborate a global dynamic picture of the inner part of the main belt, using the Spectral Analysis Method introduced by Michtchenko *et al.* (2002). The main idea of the method consists of using the information about the distribution of the test particles, which were initially placed over a perfectly rectangular grid of initial conditions, after 4.2 Myr of gravitational interactions with the Sun and five planets, from Mars to Neptune.

The final distribution of the fictitious particles will reflect the peculiarities of dynamical interactions between asteroids and the planets in the region under study. Particulary, the test particles will preserve (at least, during the time span covered by integrations) some invariable quantities of motion (e.g. proper elements), if they belong to the domains of quasi-regular motion. In contrast, these quantities will exhibit variations in the domains of dynamical instabilities, which can result even in the escape of the objects from the studied region. Therefore, the test particles can form agglomerations or, on the contrary, some voids in the specific regions of the phase space, due to the action of mean-motion and secular resonances, in such a way indicating the location of these features.

2. Dynamical maps on the representative planes

At the present, there are around 200,000 numbered objects in Bowell's catalogue of asteroid orbital elements (Bowell *et al.* 1994). About 60,000 objects are found in the range of proper semimajor axis between 2.1 AU and 2.5 AU. This region of the main belt is often referred to as the inner part.

To construct the dynamical portrait of the inner belt in the three-dimensional elements space, we introduce so-called 'representative planes' (Milani & Knežević 1992): the sample of the objects is separated in different eccentricity intervals and each one is presented on the (a, I)–plane of the semimajor axis and inclination. We have chosen arbitrarily four eccentricity intervals; inside each interval we identified one known dynamical family and associated the name of this family to the interval. We also used the values of the osculating eccentricity and angular elements of the main object of the family as the input of numerical integrations, to construct the dynamical maps.

The first interval is characterized by very low eccentricities, less than 0.075. There are neither known families nor clusters, so we select one large object from this interval, 230 Athamantis with magnitude 7.3, and associate it to this interval. The next interval

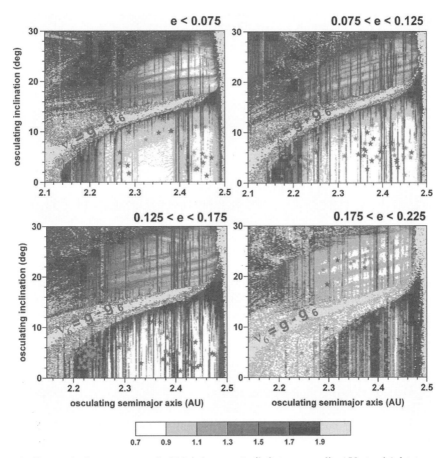

Figure 1. Dynamical maps around: 230 Athamantis (left-top panel), 4 Vesta (right-top panel), 298 Baptistina (left-bottom panel) and 163 Erigone (right-bottom panel) on the (a, I)–planes of the osculating semimajor axis and inclination. The gray color levels correlate the stochasticity of motion with the spectral number N, in logarithmic scale: lighter regions corresponds to regular motion, while darker tones indicate increasingly chaotic motion. The hatched regions correspond to initial conditions that lead to escape of objects in less than 4.2 Myr. The real objects from the corresponding eccentricity interval are superposed on each graph: red stars are objects with magnitudes less than 10, while green stars are objects with magnitudes in the range from 10 to 12.

is centered around 4 Vesta and covers low-to-moderate eccentricities within the range $0.075 \leqslant e < 0.125$. Finally, the next two intervals are associated to 298 Baptistina and 163 Erigone and cover the moderate ($0.125 \leqslant e < 0.175$) and high ($0.175 \leqslant e < 0.225$) eccentricities, respectively.

60300 massless particles were chosen in each eccentricity interval: their initial semimajor axes and inclinations were distributed over a rectangular 201×300 grid, covering the representative plane. Each particle has been integrated over 4.2 Myr, accounting for planetary perturbations from Mars to Neptune. During the integration, a low pass-band digital filter has been applied to remove the short period oscillations of the order of orbital periods.

The output series of filtered elements for each test orbit have been Fourier analyzed, in order to identify the relevant peaks in their Fourier spectra. The number of the relevant peaks in the semimajor axis variation, known as a 'spectral number N', was used to characterize the chaoticity of the orbit: for regular orbits with small values of N, a few well defined lines appear in the spectra, while for chaotic orbits the number of peaks (consequently, N) is huge. This information has been translated to a gray scale code and plotted on the (a_{osc}, I_{osc})–planes of the osculating semimajor axis and inclination.

Figure 1 shows the dynamical maps of the inner main belt plotted on planes corresponding to different eccentricity intervals. Since the variation of the asteroidal semimajor axis is strongly affected by mean-motion resonances, its time evolution was chosen as a basis for calculation of N. The calculated values of N, in the range from 1 to 80, were coded by a gray level scale that varied logarithmically from white ($\log N = 0$) to black ($\log N = 1.9$). Large values of N indicate the onset of chaos, while lighter regions on the dynamical maps correspond to regular motion, and darker tones indicate increasingly chaotic motion. The domains, where the test particles escape from the region within the time-interval of integration ($\sim$4.2 Myr), are hatched.

On the dynamical maps in Fig. 1 we superpose proper elements of some real objects from the corresponding eccentricity range. Only large objects are shown, those with absolute magnitudes lower than 10 (red stars) and between 10 and 12 (green stars). It is worth emphasizing that the dynamical map constructed over a grid of osculating orbital elements cannot be directly compared to the proper elements of the asteroids; thus, Figure 1 provide only an estimation of the dynamical distribution of real objects.

The dominant effects on the asteroidal motion are produced by the strong secular resonances, ν_6 and ν_{16}. The ν_6 secular resonance occurs in the regions where the precessional rate of the asteroid's longitude of perihelion equals the precessional rate of Saturn's perihelion. The ν_6 resonance is responsible for large-scale instabilities followed by rapid escape of the test particles inside diagonal bands crossing all maps from the lower-left to the upper-right corners in Fig. 1. The bulk of real objects is located below this band, forming the low inclination population.

The ν_{16} secular resonance occurs in the regions where the precessional rate of the asteroid's longitude of node equals the precessional rate of Saturn's node. The ν_{16} SR produces the large domains of chaotic motion in the high inclination zones and forms a natural upper boundary of the asteroidal population in the inner belt. High inclination domains of quasi-regular motion can be observed between the ν_6 and ν_{16} resonances, on each panel in Fig. 1. The curious feature is the absence of high inclined asteroids (at least, with *mag* < 12) at low and moderate eccentricities. This population is significant only at $e > 0.18$. The motion of the objects in this region is highly nonharmonic (later we will associate this property to the action of nonlinear secular resonances). Despite this fact, the 25 Phocaea, 5247 Krylov and 1660 Wood groups are found inside the high inclination zone.

A relevant feature clearly visible in Fig. 1, is the occurrence of several vertical stripes of chaotic motion, which are associated with two- and three-body mean-motion resonances with the planets from Mars to Uranus. The dominating mean-motion resonance is the 3J/1A mean-motion resonance with Jupiter, whose region is strongly chaotic, with rapid escapes of the test particles (the right border of all graphs). Thus, the devastating effects of the 3J/1A mean-motion resonance, together with ν_{16} SR, create the physical boundaries of the inner belt, while ν_6 SR delimits the domains of the low inclination orbits.

Apart the 3J/1A resonance, mean-motion resonances in the inner belt are weak: this is due to the large distance from the Jupiter–Saturn system and the small mass of Mars. Many of these resonances cut through the asteroid families and we may expect several family members to have involved in the resonances. Among the mostly relevant ones, there are 7J/2A MMR with Jupiter and 1M/2A MMR with Mars. There are also several three-body mean-motion resonances of low order, such as the $4J : -1S : -1A$, $4J : -2S : -1A$, and $5J : -4S : -1A$ resonances, where the letters J, S, A denote Jupiter, Saturn and an asteroid, respectively.

It is known that the density and the strength of the mean-motion resonances increase with increasing eccentricities; moreover, at high eccentricities, many of them may overlap creating the domains of unstable motion. The low inclination region on the bottom-right panel in Fig. 1 showing high indices of stochasticity, seems to confirm this assumption. The surprising fact is that this region is significantly populated by large objects. More-over, this population strongly contrasts with the population on the nearly circular orbits (see top-left panel in the same figure), which does not undergo the action of the overlapping mean-motion resonances, but presents low density of large objects.

Acknowledgements

This research has been supported by the Brazilian National Research Council – CNPq, the São Paulo State Science Foundation - FAPESP, and the Rio de Janeiro State Science Foundation - FAPERJ. The authors gratefully acknowledge the support of the Computation Center of the University of São Paulo (LCCA-USP) and of the Astronomy Department of the IAG/USP for the use of their facilities.

References

Bowell, E., Muinonen, K., & Wasserman, L. H. 1994, in: A. Milani, M. di Martino & A. Cellino (eds.), *Asteroid, Comets and Meteorids III*, (Dordrecht: Kluwer), p. 477
Michtchenko, T., Lazzaro, D., Ferraz-Mello, S., & Roig, F. 2002, *Icarus*, 158, 343
Milani, A. & Knežević, Z. 1992, *Icarus*, 98, 211
Milani, A. & Knežević, Z. 1992, *Icarus*, 107, 219

Icy Bodies of the Solar System
Proceedings IAU Symposium No. 263, 2009
J.A. Fernández, D. Lazzaro, D. Prialnik & R. Schulz, eds.

© International Astronomical Union 2010
doi:10.1017/S1743921310001845

Magnetite microspheric particles from bright bolide of EN171101, exploded above the Trans-Carpathians mountains on Nov. 17, 2001

Klim I. Churyumov,[1] Rudolf Ya. Belevtsev,[2] Emlen V. Sobotovich,[2] Svitlana D. Spivak,[2] Volodymyr I. Blazhko,[2] and Volodymyr I. Solonenko[3]

[1]Kyiv Shevchenko National University,
Box 04053, Observatorna str., 3, Kyiv, Ukraine
email: klim.churyumov@observ.univ.kiev.ua

[2]Institute of environmental geochemistry of NAS and MES of Ukraine
[3]Vinnytsia Kotsyubynskiy State Pedagogical University

Abstract. In 2007-2008 the authors found many magnetite microspheric particles in ground samples at the Trans-Carpathians mountains near the village of Tur'yi Remety. Their diameters are of 0.1 - 0.3 mm and they have Ni, Co and Cr as chemical composition. We think that these particles are part of the bright bolide of EN171101 which exploded above Trans-Carpathians mountains on Nov. 17, 2001.

Keywords. meteors, meteoroids, bolide, microspherule, Earth

1. Introduction

In 2007-2008 mineralogical-geochemical researches were made in the region around the fall of the bright bolide of EN171101 (near the village of Tur'yi Remety) in Trans-Carpathians. The bolide of EN171101 (Tur'yi Remety) was photographed by the Czech and Slovakia chambers of the European bolide network (EN) on November, 17, 2001 at 16:52:44 UT (Spurny & Porubchan, 2002). The length of the luminous track of the bolide reached $107km$. It began at height of $81.4km$ at 10 km to a south-west from the Ukrainian city of Dolina. The maximal brightness of -18^m absolute star magnitude of the bolide was attained at the height of 20 km. The bolide path finished at the height of 13.5 km near the village Tur'yi Remety. It is the deepest penetration of a bolide in the atmosphere of Earth, taken by a photographic chamber. Initial dynamic mass of bolides was estimated of $4500kg$ and the eventual mass, after los by ablation was $370kg$. The orbit was typical for bolides deeply penetrating into the atmosphere of Earth: $a = 1.33$ A.U., $e = 0.484$, $q = 0.684$ A.U., $\omega = 266.8°$, $\Omega = 235.4°$, $i = 7.4°$. Searches of fragments of meteoroid with general eventual mass 370 kg, made in Tur'yi Remety sat down in a district, unfortunately, did not give the positive results.

2. Search of space matter from the bright bolide of EN171101 (Tur'yi Remety)

Geologists Belevtsev R. Ya., Spivak S.D., Blazhko V.I. in 2007 made mineralogical-geochemical researches in a pool of the river Tur'ya, and in 2008 these works were continued with the participation of Churyumov K.I. and Solonenko V.I. Presence of basalt and

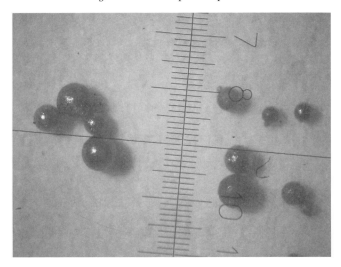

Figure 1. Magnetite microspherules that are possibly material from Bolide EN171101. The diameter of most microspherules is 0,3 mm

other volcanic minerals, similar conduit ones, complicated the authentication of possible space matter from bolide of EN171101 in the studied Earth samples. One of the obvious substance features of the bolide EN17110 is magnetite microspherules in soil samples with diameters 0.1-0.3 mm, a content of which is traced on magnetic fractions of concentrate and soil samples. These microspherules with spherical shape (and rarer elliptic or dropshaped form) were discovered with their primary concentration in the mangrove of the river Uzh between village Zarichevo and village Simer. Usually microspherules have different sizes in one sample, that allows to suppose their slow parachuting in the Earth's atmosphere after the explosion of the bolide of En171101 in windless weather (Fig. 1). Analogical microspherules were found out and investigated in the samples of soils, brought by L.A.Kulik from the region of the explosion of the Tunguska meteorite, by A.A.Yavnel in 1957, who with the help of the microchemical analysis found out the presence of nickel iron, that was a confirmation of the space origin of the material explored (Yavnel, 1957). Similar microspherules are described by E.V.Sobotovich (Sobotovich, 1976). Physical and chemical analyzes of magnetic fractions of the concentrate samples allowed to select the band of the promoted content in them Ni, Co and Cr, which is directed from Turichki-Lumshory to Simerka-Simer, i.e. approximately on 3-4 km to north-west than the published trajectory of flight of the En171101 Tur'yi Remety bolide (Spurny & Porubchan, 2002). Presently physico-chemical and spectral analysis of the magnetite microspherules discovered by us is underway in order to confirm their space origin.

References

Sobotovich, E. V. 1976, *Cosmic matter in the Earth's crust.*, 1, 159

Spurny, P., Porubchan, V. 2002, *Proceedings of Asteroids, Comets, Meteors*, (ESA-SP-500), pp. 269–272

Yavnel A. A. 1957, *Astronomicheskii Journal*, 34(5), 794

Icy Bodies of the Solar System
Proceedings IAU Symposium No. 263, 2009
J.A. Fernández, D. Lazzaro, D. Prialnik & R. Schulz, eds.

© International Astronomical Union 2010
doi:10.1017/S1743921310001857

The meteoroid above Mediterranean Sea on July, 6th 2002 was a fragment of a cometary nucleus?

Klim I. Churyumov,[1] Vitaly G. Kruchynenko,[1] Larissa S. Chubko[2], and Tatyana K. Churyumova[1]

[1] Kyiv Shevchenko National University,
Box 04053, Observatorna str., 3, Kyiv, Ukraine
email: klim.churyumov@observ.univ.kiev.ua

[2] National Aviation University,
Box 03680, Kosmonavta Komarova ave. 1, Kiev, Ukraine

Abstract. It is shown, that the explosion of the bolide above the Mediterranean sea on July, 6th 2002 with a high probability was a fragment of an icy cometary nucleus which initial weight was equal approximately to $7 \cdot 10^8$ grams. This follows from the determined energy of the outburst equal to $26kT$ of TNT (Brown *et al.* 2002). We think that this energy refers to the height of the maximal braking of the body in the Earth's atmosphere. At the speed of $20.3km/s$ accepted by authors, the weight of a body at this height is equal $5 \cdot 10^8$ gr, and at the entrance in the atmosphere it was approximately of $7 \cdot 10^8$ gr.

Keywords. meteors, meteoroids, bolide, comet, Earth

---

1. Distribution of meteoroids masses in the atmosphere of Earth

In (Brown *et al.* 2002) are given the results of the analysis of large meteoroids flashes in the Earth's atmosphere, obtained by USA geostationary satellites. For 8.5 years (from February 1994 to September 2002) 300 such phenomena were registered. On the basis of the data about optical energies of these bolides, and considering velocities at their input into the atmosphere of $20.3km/s$ and density of $3g/cm^3$, the distribution of diameters of falling bodies is obtained. The considered value of the velocity is close to $21km/s$, which follows data from catalogues (McCrosky *et al.* 1976; 1977) for bodies which initial mass is not inferior to $1kg$ (Kruchinenko, 2002). On these basis, the integral distribution of diameters of bodies is given (Brown *et al.* 2002). This distribution can be transformed in distribution of masses as:

$$logN_R = 7.146 - 0.90log\text{m}, \tag{1.1}$$

$$logN_C = 7.86 - 0.892log\text{m} \tag{1.2}$$

where N_R is the flux of bodies with masses not inferior to m (in grams) per one year on the whole Earth. The interval of the masses $6.5 \cdot 10^5$ $g \ldots 1.0 \cdot 10^{13}$ g follows from the interval of observed energies of $3.2 \cdot 10^{-2}$ $kt \ldots 5.0 \cdot 10^5$ kt. Comparison of dependence (1.1) and the generalized formula, got by (Kruchinenko, 2002), with (2.1) shows that both the curves almost do not differ except by the value of inclination (or by a parameter, characteristic of the distribution of bodies in mass), but there is a considerable shift in the ordinate. We suppose, that this difference can be related to that of the energies of flashes (explosions) which take place deeply in the atmosphere at the heights of the maximal braking of H_* and authors consider as that of their entrance in the atmosphere.

We accept the following first approach: all the bodies of a given interval of mass attained the height of H_* and exploded. All the remaining mass of the meteoroids at the height of H_* went away in a flash. This possibility only occurs in the case of comet bodies. If speed at the entrance in the atmosphere was $20.3 km/s$, at explosion it will be equal approximately $12 km/s$, and the mass of the body will become equal to 0.78 m_0. Then the corrected interval of initial energies will be: $0.13\ kt. - 2.0 \cdot 10^6\ kt$, and the interval of corrected the initial masses: $2.6 \cdot 10^6\ g \ldots 4.1 \cdot 10^{13}\ g$. In this case we will get the integral distribution of the flux of bodies in masses as:

$$lgN_R^* = 7.685 - 0.90 lgm \qquad (1.3)$$

Comparing equation (1.3) with (1.1) we note that the shift on the ordinate decreased by 4 times and substantially approaches equation (1.2). Among observed phenomena there are those which initial masses exceed $10^9 g$. From the above calculations, we assert that all flashes of meteoroids given in (Brown *et al.* 2002) corrected the masses of which are in the interval of $10^9\ g \ldots 4.1 \cdot 10^{13}\ g$, are caused the comet bodies of low density and that fully split at heights H_*. In other words, all flashes energies which exceed $20kt$ are generated by comet bodies. According to this results, in this selection there are not more than two. At least, the bolide registered on July, 6 2002 (Fig. 1) above Mediterranean Sea and with an estimated energy equal to $26kt$, was generated by a cometary body. Accordingly, the density of the body would be of 1 g/cm^3 and having a high initial velocity ($\sim 30 km/s$).

2. Conclusion

Thus, we believe that the explosion of the meteoroid above Mediterranean, with high probability, was generated by a cometary body the initial mass of which was equal approximately to $7 \cdot 10^8\ g$. This follows from the energy of the flash, equal to $26kt$ TNT (Brown *et al.* 2002). We consider that this energy belongs to the height of the maximal breaking. At the velocity of $20.3 km/s$, accepted by the authors, the mass of body at this height was equal to $5 \cdot 10^8\ g$ and of approximately $7 \cdot 10^8\ g$ at its entrance into the atmosphere.

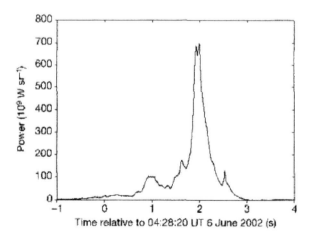

Figure 1. Optical curve of the bolide above the Mediterranean Sea, on 6 June 2002.

References

Brown, P., Spalding, R. E., ReVelle, D. O. *et al.* 2002, *Nature*, 420, 314

Kruchynenko, V. G. 2004, *Kinematics and physics of celest. bodies.*, 20(3), 269

McCrosky, R. E., Shao, C.- Y., & Posen, A. 1976, *Center for Astrophysics. Prep. Ser.*, 665, 13

McCrosky, R. E., Shao, C.- Y., & Posen, A. 1977, *Center for Astrophysics. Prep. Ser.*, 721, 61

Icy Bodies of the Solar System
Proceedings IAU Symposium No. 263, 2009
J.A. Fernández, D. Lazzaro, D. Prialnik & R. Schulz, eds.

© International Astronomical Union 2010
doi:10.1017/S1743921310001869

The solar cycle effect on the atmosphere as a scintillator for meteor observations

Asta Pellinen-Wannberg[1], Edmond Murad[2], Noah Brosch[3], Ingemar Häggström[4], and Timur Khayrov[5]

[1] Umeå University and Swedish Institute of Space Physics,
Box 812, SE-98128 Kiruna, Sweden
email: `asta@irf.se`

[2] Retired AFRL,
20 Kenrick Terrace, Newton, MA 02458 USA
email: `emurad@verizon.net`

[3] The Wise Observatory and Tel Aviv University,
Tel Aviv 69978, Israel
email: `noah@wise.tau.ac.il`

[4] EISCAT Headquarters,
Box 812, SE-98128 Kiruna, Sweden
email: `ingemar@eiscat.se`

[5] Luleå University of Technology and Julius Maximilians Universität Würzburg,
Box 812, SE-98128 Kiruna, Sweden
email: `timkha-6@student.ltu.se`

Abstract. We discuss using high solar cycle atmospheric conditions as sensors for observing meteors and their properties. High altitude meteor trails (HAMTs) have sometimes been observed with HPLA (High Power Large Aperture) radars. At other times they are not seen. In the absence of systematic studies on this topic, we surmise that the reason might be differing atmospheric conditions during the observations. At EISCAT HAMTs were observed in 1990 and 1991. Very high meteor trails were observed with Israeli L-band radars in 1998, 1999 and 2001. Through the Leonid activity, around the latest perihelion passage of comet Tempel-Tuttle, optical meteors as high as 200 km were reported. This was partly due to new and better observing methods. However, all the reported periods of high altitude meteors seem to correlate with solar cycle maximum. The enhanced atmospheric and ionospheric densities extend the meteoroid interaction range with the atmosphere along its path, offering a better possibility to distinguish differential ablation of the various meteoric constituents. This should be studied during the next solar maximum, due within a few years.

Keywords. Sun: activity, solar-terrestrial relations, atmospheric effects, meteors, meteoroids

1. Introduction

Until quite recently it was considered that meteoroid ablation starts below 130 km. At the time of the last apparition of comet 55P/Tempel-Tuttle, parent of the Leonids, however, many high altitude optical meteors were reported. This might be due to new and improved methods such as intensified TV observations, which have recorded meteors up to 160 km altitude (Fujiwara *et al.* 1998). Koten *et al.* (2006) reported 164 meteors with beginning heights above 130 km from Leonids in 1998–2001. The altitude 130 km is often given as the limit above which the dominating process of sputtering from the meteoroid surface due to the neutral and ionized atmospheric constituents switches over to ablation.

Table 1. High altitude meteor trails

Date	Time [UT]	Altitude [km]
1990-12-12	19 : 24 : 52	139
1990-12-12	19 : 24 : 52	154
1990-12-12	20 : 14 : 46	160
1990-12-13	01 : 11 : 17	160
1990-12-13	01 : 11 : 54	160
1990-12-13	02 : 13 : 42	147
1990-12-14	19 : 10 : 32	133
1990-12-14	22 : 00 : 24	111
1990-12-14	22 : 33 : 44	141
1990-12-14	23 : 19 : 10	223
1990-12-14	23 : 28 : 14	214
1990-12-15	00 : 01 : 54	138
1990-12-15	02 : 44 : 12	142
1990-12-15	02 : 54 : 10	134

It is known that the altitude at which meteor phenomena are observed depends on the characteristics of the meteoroid, for example mass, density, entry velocity, impact angle, structure and composition. In addition to the observation method, variations in the atmospheric and ionospheric altitude density profiles contribute to the altitude profiles of meteors. Thus the meteor beginning altitudes vary between high and low latitudes, time of day and season. The dominant effect might, however, be due to the 11-year solar cycle which decreases and increase the comparable densities of the Earth's atmosphere. An essential issue here is that the recent Leonid activity reached its maximum simultaneously with the latest solar cycle 23 around 1999-2002. This feature might contribute to the many reports on high altitude meteors. In this paper we discuss the impact of the solar cycle variations on the atmospheric density profiles in some radar and visual meteor observations.

2. Meteor altitudes

The first high altitude meteor trails (HAMTs) were recorded with the EISCAT UHF radar during the initiation of meteor observations at the facility in 1990 and 1991. Typically these events lasted a few seconds, but events lasting up to 14 seconds were also recorded. They were interpreted as trails drifting through the antenna beam (Pellinen-Wannberg and Wannberg, 1996). Table 1 summarizes the observations.

The first published HAMTs extended up to 400 km during the 1998 and 1999 Leonids (Brosch *et al.* 2001). The measurements were performed with the L-band Israeli phased array radar facility. In a later search through 2005-2006 EISCAT common program data no HAMTs were found (Khayrov 2008). Neither were they seen in a dedicated EISCAT TNA campaign by Brosch *et al.* (2009).

The common item with the EISCAT 1990 and 1991 measurements and the Israeli 1998 and 1999 observations is that they both occured during high solar activity, while the later reconstructions were performed close to the present exceptionally quiet solar minimum. At EISCAT there were no common programs optimized for such searches before 2005. The 1990 and 1991 observations were highly focused special programs, which were not run later when studying head echoes.

Head echo distributions also show altitude variations, though on a smaller scale. Westman (1997) proposed that there is a seasonal variation in meteor head echo altitude distributions, which also depends on the frequency. The concept of a "head echo height ceiling" was introduced for the altitude below which 90% of the head echoes occurred. The wintertime distributions from December 1990 and 1991 are 2-4 km higher up than the summertime one from August 1993. This feature was associated with the seasonal variation of the atmospheric density.

Sparks and Janches (2009) have published diurnal as well as seasonal variations in micrometeor (i.e. head echo) altitude distribution peaks from Arecibo at 18.3°N and PFISR at 65.1°N. The seasonal variation is small in the tropics, but the diurnal variation can be as large as 10 km. The high latitude observations show the opposite, the diurnal variation is smaller and the winter distributions are about six km higher than the summer ones. The authors correlated the results with the astronomical properties of the meteoric flux.

Khayrov (2008) compared the Westman (1997) results with EISCAT common program data from 2005 and 2006. These were 3–5 km lower down than the distributions from the early 90's, but the wintertime distribution was still 3-4 km higher up than the summer ones, as in the earlier data. The 1990/91 observations were performed during high and the 2005/06 observations during low solar activity. Thus these altitude differences might be related to the variations in atmospheric densities due to the solar cycle.

During the 2002 meteor shower a Leonid starting at 145 km was observed with the ALIS optical network operating CCD imagers in Northern Sweden (Pellinen-Wannberg *et al.* 2004). The signal was recorded in two imagers. One of them had a sodium filter allowing emissions at 589.3 ± 10.0 nm and the second one had a calcium filter 422.7 ± 14.0 nm also covering iron and Balmer hydrogen gamma emissions. The Leonid trail through the second filter was very strong and started at 145 km altitude, while the one in the usually stronger sodium line was much weaker and started at 110 km. Thus it was assumed that the strong emission came from the hyper-thermal collisions between a hydrogen compound in the meteoroid with the atmospheric constituents. This Leonid came from the 55P/Tempel-Tuttle 1767 trail and had traveled seven orbits after its ejection from the comet. Thus it was possible that it still contained ice at entry into Earth's atmosphere if a fast rotating grain with albedo 0.1 is assumed.

3. The expanding atmosphere

The enhanced solar activity causes the atmosphere, as well as the ionosphere, to expand to higher altitudes. This is known in connection with the enhanced drag on satellites, and accelerated orbital decay. The atmospheric density profiles can be estimated from

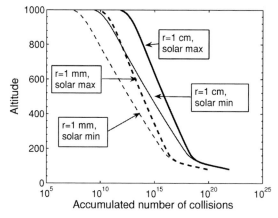

Figure 1. The accumulated number of collisions of a 1 cm (solid line) and a 1 mm (dashed) radius meteoroid with the atmospheric constituents during solar maximum (bold) and minimum conditions at 60°N and 20°E

the MSIS model (Hedin 1991) for given latitude and time, which includes the time of day, season and solar cycle. For example, at 200 km during the solar cycle maximum the atomic O density is on average as high as the density about 20 km lower down during the minimum. At 150 km the difference is about 5 km. At 400 km this altitude difference is close to 100 km. For the ionospheric part of the atmosphere the situation is more complicated and the altitude differences can be even larger.

For studying the meteoroids' interaction with the atmosphere, another parameter might be more relevant. The meteoroids' accumulated collisions with the atmospheric constituents is shown in Fig. 1. The number of collisions of a 1 cm and a 1 mm size meteoroid is the same at 185 km at solar maximum as it is at 150 km at the minimum; at 130 km the difference is only about 5 km. At 400 km the difference is again about 100 km. Since the meteoroid temperature is related to the start of the observable meteor process and the body is heated up due to every collision, comparable meteors can flame up earlier during solar maximum than minimum conditions.

4. Summary

The atmospheric density has an impact on both meteor trail and head echo beginning heights. The extreme variations of atmospheric densities differ from a few kilometers to 100 km for solar min and max conditions. This might contribute to the explanation of appearance of the HAMTs in 1990 and 1991 at EISCAT and many of the high altitude Leonids around the latest perihelion passage during solar cycle maximum. The prevalent atmospheric density profile of the time and location should be related to the micrometeor altitude distributions as well as trail echo observations. The condition of the expanded atmosphere during the coming solar cycle maximum might systematically be used as a "scintillator" for detecting various features of impacting meteors within an essentially longer path than during solar minimum. Appearance of various type of constituents, which under low activity disintegrate together below 130 km altitude, can occur differentially at higher altitudes in solar maximum. Even volatiles locked in the meteoroid structures can be released at higher altitudes, which might explain the Leonid observed in 2002 (Pellinen-Wannberg *et al.* 2004) and other HAMTs during the past Leonids. A in-depth study has been initiated to investigate the solar cycle effect on meteor properties including chemical reactions between the meteoroid and atmospheric constituents.

References

Brosch, N., Schijvarg, N., Podolak, M., & Rosenkrantz, M. R. 2001, in: B. Warmbein (ed.), *Proceedings of the Meteoroids 2001 Conference* (ESA Publications Division), pp. 165–173

Brosch, N., Häggström, I., Pellinen-Wannberg, A., & Westman, A. 2009, *MNRAS*, in press

Fujiwara, Y., Ueda, M., Shiba, Y., Sugimoto, M., Kinoshita, M, & Shimoda, C. 1998, *J. Geophys. Res.*, 25, 285

Hedin, A. E. 1991, *J. Geophys. Res.*, 96, 1159

Khayrov, T. 2008, MSc Thesis *LTU*, ISSN 1653-0187, ISNB LTU-PB-EX–09/016–SE

Koten, P., Spurný, P., Borovička, J. Evans, S., Elliott, A., Betlem, H., Štork, R., & Jobse, K. 2006, *Meteorit. Planet. Sci.*, 40, 1635

Pellinen-Wannberg, A., & Wannberg, G. 1996, *J. Atmos. Terr. Phys.*, 58, 495

Pellinen-Wannberg, A., Murad, E., Gustavsson, B., Brändström, U., Enell, C.-F., Roth, C., Williams, I. P., & Steen, Å. 2004, *Geophys. Res. Lett.*, 31, doi: 10.1029/2003GL018785

Sparks, J. J., & Janches, D. 2009, *Geophys. Res. Lett.*, 36, doi: 10.1029/2009GL038485

Westman, A. 1997, PhD Thesis *IRF Scientific Report*, ISSN 0284-1703, ISNB 91-7191-351-3

Icy Bodies of the Solar System
Proceedings IAU Symposium No. 263, 2009   © International Astronomical Union 2010
J.A. Fernández, D. Lazzaro, D. Prialnik & R. Schulz, eds.   doi:10.1017/S1743921310001870

Study of meteoroid stream identification methods

Regina Rudawska[1] and Tadeusz J. Jopek[2]

Astronomical Observatory of Adam Mickiewicz University, Poznan, Poland
[1]email: reginka@amu.edu.pl [2]email: jopek@amu.edu.pl

Abstract. We have tested the reliability of various meteoroid streams identification methods. We used a numerically generated set of meteoroid orbits (a stream component and a sporadic background) that were searched for streams using several methods.

Keywords. meteoroids, data analysis

1. Introduction

The meteoroid stream identification methods are based on three components: dynamical similarity (distance) function, similarity threshold D_c and cluster analysis algorithm – which define the stream itself. Several similarity functions have been proposed: D_{SH} the orbital D-criterion introduced by Southworth and Hawkins (1963), its modifications by Drummond (1981), Jopek (1993), Valsecchi *et al.* (1999), and Jenniskens (2008) and the D_V function defined in the domain of the vectorial heliocentric orbital elements (Jopek *et al.* (2008)). The threshold, D_c is used to test the similarity among two orbits $\mathbf{O}_i, \mathbf{O}_j$ – these two orbits are associated if $D(\mathbf{O}_i, \mathbf{O}_j) < D_c$. Having the distance function and the similarity threshold a meteoroid stream can be detected by a cluster analysis algorithm, one can use e.g.: an iterative methods proposed in (Sekanina (1976), Welch (2001)), a single neighbour linking technique (Southworth and Hawkins (1963), Lindblad (1971)), method of indices (Svoreň *et al.* (2000)) or the wavelet transform technique (Galligan and Baggaley (2002), Brown *et al.* (2008)).

2. Meteoroid data sample preparation

We searched for streams among the numerically generated orbits. For selected NEOs (see table 1), fifteen sets of genetically associated particles were generated; the motion of all stream particles was integrated numerically over 40KA with the Newtonian force model of the Planetary System. At each of six intermediate epochs (see table 2) the stream component was completed by a set of sporadic orbits generated with the distribution of orbital elements that were statistically similar to the distribution of the background meteors taken from the IAU MDC photographic catalogue, Lindblad *et al.* (2003).

3. Searching methods tested in this study

The orbital samples were searched by the following methods:
- *W1DSH* – a simplified version of Welch's method (Welch (2001)) equipped with the D_{SH} function. In this method, the density at mean orbit of a stream $\mathbf{O}_A$ in orbital

Parent body	q	a	e	i	Ω	ω	Q	P	Associated meteoroid stream
1998 SH$_2$	0.743	2.686	0.723	2.5	13.1	261.2	4.629	4.40	α Virginids
2004 BZ$_{74}$	0.330	3.048	0.892	16.6	233.9	121.3	5.767	5.32	α Scorpiids
2004 HW	0.976	2.688	0.637	0.8	220.4	62.3	4.401	4.41	Corvids
2004 TG$_{10}$	0.315	2.242	0.859	3.7	212.3	310.0	4.169	3.36	N. Taurids, Dayt. β Taurids
2005 NZ$_6$	0.248	1.834	0.865	8.5	39.7	48.0	3.419	2.48	Dayt. April Piscids
3200 Phaethon	0.140	1.271	0.890	22.2	265.4	322.0	2.403	1.43	Geminids
1P/Halley	0.587	17.942	0.967	162.2	58.9	111.9	35.296	76.00	Orionids, η Aquariids
2P/Encke	0.331	2.209	0.850	11.9	334.7	186.2	4.087	3.28	S. Taurids, Dayt. ζ Perseids
7P/Pons-Winnecke	1.256	3.435	0.634	22.3	93.4	172.3	5.615	6.37	June Bootids
8P/Tuttle	0.998	5.672	0.824	54.7	270.5	206.7	10.346	13.50	Ursids
21P/Giacobinni-Zimmer	1.034	3.522	0.706	31.8	195.4	172.5	6.010	6.61	Draconids
26P/Grigg-Skjellerup	0.997	2.965	0.664	21.1	213.3	359.3	4.933	5.11	π Puppids
55P/Temple-Tuttle	0.977	10.337	0.905	162.5	235.3	172.5	19.698	3.28	Leonids
73P/Schwassmann-Wachmann3	0.941	3.063	0.693	11.4	69.9	198.8	5.186	5.36	τ Herculids
109P/Swift-Tuttle	0.958	26.317	0.963	113.4	139.4	153.0	51.675	135.00	Perseids

Table 1. Orbital elements of 15 NEOs for which the responding, artificial streams have been generated. The orbits were gathered from Marsden and Williams (2003), NeoDys (2007).

element space, operating on a set of orbits $\mathbf{O}_i, i = 1, ..., N$, was given by

$$\rho(\mathbf{O}_A) = \sum_{i=1}^{N} \left(1 - \frac{D^2(\mathbf{O}_i, \mathbf{O}_A)}{D_c^2}\right) \tag{3.1}$$

- $W2DSH$ - a simplified Welch's method with D_{SH} function, however the density at mean orbit of a stream was calculated from

$$\rho(\mathbf{O}_A) = \sum_{i=1}^{N} \left(1 - \frac{D(\mathbf{O}_i, \mathbf{O}_A)}{D_c}\right)^2 \tag{3.2}$$

- $W1DV$ - similarly to $W1DSH$, but with D_V function as described in Jopek *et al.* (2008),
- $W2DV$ - similarly to $W2DSH$, but with D_V function,
- MI - method of indices (Svoreň *et al.* (2000)),
- $SLDSH$ - single neighbour linking technique with D_{SH} function,
- $SLDV$ - as above, but with D_V function.

All methods, except for the last three, were applied with the values of D_c corresponding to their largest reliability. In case of $SLDSH$ and $SLDV$ methods, the constant threshold $D_c = 0.02$ and $D_c = 0.01 \cdot 10^{-2}$ were adopted. MI method was used in the form described by Svoreň *et al.* (2000).

To evaluate the reliability level of the result obtained for a given stream with the applied method, we introduced two parameters, S_1 and S_2, defined as

$$S_1 = \frac{N_p}{N_{max}} \cdot 100\%, \qquad S_2 = \frac{N_i}{N_p + N_i} \cdot 100\%$$

where N_p – a number of correctly identified members of a stream, N_{max} – a total amount of particles in the stream, and N_i – a number of interlopers i.e. sporadic meteoroids and meteoroids belonging to another streams.

In addition, for all streams identified at the same epoch, a general reliability parameter was evaluated with

$$SS_1 = \sum_{k=1}^{N} N_{kp} \cdot \left[\sum_{k=1}^{N} N_{k\,max}\right]^{-1} \cdot 100\%$$

where $k = 1, \ldots, N = 15$, denotes all the identified streams.

4. Results

At each epoch, for each stream, the values of the above parameters have been calculated. The results of the stream identification were accepted as reliable, only if $S_2 < 10\%$. Next, a ranking of all reliable results was accomplished, and the results obtained are given in table 2. We can see that the results with highest reliability were most often obtained using D_V function and the Welch iterative or single linkage cluster analysis algorithm. When the same algorithms were used with D_{SH} function, the results were considerably worse.

The ranking, in the way it was carried out, informs us only which of the methods was better. To illustrate the relative differences between the results obtained with various methods we need another measure. For this purpose we used a general reliability level SS_1. Close to the starting epoch SS_1 was above 90% for all the methods (Fig. 1), while as the stream dispersion proceeded in time the reliability of methods decreased. For $W1DV$, $W2DV$ and $SLDV$ methods decrease was approximately linear, while for $W1DSH$, $W2DSH$, MI and $SLDSH$ methods the decrease of reliability was faster, with distinct fluctuation. In the first group the most reliable results were obtained more often with $W1DV$ method (from 100% to 60%), and the reliability of $W2DV$ and $SLDV$ was equivalent. At the beginning, $SLDV$ was slightly more effective, while the $W2DV$ method gained the advantage in the later epochs. In the case of the second group of methods, their initial high reliability distinctly decreased with time, and finally reached values below 50–40%. The lowest reliability, about 20%, was obtained by $W2DSH$ method.

Table 2. Final score of the ranking of the meteoroid stream identification. In each column we see how many times a given method achieved the best result, e.g. at starting epoch, using $W1DV$ method three streams have been identified with the highest reliability level. Sometimes a few methods achieved exactly the same reliability for a given stream. In such cases each method scored one point.

Epoch\Method	W1DSH	W2DSH	W1DV	W2DV	MI	SLDSH	SLDV
00000	2	2	3	3	10	8	4
01200	2	1	1	2	3	1	8
07200	0	0	7	3	2	0	5
15200	0	0	10	3	0	0	3
22000	0	0	2	5	2	0	4
30000	0	0	10	6	0	0	0
Total	4	3	33	22	17	9	24

5. Conclusion

Our survey was the first step in the assessment of reliability of meteoroid stream identification methods. The obtained results let us state that identification methods based on D_V function clearly distinguish themselves from others. Cluster analysis algorithms: simplified Welch's algorithm and single linking technique with D_V function, most often appeared to be the most effective, whereas methods with D_{SH} criterion, i.e $W1DSH$, $W2DSH$ and $SLDSH$ were less effective.

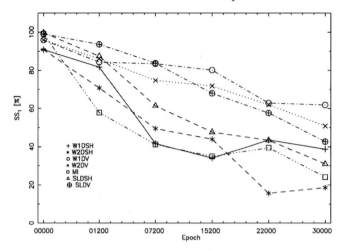

Figure 1. The general reliability parameter SS_1 obtained for all identified streams. The value of SS_1 was calculated by formulae 3 for each epoch.

Acknowledgements

The work was supported by the KBN Project N N203 2006 33. The majority of calculations were done at Poznań Supercomputing and Networking Centre. The authors wish to thank Dr. Anna Marciniak for improving English in the manuscript.

References

Brown, P., Weryk, R. J., Wong, D. K., & Jones, J. 2008, *Icarus*, 195, 317

Drummond, J. D. 1981, *Icarus*, 45, 545

Galligan, D. P. & Baggaley, W. J. 2002, *IAU Colloq.181: Dust in the Solar System and Other Planetary Systems*, 42

Jenniskens, P. 2008, *Icarus*, 194, 13

Jopek, T. J. 1993, *Icarus*, 106, 603

Jopek, T. J., Rudawska, R., & Pretka-Ziomek, H. 2006, *MNRAS*, 371, 1367

Jopek, T. J., Rudawska, R., & Bartczak, P. 2008, *Earth, Moon and Planets*, 102, 73

Lindblad, B. A., Neslušan, L., Porubčan, V., & Svoreň, J. 2003, `http://www.ta3.sk/ne/IAUMDC/Ph2003/`

Lindblad, B. A. 1971, *Smithson. Contr. Astrophys*, 12, 1

Marsden, B. G. & Wiliams, G. V. (2003) *Catalog of Cometary Orbits 2003*, 15th edn, Smithsonian Astrophysical Observatory, Cambridge,MA

NEO Dynamic Site: 2007, `http://newton.dm.unipi.it/neodys/`

Southworth, R. B. & Hawkins, G. S. 1963, *Smithson. Contr. Astrophys*, 7, 261

Sekanina, Z. 1976, *Icarus*, 27, 265

Svoreň, J., Neslušan, L., & Porubčan,V. 2000, *Planet. Space Sci.*, 48, 933

Valsecchi, G. B., Jopek, T. J., & Froeschle, C l. 1999, *MNRAS*, 304, 743

Welch, P. G. 2001, *MNRAS*, 328, 101

Part VII

Physical Processes in Comets

Part VII

Physical Processes in Comets

Icy Bodies of the Solar System
Proceedings IAU Symposium No. 263, 2009
J.A. Fernández, D. Lazzaro, D. Prialnik & R. Schulz, eds.

© International Astronomical Union 2010
doi:10.1017/S1743921310001882

Recent polarimetric observations of comet 67 P/Churyumov-Gerasimenko

Anny-Chantal Levasseur-Regourd[1], Edith Hadamcik[1], Asoke K. Sen[2], Ranjan Gupta[3], and Jeremie Lasue[4,5]

[1] UPMC Univ. Paris 06 / LATMOS-IPSL, BP 3, 91371 Verrieres, France
email: aclr@latmos.ipsl.fr
email: edith.hadamcik@latmos.ipsl.fr

[2] Assam Univ., Silchar 788001, India
email: asokesen@yahoo.com

[3] IUCAA, Post Bag 4, Ganeshkhind, Pune-411007, India
email: rag@iucaa.ernet.in

[4] LPI, 3600 Bay Area Blvd, Houston, TX 77058-1113 USA
[5] LANL, Space Science and Applications, ISR-1, Mail Stop D-466, Los Alamos, NM 87545 USA
email: lasue@lpi.usra.edu

Abstract. Remote observations of solar light scattered by dust in comet 67P/Churyumov-Gerasimenko coma are of major importance to assess the properties of the dust and thus to prepare the rendezvous of the Rosetta spacecraft with comet 67P/Churyumov-Gerasimenko. We present polarimetric data obtained from India in December 2008 and France in March 2009. Compared with previous observations of this comet and of other Jupiter family comets, they confirm that it is dust-poor, although it may exhibit outbursts leading to the ejection of dust particles from its subsurface, especially after its perihelion passage.

Keywords. Techniques, polarimetric, comets, individual (67P/Churyumov-Gerasimenko)

1. Introduction

Evaporation of ices from cometary nuclei releases gases that carry out solid particles of dust. Together, they contribute to the bright coma, visible when a comet is not too far from the Sun on its elongated orbit. Understanding the properties of such dust particles is of major interest, not only to assess the diversity of comets and understand the processes that allowed their formation in the early solar system, but also to better describe the environment of the nucleus. This latter issue is of major importance to prepare the ESA Rosetta mission that will rendezvous in 2014-2015 with comet 67P/Churyumov-Gerasimenko, observe the surface of the nucleus, deploy a lander onto the nucleus and follow the evolution of the coma (Glassmeier *et al.* 2007).

Remote observations of solar light scattered by dust, and more specifically measurements of its partial linear polarization, provide unique opportunities to estimate some of its properties, i.e. size distribution, morphology and complex refractive indices (see e.g. Levasseur-Regourd 1999, Kolokolova *et al.* 2004). Intensity and polarization images are derived from images of four polarized components with fast axis at 45° from one another, thereafter called Z_0, Z_{45}, Z_{90} and Z_{135}. Properties are evaluated through a comparison of the dependence of the polarization upon phase angle and wavelength of the observations, location within the coma and temporal evolution, with results from light scattering experimental and numerical models (e.g. Hadamcik & Levasseur-Regourd 2003; Hadamcik *et al.* 2007a, Levasseur-Regourd *et al.* 2008, Lasue *et al.* 2009).

Table 1. Log of the observations, including date, number of days before or after perihelion, geocentric distance Δ, heliocentric distance R, predicted visual magnitude, phase angle α, Sun position angle and code of the filters.

Date	Days to perihelion	Δ (AU)	R (AU)	m_v (predicted)	α (°)	Sun PA (°)	Filter (code)
2008, India							
25 Dec.	-65	1.67	1.45	14.2	35.8	70.5	CR
26 Dec.	-64	1.67	1.46	14.2	35.8	70.4	CR
27 Dec.	-63	1.67	1.46	14.2	35.8	70.3	CR & CB
2009, France							
17 March	+17	1.72	1.26	13.5	34.7	69.9	R
18 March	+18	1.73	1.27	13.5	34.7	70.0	R
19 March	+19	1.73	1.27	13.5	34.7	70.2	R

A joint effort between French and Indian teams, already familiar with polarimetric measurements, has been made in 2008-2009, to provide remote observations of the dust coma of 67P/Churyumov-Gerasimenko. It was indeed the last apparition of this comet, the period of which equals 6.57 years, before Rosetta rendezvous.

2. Instrumental conditions

In December 2008, observations were performed from IUCAA observatory at Girawali, near Pune, India, with a 2 m Cassegrain telescope and an Imaging Polarimeter (achromatic rotating half-wave plate; Wollaston prism as analyser; 2x2 binning for present observations; see e.g. Ramprakash *et al.* 1998). In March 2009, soon after perihelion, observations were performed from OHP observatory near Marseille, France with a 0.8 m Cassegrain telescope (4 polaroid filters on a rotating wheel; resolution 0.21 arcsec, in the present observations a 4x4 binning for present observations; see e.g. Hadamcik *et al.* 2007b). To minimize the contribution of cometary gaseous emissions, observations were performed, as previously in France, with a special red filter (R for 650 nm ± 45 nm). In India, ESA narrowband filters were used in the blue (CB for 443 nm ± 2 nm) and red (CR for 684 nm ± 4.5 nm) domains. Table 1 summarizes the conditions of the observations.

3. Results and discussion

Intensity. Images are obtained by adding two polarized images corresponding respectively to ($Z_0 + Z_{90}$) and to ($Z_{45} + Z_{135}$), once the sky background is subtracted and the images are centered on the photometric center. The validity of the results is assessed by ensuring that ($Z_0 + Z_{90}$) and ($Z_{45} + Z_{135}$) be similar. As noticed in Fig. 1, the coma was elongated in the antisolar direction, with possibly a jet-like feature towards the North in late December 2008; bright structures were present in the solar and antisolar directions, with a fan-type structure suspected towards the South in mid-March 2009. A radial decrease in intensity is computed in order to estimate the deviation from an isotropic coma. As expected, its slope in log-log scale is equal to -1 on average. However, it is found to be steeper for photometric center distances larger than 2000 km in December and larger than 4000 km in March.

Polarization. Values are obtained by combining four polarized intensities, with the linear polarization (in percent) equal to 200 $[(Z_0 + Z_{90})^2 + (Z_{45} + Z_{135})^2]^{0.5}$ / ($Z_0 + Z_{90} + Z_{45} + Z_{135}$). Although the phase angles are similar in December and in March (of the order of 35°), the polarization appears to be about 1% higher in March than in December, while the error bars remain below 0.4 %.

Discussion. The intensity was higher than expected from ephemeris, by about 2 magnitudes, in mid-March 2009; this result agrees with Yoshida observations (2009). Intensity and polarization observations had already been obtained during previous apparitions, as reviewed by Levasseur-Regourd *et al.* (2004). Various authors had reported post-perihelion increases in intensity (see e.g. Marsden 1986; Kidger 2003; Ferrin 2007). As pointed out by Kidger (2003), this phenomenon seems typical of comets that have recently suffered a drop in perihelion distance and could originate in post-perihelion outbursts. Variability of the mean radial profile, together with a large pre/post-perihelion asymmetry, was also reported by Schleicher (2006) for the apparitions 1982-1983 and 1995-1996 apparitions.

Finally, a few polarimetric data had been obtained during the 1982-1983 apparition in the 12° to 38° phase angle range. (Myers & Norsieck 1984, Chernova *et al.* 1993). Although the phase angles were below 40°, the results, together with ours, suggest that 67P/Churyumov-Gerasimenko belongs to the low polarization class of comets. This is in agreement with the absence of any silicate emission feature in the near infrared (Hanner *et al.* 1995) and with the fact that Jupiter family comets are typically dust-poor comets (Hadamcik and Levasseur-Regourd 2009). The slight increase in polarization detected in March 2009, at a time when a sudden increase in intensity took place, fairly agrees with the fact that outbursts seem to be correlated with increases in polarization, at least for phase angles above about 35°, as typically noticed for 47P/Ashbrook-Jackson or C/1996 B2 Hyakutake (Renard *et al.* 1996, Desvoivres *et al.* 2000, Tozzi *et al.* 1997). More detailed results will be presented in Hadamcik *et al.* (in preparation).

4. Conclusions

Light scattering observations of 67P/Churyumov-Gerasimenko have been obtained in late December 2008 and mid-March 2009, during the last apparition of this comet before its encounter with the Rosetta spacecraft. A significant increase in intensity, possibly induced by a significant outburst activity, is noticed after perihelion, as for previous apparitions; the pre/post perihelion asymmetry leads us to emphasize the fact that it would be of great interest to tentatively extend the Rosetta mission after the perihelion passage. Our observations confirm that the comet behaves like most Jupiter family comet (i.e. is rather dust-poor) with relatively large dust particles in its coma.

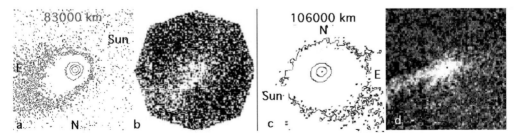

Figure 1. Typical intensity images of the dust coma of comet 67P/Churyumov-Gerasimenko, as obtained from IUCAA observatory on 27 December 2008 (average time 14:38 UT) and from OHP observatory on 18 March 2009 (average time 19:06 UT). Images a and c are isophotes in Log intensity scale. Images b and d (emphasized by rotational gradient technique) allow the detection of jet-like and fan-type structures.

The outburst, suspected in intensity and polarization data in March 2009, suggests that the coma may exhibit some drastic changes, with possible ejection of smaller grains from the subsurface. Last but not least, it is important to keep in mind that the resolution obtained for remote line-of-sight observations does allow the detection of any local variations in polarization in the innermost coma. The detection of changes in the physical properties of the dust particles at nucleus distances smaller than about 3000 km, e.g. with some dust particles consisting of large icy conglomerates, will only be possible from the Rosetta spacecraft after its rendezvous with the comet.

Acknowledgements

We gratefully acknowledge ESA for providing narrowband cometary filters for India. We acknowledge PNP for the funding of the OHP observations and IUCAA, Pune India for making telescope time available.

References

Chernova, E. P., Kiselev, N. N., & Jockers, K. 1993, *Icarus*, 103, 144

Desvoivres, E., Klinger, J., & Levasseur-Regourd, A. C. 2000, *Icarus*, 144, 172

Ferrin I. 2007, *Icarus*, 191, 22

Glassmeier, K. H., Boehnhardt, H., Koschny, D., Kuhrt, E., & Richter, I. 2007, *Space Sci. Revs*, 128, 1

Hadamcik, E. & Levasseur-Regourd, A. C. 2003, *J. Quant. Spectros. Radiat. Transfer*, 79, 661

Hadamcik, E., Renard, J.-B., Rietmeijer, F. J. M., & Levasseur-Regourd, A. C. *et al.* 2007a, *Icarus*, 190, 660

Hadamcik, E., Levasseur-Regourd, A. C., Leroi, V., & Bardin, D. 2007b, *Icarus*, 190, 459

Hadamcik, E. & Levasseur-Regourd, A. C. 2009, *Planet. Space Sci.*, 57, 1118

Hanner, M. S., Tedesco, M. S., & Tokunaga, A. T., *et al.* 1985, *Icarus*, 64, 11

Kidger, M. R. 2003, *A&A*, 408, 767

Kolokolova, L., Hanner, M. S., Levasseur-Regourd, A. C., & Gustafson, B. 2004, in: M. Festou, H. U. Keller & H. A. Weaver (eds.), *Comets II* (Tucson: Univ. Arizona Press), p. 577

Lasue, J., Levasseur-Regourd, A. C., Hadamcik, E., & Alcouffe, G. 2009, *Icarus*, 199, 129

Levasseur-Regourd, A. C. 1999, *Space Sci. Revs*, 90, 163

Levasseur-Regourd, A. C., Hadamcik, E., Lasue, J, Renard, J. B., & Worms, J. C. 2004, in: L. Colangeli, E. Mazzotta, E. Epifani, & P. Palumbo (eds.), *The new Rosetta targets* (Dordrecht: Kluwer), p. 111

Levasseur-Regourd, A. C., Zolensky, M., & Lasue, J. 2008, *Planet. Space Sci.*, 56, 1719

Myers, R. V. & Nordsiek, K. H. 1984, *Icarus* , 58, 431

Marsden, B. 1983, *QJRAS* , 27, 102

Ramprakash, A. N., Gupta, R., Sen, A. K., & Tandon, S. N. 1998, *A&AS*, 128, 369

Renard, J. B., Hadamcik, E., & Levasseur-Regourd, A. C. 1996, *A&A*, 316, 263

Tozzi, G. P., Cimatti, A., di Serego Alighieri, S., & Cellino, A. 1997, *Planet. Space Sci.*, 45, 535

Schleicher, D. G. 2006, *Icarus* , 181, 442

Sen, A. K., Deshpande, M. R., Joshi, U. C., Rao, N. K., & Raveendran, A. V. 1991, *A&A*, 242, 496

Yoshida, S. 2009, see magnitude data in www.aerith.net/index.html

Icy Bodies of the Solar System
Proceedings IAU Symposium No. 263, 2009
J.A. Fernández, D. Lazzaro, D. Prialnik & R. Schulz, eds.
© International Astronomical Union 2010
doi:10.1017/S1743921310001894

Secular light curves of comets

Ignacio Ferrín

Center for Fundamental Physics, University of the Andes,
Merida, Venezuela
email: ferrin@ula.ve

Abstract. We present a resume of the fundamental ideas developed in the *Atlas of Secular Light Curves of Comets* (Ferrín, 2009).

Keywords. comets: general, 2P/Encke

1. Introduction

Recently I made some advances on the subject of the secular light curves of comets (SLCs) (Ferrín, 2005a, 2005b, 2006, 2007, 2008, 2009, from now on, papers I to VI). In particular the *Atlas of Secular Light Curves of Comets (Paper VI)* will appear in Planetary and Space Science, and is also available in the Astro-ph pre-print server (http://arxiv.org/ftp/arxiv/papers/0909/0909.3498.pdf). The *Atlas* contains 27 comets, 54 SLCs, and 70 plots. We are going to resume some of the knowledge derived in the Atlas.

2. Secular Light Curves

The magnitude at Δ, R,α, is denoted by m1(Δ ,R) for visual observations and V(Δ,R,α) for instrumental magnitudes, where Δ is the comet-Earth distance, R is the Sun-comet distance, and α is the phase angle (Earth-comet-Sun). The brightness of the comet is presented in two plots, the Log R plot and the time plot. The reason to select the two plots is because they give independent and different parameters. The Log R plot may be reflected at R= 1 AU or may be reflected at q, the perihelion distance (see Figure 1).

The 'reflected double-Log R' plot presents the reduced magnitude vs Log R reflected at R = 1 AU. Reduced means reduced to Δ = 1 AU, m(1,R)= m(Δ,R) −5 LogΔ . In this plot time runs horizontally from left to right, although non linearly. Negative Logs indicate observations pre-perihelion. *The value of the reflected double-Log R diagram is that power laws on R (R^{+n}) plot as straight lines.* The slope 5 line at the bottom of the plot in the form of a pyramid is due to the atmosphereless nucleus (Figure 1). Additionally the inclusion of the R= 1 AU line allows the determination of the absolute magnitude by extrapolation (actually by interpolation of the pre and post-perihelion intervals).

The 'time plot' (Figure 2) presents the reduced magnitude vs time to perihelion. This is the most basic and simple plot. Negative times are pre-perihelion. The advantage of this plot is that time runs horizontally linearly from left to right thus showing the time history of the object.

3. Envelope

A visual observation of a comet is affected by several effects, all of which decrease the perceived brightness of the object: moon light, twilight, cirrus clouds, dirty optics (dust on the mirror), dirty atmosphere (pollution), low altitude (haze), excess magnification, the Delta-effect, etc. *Thus it is a fundamental premise of The Atlas that the envelope of the observations defines the secular light curve, since there is no known physical effect that could increase+ the perceived brightness of a comet measured by two different observers, at the same instant of time.*

4. Parameters Measured From the Plots

The Log R and the time plots provide a wealth of new information. There are ~ 40 parameters listed, of which ~ 20 are new and measured from the plots (Paper I).

Log R Plots

The symbol legend for the log R plot can be found on the time plot (see Figures 1 and 2). The title of each plot identifies the comet in the new and old nomenclature system

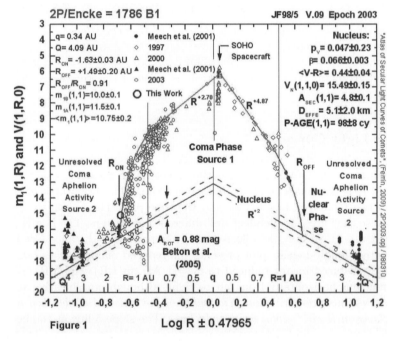

Figure 1. Secular light curve of comet 2P/Encke, Log plot. The symbols correspond to different apparitions identified in the time plot. At left and right are listed some orbital information and the parameters derived from this plot. The negative sign before logs and Rs, is a label, not a mathematical sign, and indicates only observations pre-perihelion. Since this comet has q $< 1AU$, Log $0.34 = 0.47965$ has been subtracted before perihelion and added after perihelion to the horizontal axis to make room for the observations inside the Earth's orbit. There are three phases of the secular light curve, the nuclear phase, the coma phase and the nuclear phase again. On log plots power laws plot as straight lines, therefore the nucleus makes a pyramidal line at the bottom, since it follows a R^{+2} law. Most nuclear magnitudes lie inside the amplitude of the rotational light curve range (A_{ROT}). The turn-on and turn-off points are very sudden affairs, so it is easy to decide what is a nuclear magnitude and what observations are coma contaminated. The coma reaches a pointed sharp maximum after perihelion, and turns off with a steep descent. The secular light curve is very asymmetric with $R_{OFF}/R_{ON} = 0.91$.

to allow going back to historical references. The first designation of the comet gives the first apparition (the discovery year). The label at the top right indicates if the comet belongs to the Jupiter family (JF), to the Saturn family (SF), to the Oort Cloud (OC), to the Asteroidal Belt (ABC), to the Encke type (ET), if it is a Centaur (CEN) or if it belongs to the Halley type (HT). This classification follows closely the one given by Levison (1996), except that Chiron type and Centaur are synonymous, ABC is included in his ecliptic comets and finally he did not consider the existence of Saturn family comets (28P, 85P). The two numbers following the family are the photometric age, P-AGE (1,1) later, measured in units of comet years (cy) and the diameter measured in km. The reason for using P-AGE as a label here is that the definition of P-AGE is robust, as is demonstrated in Paper I.

When reading these numbers it is important to keep in mind that $\sim 95\%$ of comets have $0 < P - AGE < 100$ cy and $0 < DEFFE < 10$ km. Any object above 100/10 or near 0/0 is exceptional (very old, very large, very young, very small).

Next is the Version of the plot. All plots are identified as Version V.09 (the year of data completion). The upper left hand side of each plot gives the perihelion distance, q, the aphelion distance for that epoch, Q and Log Q to identify the extent of the plot. Q and q can be located at the bottom of the plots and come in units of AU. At the extreme lower right of the plots, the date line tilted 90^0 is the last update of observations.

The Epoch label identifies the apparition that has contributed most significantly to the definition of the envelope. *The importance of this label is that each apparition of the*

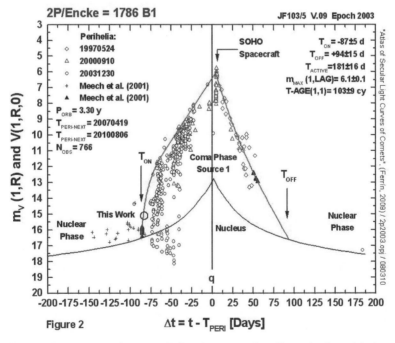

Figure 2. Secular light curve of comet 2P/Encke, time plot. P_{ORB} is the orbital period of the comet around the sun, and the perihelia selected are indicated. The turn-off of activity is a sudden event. The comet was active for 181 days. The time-age is $T - AGE = 103 \pm 9$ comet years, indicating that this is a methuselah comet. The dates are in the format YYYYMMDD.

comet is a frame of a movie. The next apparition will have a new Epoch label. After many apparitions it will be possible to create a movie of the photometric history of the comet. The *Atlas* (Paper VI) contains the first frames of those movies.

Due to space limitations, we are going to describe only four parameters in the Log R plot, the turn on point, R_{ON}, the turn off point, R_{OFF}, the amplitude of the secular light curve, A_{SEC}, and the photometric age, P-AGE (Figure 1).

1. R_{ON} [AU]. The turn-on distance of the coma. The negative sign in this parameter in R_{ON} and in Log R in the plots indicates values before perihelion but it does not mean that the log is negative. Physically R_{ON} corresponds to the onset of *steady* activity. It is the interception of the nuclear line and the coma envelope. Browsing the secular light curves of the Atlas, it can be seen that the turn-on and turn-off points are very sudden affairs. When there are enough data points these parameters can be measured easily and accurately because of the sharp change of slope. Usually R_{ON} takes place before perihelion, but for comets 107P and 133P it takes place *after* perihelion.

2. R_{OFF} [AU] The turn-off distance of the coma, usually larger than R_{ON}. This is the interception of the coma envelope and the nuclear line post-perihelion. R_{OFF} takes place *after* perihelion.

3. A_{SEC} (1,1) = V_N (1,1,0) $-m1(1,1)$ = amplitude of the secular light curve above the nuclear magnitude measured at 1 AU from the Earth, and 1 AU from the sun. $A_{SEC}(1,1)$ is a measure of the activity of the nucleus and thus of age. The value is calculated at (1,1) to allow comparison to different comets. Do not confuse with A_{ROT} the amplitude of the rotational *light* curve. For comets with q < 1 AU, there are two values, A_{SEC} (1,-1) and $A_{SEC}(1,+1)$. The listed value is the mean of the pre and post-values.

4. P-AGE (1,1) = Photometric Age measured at 1 AU from the Earth, 1 AU from the sun. It is an objective of the *Atlas* to be able to define a parameter that measures the age of a comet solely from the secular light curves. Although it is not possible in most cases to assign an absolute physical age (exception 2P/Encke), it is nevertheless possible to define a parameter related to activity that ranks the comets by age. We call it P-AGE to distinguish it from a real age. It should be emphasized that P-AGE is related to the loss of volatiles as a proxy for age. The ability to order comets according to their relative ages could be a useful tool to understand a number of cometary events.

Consider the three parameters A_{SEC}, R_{ON} and $-R_{ON} + R_{OFF}$. As a comet ages, the amplitude of the secular light curve, A_{SEC}, must decrease. In fact A_{SEC} must be zero for an inert nucleus. Thus A_{SEC} must be related to activity and age. In the Atlas both are taken as synonymous. In fact, activity is a proxy for age. R_{ON} is also related to age. As the comet ages, the crust on the nucleus increases in depth, sublimating ices must recede inside the nucleus, sustained sublimation is quenched, and the comet needs to get nearer to the Sun to be activated (Yabushita & Wada, 1988; Meech 2000). Thus R_{ON} decreases with age. On the other hand, $R_{SUM} = -R_{ON} + R_{OFF}$ measures the total space of activity of the comet. Comets that have exhausted their CO and CO2, must get nearer to the Sun to be active. Comets whose activity is dominated by water ice become active much nearer to the Sun than CO or CO_2 dominated comets (Meech 2000). Thus a parameter that measures age and activity at the same time, and that includes the three above quantities could be $A_{SEC}* (-R_{ON} + R_{OFF})$. This value defines the area of a rectangle in the phase space A_{SEC} vs R_{SUM}.

So defined, P-AGE would give small values for old comets and large values for new comets, inverted from what we would like. It would be interesting to scale these values to human ages. We will call these 'comet years (cy)' to reflect the fact that they have not yet been scaled to Earth's years. To calibrate the scale, we will arbitrarily set to 28P/Neujmin 1 an age of 100 cy. With this calibration we define P-AGE thus:

$$P - AGE(1,1) = 1440/[A_{SEC}(1,1) * (-R_{ON} + R_{OFF})] \; comet \; years \; (cy) \qquad (4.1)$$

The value of the constant has been chosen to force comet 28P to an age of 100 cy. Scaling to human ages may seem naïve and unorthodox. However it places the comets in perspective and provides a scale for comparison. This enhances the usefulness of P-AGE, and when the evolution of A_{SEC}, R_{ON} and R_{OFF} with time is studied and calibrated with a suitable physical model, it will be possible to convert these values to a real physical age, thus achieving the objective we have set in this paper. This parameter classifies the secular light curves by shape, a proof of its validity (see Paper VI).

Notice that the photometric age, P-AGE, is measured at Δ =1 AU, R=1 AU, thus the correct notation is P-AGE(1,1). This is done to be able to compare comets with different q. For comparison with the definition of absolute magnitude m(1,1), P-AGE (1,1) could be interpreted as the *absolute photometric age*, and its main value is that it allows the comparison of comets with different qs.

Time plots

Due to space limitations, we are going to describe only three parameters in the Log R plot, the turn on point, T_{ON}, the turn off point, T_{OFF} and the time-age, T-AGE(1,1) (Figure 1). The symbol legend for the plots can be found on the time plot.

1. T_{ON} [days] = the time at which the nucleus turns on. The negative sign in this parameter indicates pre-perihelion quantities. It corresponds to RON but in the time domain.

2. T_{OFF}= the time after perihelion at which the nucleus turns off.

3. T-AGE (1,1)= time-age. It is possible to define an age from the time plot in the same way we did for P-AGE:

$$T - AGE(1,1) = 90240/[A_{SEC}(1,1) * (-T_{ON} + T_{OFF})] \; comet \; years \; (cy) \qquad (4.2)$$

The value of the constant has be chosen to force comet 28P to an age of 100 cy.

5. Example

Due to space limitations, we can present only the SLC of one comet, 2P/Encke in 2003. It can be seen in Figures 1 and 2. See the Atlas (paper VI) for 26 additional comets.

Acknowledgements

To FUNDACITE-Mérida and CDHT-ULA for their support through several grants.

References

Ferrín, I., 2005a, *Icarus*, 178, 493
Ferrín, I., 2005b, *ICQ*, 27, 249
Ferrín, I., 2006, *Icarus*, 185, 523

Ferrín, I., 2007, *Icarus*, 187, 326

Ferrín, I., 2008, *Icarus*, 197, 169

Ferrín, I., 2009, *Planetary and Space Science*, Paper VI, in press

Levison, H., 1999, *ASP Conference Series.*, 107, 173

Meech, K. J., 2000, *ASP Conference Series*, 213, 207

Yabushita, S., & Wada, K., 1988, *EMP*, 40, 303

Icy Bodies of the Solar System
Proceedings IAU Symposium No. 263, 2009
J.A. Fernández, D. Lazzaro, D. Prialnik & R. Schulz, eds.
© International Astronomical Union 2010
doi:10.1017/S1743921310001900

Cometary nature of the 1908 Tunguska cosmic body

F. S. Ibodov[1], S. S. Grigorian[1], and S. Ibadov[2]

[1]Moscow State University, Moscow, Russia,
emails mshtf@sai.msu.ru, firuz@pochta.ru, grigor@imec.msu.ru

[2]Institute of Astrophysics, Dushanbe, Tajikistan
email: ibadovsu@yandex.ru

Abstract. The cometary nature of the 1908 Tunguska cosmic body is compatible with the predictions of an analytical theory of the 1908 Tunguska explosion developed in 1976–1979. The theory takes into account the three simultaneously occurring processes, namely aerodynamic destruction of the cosmic body in the Earth's atmosphere, transversal expansion of the crushed mass under the action of pressure gradient on the frontal surface of the body, and an aerodynamic deceleration of crushed expanding mass. The use, for the mechanical parameters of the Tunguska cosmic body, of the characteristics of a cometary nuclei such as that of comet Halley 1986 III and comet Shoemaker – Levy 9 1994, gives parameters of the Tunguska explosion derived from observations of Tunguska event in the Siberian taiga in 1908.

Keywords. comets: general; flares: comet nuclei, planetary atmospheres, aerodynamic destruction, transversal expansion, Tunguska explosion

1. Introduction

The nature of the Tunguska explosion occurred in Siberian taiga, Russia, on June 30, 1908, that produced catastrophic destruction in the taiga, remained mysterious during many decades, and scientific investigations were carried out to accumulate data on the parameters of this event. The estimation of the initial mass of the exploded cosmic body M_o made on the basis of observations of reduction of solar radiation due to atmospheric opacity in California, USA, is of the order of 10^6 tons (Fesenkov 1949). From the analysis of crushing the taiga, the derived energy of explosion was of the order of $E_e = 4x10^{24}$ erg or 10 megatons in TNT equivalent, i.e., 1000 times the energy of the Hiroshima atomic explosion. The geocentric velocity of the bolide being considered as a fragment of comet Encke, is about 30 km/s, the inclination angle of his flight trajectory to the horizon is $\alpha = 30 - 45^o$, the altitude of explosion of the bolide above the Earth's surface is estimated as $h_e = 5 - 10$ km (Korobeinikov *et al.* 1991, and references therein).

2. Analytical theories of explosion of Tunguska cosmic body

The hypothesis of the Tunguska body as a cosmic body of anomaly small density (less than 0.01 g/cm^3), for which the generation of a strong shock wave above the Earth's surface becomes possible due to the sharp aerodynamic deceleration of the evaporating body and the fully dissipation of its large initial kinetic energy in the atmosphere, was theoretically considered in 1975. However, the theory was not applicable because of unacceptability of the starting hypothesis on the extremely low density of cosmic body (Petrov & Stulov 1975; Grigorian 1979; Surdin *et al.* 1982).

An analytical theory of the Tunguska explosion, taking into account an aerodynamic destruction of cosmic body, transverse expansion of the crushed mass under the action of pressure gradient on the frontal surface of the body and aerodynamic deceleration of the fragmented flattened expanding hypervelocity mass, was developed by Grigorian (1976, 1979).

The theory was used, applying numerical methods of integration and mathematical modeling, also for a quantitative estimation of events, occurring at the entry of fragments of the nucleus of comet Shoemaker-Levy 9 into the atmosphere of Jupiter on July 16-22, 1994 (Grigorian 1994; Fortov et al. 1996).

In 2008 we modified the theory of Grigorian (1979) to find the law of variation of the kinetic energy of the fragmented mass in the explosion zone in an explicit form, that is necessary for deriving the position of the point of maximal deceleration - maximal energy release, and solving thus the problem of analytically finding the altitude and other parameters of "explosion" of cosmic bodies like cometary nuclei, having usually small density and material strength, in the atmospheres of planets (Ibadov et al. 2008a).

The solution of the above problem is of interest also to investigate explosion of sungrazing comets in the atmosphere of the Sun and comet-like bodies, as well as stargrazers, in the atmospheres of young stars in order to study flare mechanisms of such stellar objects (Ibadov 2007; 2008b).

3. Nature of the Tunguska cosmic body

The basic parameter of the Tunguska event, the altitude of "explosion" of bolide in the Earth's atmosphere where the maximal energy release due to sharp aerodynamic deceleration of the crushed and transversally expanding mass occurs, is determined as

$$z_m = z_e = z_* - H \ln \left(1 + \frac{\sqrt{3C_x}C}{2b} \right) = H \ln \left(\frac{\rho_o V_o^2}{(1 + \frac{\sqrt{3C_x}C}{2b})\sigma_*} \right) = H \ln \left(\frac{2b\rho_o V_o^2}{\sqrt{3C_x}C\sigma_*} \right),$$
(3.1)

where

$$z_* = H \ln \left[\left(\frac{\rho_o V_o^2}{\sigma_*} \right) \right], \quad b = \nu \exp \left(-\frac{z_*}{H} \right),$$
(3.2)

$$\nu = \frac{3C_x\rho_o H}{4\rho_b R_o \sin \alpha}, \quad C = \left(\frac{3C_x R_o \sin \alpha}{8H} \right)^{1/2};$$
(3.3)

z_* is the altitude corresponding to the onset of the cosmic body aerodynamic destruction in the Earth's atmosphere, H is the height scale of the atmosphere, R_o is the body initial radius, V_o is the initial entry velocity of the body to the atmosphere, ρ_b and σ_* are the density and mechanical strength of the body material.

Assuming for the Tunguska cosmic body $M_o = 4x10^{12}g$ (Fesenkov 1949), $\rho_b = 1$ g/cm^3 (Marov 1994) and $R_o = 10^4$ cm, $V_o = 3x10^6$ cm/s, $\alpha = 30^0$, $\sigma_* = 10^7$ dynes/cm^2, $C_x = 1$, $\rho_o = 1.3x10^{-3}$ g/cm^3, $H = 7x10^5$ cm according to (1), (3.2) and (3.3) we obtain $\nu = 0.14$, $b = 1.2x10^{-4}$, $C = 5.1x10^{-2}$, $z_* = 48.3x10^5$ cm = 48.3 km, $z_m = 7.3x10^5$ cm = 7.3 km.

It should be noted that the theoretical values of "explosion" altitude of the cosmic body z_m, computed with (1) will correspond to the altitude of explosion h_e obtained from observations of the Tunguska phenomenon as well as to the mass and the radius of the cosmic body M_o and R_o of the order of 10^{12} g and 100 m, respectively, to the initial kinetic energy of the body E_o in the range of $10^{24} - 10^{25}$ erg in the case if the density of the body is $\rho_b = 0.5 - 1$ g/cm^3.

The obtained density of the exploded Tunguska cosmic body is characteristic for the density of nuclei of comets (Reinhard 1986; Sagdeev 1986; Fortov *et al.* 1996).

4. Conclusions

The theory of aerodynamic destruction of cosmic bodies and transversal expansion of fragmented mass in the Earth's atmosphere can adequately explain the 1908 Tunguska explosion for mechanical parameters of the body like those for comet Halley 1986 III and comet Shoemaker – Levy 1994. It indicates, along with data of observations of the phenomena, that the 1908 Tunguska cosmic body had cometary nature.

Acknowledgements

The authors are grateful to Academician A. M. Cherepashchuk for his interest and stimulating remarks and also to Drs. V.F. Esipov and G.M. Rudnitskij, Sternberg Astronomical Institute of Moscow State University for their promotion in fulfilling the research. The work is supported by the Russian Foundation for Basic Research, Project RFBR 08-02-11003-ano.

References

Fesenkov, V. G. 1949, *Meteoritika*, 6, 8

Grigorian, S. S. 1976, *Dokl. Akad. Nauk SSSR*, 231, 57

Grigorian, S. S. 1979, *Kosmich. Issled.*, 17, 875

Grigorian, S. S. 1994, *Dokl. Akad. Nauk*, 338, 752

Ibadov, S., Ibodov, F. S., & Grigorian, S. S. 2008a, in: *Intern. Conf. "100 Years Since Tunguska Phenomenon: Past, Present and Future"* // tunguska.sai.msu.ru/index.php?q=present

Ibadov, S., Ibodov, F. S. & Grigorian, S. S. 2007, in: *Star-Disk Interaction in Young Stars*, Proc. IAU Symp. No. 243, Grenoble, France, p. VI.4 // www.iaus243.org

Ibadov, S., Ibodov, F. S. & Grigorian, S. S. 2008b, in: *Universal Heliophysical Processes*, Proc. IAU Symp. No. 257, Cambridge University Press, 341

Korobeinikov, V. P., Chushkin, P. I., & Shurshalov, L. V. 1991, *Astron. Vestnik*, 25, 327

Marov, M. Ya. 1994, *Astron. Vestnik*, 28, 5

Petrov, G. I. & Stulov, V. P. 1975, *Kosmich. Issled.* (Letters), 13, 587

Reinhard, R. 1986, *Nature*, 321, 313

Sagdeev, R. Z., Blamont, J., Galeev, A. A., Moroz, V. I., Shapiro, V. D., Shevchenko, V. I., & Szego, K. 1986, *Nature*, 321, 259

Surdin, V. G, Romeiko, V. A, & Koval, V. I. 1982, *Astron. Circular USSR*, 1206, 1

Icy Bodies of the Solar System
Proceedings IAU Symposium No. 263, 2009
J.A. Fernández, D. Lazzaro, D. Prialnik & R. Schulz, eds.

© International Astronomical Union 2010
doi:10.1017/S1743921310001912

On the relationship between gas and dust in 15 comets: an application to Comet 103P/Hartley 2 target of the NASA *EPOXI* mission of opportunity

G. C. Sanzovo[1], D. Trevisan Sanzovo[1], and A.A. de Almeida[2]

[1]Department of Physics, State University of Londrina, Londrina, PR, Brazil
email: gsanzovo@uel.br

[2]Department of Astronomy, University of São Paulo, São Paulo, SP, Brazil
email: amaury@astro.iag.usp.br

Abstract. After the success of Deep Impact mission to hit the nucleus of Comet 9P/Tempel 1 with an impactor, the concerns are turned now to the possible reutilization of this dormant flyby spacecraft in the study of another comet, for only about 10% of the cost of the original mission. Comet 103P/Hartley 2 on UT 2010 October 11 is the most attractive target in terms of available fuel at rendezvous and arrival time at the comet. In addition, the comet has a low inclination so that major orbital plane changes in the spacecraft trajectory are unnecessary. In an effort to provide information concerning the planning of this new NASA *EPOXI* space mission of opportunity, we use in this work, visual magnitudes measurements available from International Comet Quarterly (ICQ) to obtain, applying the Semi-Empirical Method of Visual Magnitudes - SEMVM (de Almeida, Singh, & Huebner 1997), the water production rates (in molecules/s) related to its perihelion passage of 1997. When associated to the water vaporization theory of Delsemme (1982), these rates allowed the acquisition of the minimum dimension for the effective nuclear radius of the comet. The water production rates were then converted into gas production rates (in g/s) so that, with the help of the strong correlation between gas and dust found for 12 periodic comets and 3 non-period comets (Trevisan Sanzovo 2006), we obtained the dust loss rates (in g/s), its behavior with the heliocentric distance and the dust-to-gas ratios in this physically attractive rendezvous target-comet to Deep Impact spacecraft at a closest approach of 700 km.

Keywords. Gas release rates - Dust release rates - Dust-to-gas ratios - Short-period comets - Comet 103P/Hartley 2

1. Introduction

Comet 103P/Hartley 2 was discovered on 1984 November 28 in Australia, by Malcolm Hartley, having estimated visual magnitude 17-18 (Hartley 1984). For its 1997 return, ICQ makes available a set of 857 visual magnitudes measurements, obtained by several observers, taken between 1997 March 3.16 (r = 2.803 AU) and 1998 May 19.48 (r = 2.104 AU). In the present work, we apply the Semi-Empirical Method of Visual Magnitudes (SEMVM) to this visual magnitude data to find water, and hence gas production rates. Combining our derived water release rates with the vaporization theory of Delsemme (1982), provided the determination of the comet's nuclear dimensions and active areas. We also used the gas-to-dust correlation found by Trevisan Sanzovo (2006) for 12 Jupiter Family (JF) comets with the purpose to estimate the corresponding dust loss rates for Comet 103P/Hartley 2.

2. Theoretical Considerations

2.1. *The Semi-Empirical Method of Visual Magnitudes (SEMVM) and the Gas-to-Dust correlation*

If $m_{v'}$ is the total visual magnitude observed from the coma of a comet, reduced to the standard diameter of 6.78 cm when the observer is placed at a standard geocentric distance $\Delta = 1$ AU, (Morris (1973)) and the comet at a heliocentric distance r (in AU), the water production rate (in molecules/s) is given by (de Almeida, Singh, & Huebner 1997; Sanzovo *et al.* 2001)

$$Q\left(H_2O\right) = \left\{ \frac{r^2 . 10^{[0.4(-26.8-m_{v'})]} - p.R_N^2.\Phi_N}{R.l_r.\left[1 + \delta\left(r, \theta\right)\right]} \right\}^{0.825}, \quad (2.1)$$

where $p = p(\lambda)$ is the geometric visual albedo of the comet's nucleus with radius R_N, and the other parameters of Equation (2.1) and framework of SEMVM are reported in details elsewhere (Sanzovo *et al.* 1996, and de Almeida, Singh, & Huebner 1997). Once the water production rates are obtained, its conversion into gas loss rates can be accomplished, considering a gaseous mixture of $\sim 77\%$ H_2O, $\sim 13\%$ CO, and $\sim 10\%$ of other molecular species with average molecular weight ~ 30 amu (Sanzovo *et al.* 1996; de Almeida, Singh, & Huebner 1997). Active surface areas and nuclear dimensions were inferred combining SEMVM and Delsemme's theory (Delsemme 1982). In view of lack of information in literature about continuum observations refering to Comet 103P/Hartley 2, we used for the estimate of the dust loss rates, q_d, the gas-to-dust correlation found by Trevisan Sanzovo (2006) for 12 periodic comets determined for $\lambda = 4845$ and 4770 Å, and dust particle density $\rho_d = 0.5$ g/cm^3. The strong gas-to-dust correlation is given by $\log(q_g) = (2.796 \pm 0.161) + (0.599 \pm 0.025) \times \log(q_d)$. Trevisan Sanzovo (2006) also provide other dust-to-gas correlation $[\log(q_g) = (-0.200 \pm 0.203) + (1.067 \pm 0.031) \times \log(q_d)]$ valid for Comets C/Hale-Bopp, C/Hyakutake, C/Levy, and also including 1P/Halley. The use of the gas-to-dust correlation for our JF cometary sample made possible the estimate of the dust loss rates, as well as the dust-to-gas ratios for the 1997 return of Comet 103P/Hartley 2.

3. Results and Discussion

Our analysis is based on Figure 1 and Table I, which summarize main results. There, in addition to the nomenclature, the first column shows the perihelion passages corresponding to the gas and dust analysis quoted between parenthesis and brackets, respectively. In column 2, we have the orbital period (in years) of the objects, while the nuclear radius (in km), the fraction of active area (in %), and the active surface area (in (km^2) is shown in column 3 of the same table.

3.1. *Nuclear Dimensions and Masses*

We fixed the active surface area in 20% for the nuclear hemisphere lit by the sun, and the water production rates obtained through the application of SEMVM were combined with the vaporization rates of Delsemme (1982), resulting in an effective nuclear radius of $\sim$ 1.8 km for Comet 103P/Hartley 2. If $f_{AA} = 1.0$, the water vaporization and production rates will be compatible with a minimum nuclear radius of ~ 0.8 km. We verify therefore, that Comet 103P/Hartley 2 has, amongst the JF comets of the sample, about the same dimension of Comet 21P/GZ, whose effective nuclear radius is ~ 1.7 km. For an activity of 100%, Comet 103P/Hartley 2 has about the same dimension of Comet C/Hyakutake (see Table I). We also make an estimate of the nuclear masses which are shown in column

Table 1. Gas and dust properties for Comet 103P/Hartley 2 and other 15 comets.

Comet	P (years)	R_N - f_{AA} - A_A (km - % - km^2)	$M^{(a)}$ (g)	q_g (g/s)	q_d (g/s)	$Q_T^{(b)}$ (kg/s)	χ	$R_N^{(c)}$ (km)	References
1P/Halley (1986);[1986]	~ 76	1.5 - 100 - 14 / 5.0 - 10 - 157	2.6(17)	$\propto r^{-4.74}$* / $\propto r^{-3.15}$◊	$\propto r^{-3.18}$●	~2670* / ~6100◊	0.18 - 2.17	5.0	WEA97
9P/Tempel 1 (1972,83,89,94, 00,05); [1983]	~ 5.5	1.5 - 100 - 14 / 2.0 - 10 - 25	1.7(16)	$\propto r^{-8.79}$* / $\propto r^{-6.47}$◊	$\propto r^{-4.21}$	~340*	0.02 - 0.38	2.3	TAN00
10P/Tempel 2 (1983,88,94,99); [1983,88,99]	~ 5.5	1.2 - 100 - 9 / 2.7 - 20 - 46	4.1(16)	$\propto r^{-7.56}$* / $\propto r^{-7.62}$◊	$\propto r^{-6.02}$●	~176* / ~532◊	0.01 - 0.47	2.9	TAN00
21P/GZ (1985,98);[1985]	~ 6.6	0.9 - 100 - 5 / 1.7 - 20 - 18	1.0(16)	$\propto r^{-4.75}$* / $\propto r^{-3.23}$◊	$\propto r^{-1.69}$●	~244* / ~247◊	0.08 - 0.36	1.0	TAN00
22P/Kopff (1983,90,96);[1983]	~ 6.5	1.9 - 100 - 23 / 4.3 - 20 - 116	1.7(17)	$\propto r^{-5.85}$* / $\propto r^{-4.34}$◊	$\propto r^{-3.60}$●	~895* / ~714◊	0.06 -0.34	1.8	TAN00
24P/Schaumasse (1984,93);[1984]	~ 8.2	0.4 - 100 - 1 / 0.9 - 20 - 5	1.5(15)	$\propto r^{-12.42}$* / $\propto r^{-8.20}$◊	-	≥ 23* / ≥ 82◊	0.03 - 0.85	0.8	TAN00
26P/GS (1982,87,92); [1977,82]	~ 5.1	0.3 - 100 - 1 / 1.0 - 10 - 6	2.1(15)	$\propto r^{-7.56}$* / $\propto r^{-5.37}$◊	$\propto r^{-7.27}$◊	~ 19	0.02 - 0.25	1.3	TAN00
46P/Wirtanen (1986,91,97); [1991,97]	~ 5.5	0.7 - 100 - 3 / 1.5 - 20 - 14	7.1(15)	$\propto r^{-8.06}$* / $\propto r^{-6.28}$◊	$\propto r^{-4.94}$●	~80* / ~163◊	0.02 - 0.08	0.7	TAN00
62P/Tsuchinshan 1 (1985,98);[1985]	~ 6.6	0.7 - 100 - 3 / 1.5 - 20 - 16	7.1(15)	$\propto r^{-10.08}$●	$\propto r^{-8.67}$●	~64●	0.03 - 0.14	0.8	TAN00
67P/CG (1982,96);[1982,96]	~ 6.6	0.7 - 100 - 3 / 2.1 - 10 - 28	1.9(16)	$\propto r^{-4.46}$* / $\propto r^{-4.29}$◊	$\propto r^{-4.11}$●	~ 75* / ~ 132◊	0.04 - 1.00	2.5	TAN00
81P/Wild 2 (1990,97,030; [1978,84,97]	~ 6.4	1.5 - 100 - 14 / 3.9 - 20 -96	2.5(16)	$\propto r^{-5.22}$* / $\propto r^{-6.59}$◊	$\propto r^{-2.77}$●	~ 868* / ~ 690◊	0.07 - 1.15	2.2	TAN00
C/Hyakutake (1996);[1996]	> 200	1.0 - 100 - 6 / 2.4 - 20 - 36	2.9(16)	$\propto r^{-2.73}$* / $\propto r^{-2.99}$◊	$\propto r^{-1.85}$*	3340*	0.15 - 1.47	2.4	LIS99
C/Levy (1990);[1990]	> 200	1.6 - 100 - 16 / 3.4 - 10 - 73	8.2(16)	$\propto r^{-2.80}$* / $\propto r^{-4.01}$◊	$\propto r^{-2.10}$●	~ 12240* / ~9560◊	0.19 - 1.53	2.4-3.4	SCH91
C/Hale-Bopp (1997);[1997]	> 200	16.8 - 100 - 1774 / 26.7 - 40 - 4480	4.0(19)	$\propto r^{-3.53}$* / $\propto r^{-3.75}$◊	$\propto r^{-1.00}$●	~ 47200* / ~ 50500◊	0.28 - 27.68	30.0	FER03
85P/Boethin (1986)	~ 11.2	0.8 - 100 - 4 / 1.9 - 20 - 23	1.4(16)	$\propto r^{-5.21}$* / $\propto r^{-5.34}$◊	$\propto r^{-6.90}$● / $\propto r^{-7.08}$◊	~ 419.5* / ~ 388.5◊	0.10 - 0.40	0.7	AHE95
103P/Hartley 2 (1997)	~ 6.4	0.8 - 100 - 4 / 1.8 - 20 - 20	1.2(16)	$\propto r^{-7.42}$● / $\propto r^{-8.81}$◊	$\propto r^{-9.83}$● / $\propto r^{-5.14}$◊	~ 51.5* / ~ 52.3◊	0.02 - 0.25	0.7 / 0.57	GRO04 / LIS09

Notes:
$^{(a)}$2.6×10^{17}; $^{(b)}$ For r= 1.6 AU; $^{(c)}$ Data from Literature for comparison;*for pre-perihelic phase; ◊ for post-perihelic phase, and ● for pre- and post-perihelic phases; WEA97 = Weaver et al. (1997); TAN00 = Tancredi et al. (2000); LIS99 = Lisse et al. (1999); SCH91 = Schleicher et al. (1991); AHE95 = A'Hearn et al. (1995); FER03 = Fernández et al. (2003); GRO04 = Groussin et al. (2004), and LIS09 = Lisse et al. (2009).

4 of Table I. For this calculation, we considered a spherical nucleus and adopted a mean nuclear density $\rho_N = 0.5$ g/cm. With this procedure we verify that Hartley 2 is a JF comet with intermediary mass (M $\sim 1.2 \times 10^{16}$ g), being comparable to 21P/GZ, and $\sim$ 22 times less massive than Comet Kopff.

3.2. *Gas, Dust, Dust-to-Gas Ratios and Productivity*

The application of SEMVM to Comet 103P/Hartley 2 yield average gas production rates (in g/s) which vary with the heliocentric distance according with the power-law $q_g = (1.572 \pm 0.045) \times 10^6 . r^{-7.43(\pm 0.16)}$ in the pre-perihelion phase interval $2.803 \leqslant r(\mathrm{AU}) \leqslant 1.032$, and $q_g = (1.328 \pm 0.221) \times 10^6 . r^{-8.81(\pm 1.00)}$ in the post-perihelion phase comprehended between r = 1.032 and 2.104 AU. We find that, at perihelion (r = 1.032 AU), Comet 103P/Hartley 2 lost gas at an average rate of $\sim 1.2 \times 10^6$ g/s, with a maximum of gas production rate estimated as $q_g \sim 3.5 \times 10^6$ g/s. At perihelion and r = 1.04 AU, A'Hearn et al. (1995), Crovisier et al. (1999) and Colangeli et al. (1999) found Q(H$_2$O) = 3.0× 10^{28}, 1.24×10^{28} and 3.1×10^{28} molecules/s, which corresponds to q_g=1.03×10^6, 4.24×10^5 and 1.06×10^6 g/s, respectively. Hence, our results are in reasonable agreement with those authors. Inspection of Table I show that Comet 103P/Hartley 2 is nearly two orders of magnitude less active than Comet 1P/Halley, and about three orders of magnitude less productive than Comets C/Hyakutake and C/Hale-Bopp. Besides the orbital period, dimensions and nuclear masses, Table I also presents in columns 5, 6, and 8 the dependencies with heliocentric distance of the gas and dust loss rates, and dust-to-gas mass ratios, respectively. In the last two columns of the same Table I we include nuclear radii and respective references found in literature, for comparison. The total mass loss rates, Q_T (in kg/s), in the form of gas and dust, are shown in column 7 of the Table I, being obtained fixing r = 1.6 AU. The results indicate that Comets 22P/Kopff and 81P/Wild 2 in 1996 and 1997 apparitions, respectively, lost $\sim$ 17 times more mass than Comet 103P/Hartley 2 which, in the pre-perihelic phase was more productive than Comets 24P/Schaumasse and 26P/GS. Comparing dust and gas mass release rates obtained in this work, we conclude that Comet Hartley 2 has predominantly $0.02 < \chi < 0.25$, being classified as belonging to the family of comets with intermediate dust-to-gas mass ratios (Sanzovo *et al.* 1996).

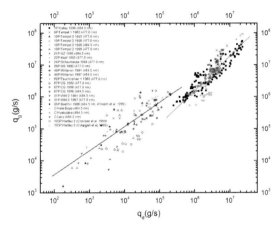

Figure 1. Correlation between gas and dust mass rates in 15 comets (after Trevisan Sanzovo 2006.)

Acknowledgements

This work was partially supported by FAPESP (São Paulo, SP, Brazil) under grant No. 09/50626-8. GCS is grateful to PROPPG-UEL (Londrina, PR, Brazil) for financial help.

References

A'Hearn, M. F., Millis, R. L., Schleicher D. G. *et al.* 1995, *Icarus* 118, 223
de Almeida, A. A., Singh, P. D., & Huebner, W. F. 1997, *Planet. Space Sci.* 45, 681 *A&A* 45, 681
Colangeli, L., Epifani, E., Brucato, J. R. *et al.* 1999, *A&A* 343, L87
Crovisier, J., Encrenaz, Th., Lellouch, E. *et al.* 1999, *ESASP* 427, 161
Delsemme, A. H. 1982, in: L. L. Wilkening (ed.) *Comets* University of Arizona Press, Tucson, p.85
Groussin, O., Lamy, P., Jorda, L. *et al.* 2004, *A&A* 419, 375
Hartley, M. 1984, *IAU Circ.* 4015, 1
Lisse, C. M., Fernández, Y. R., Kundu, A. *et al.* 1999, *Icarus* 140, 189
Lisse, C. M., Fernández, Y. R., Reach, W. T. *et al.* 2009, *PASP* (in Press)
Morris, C. S. 1973, *PASP* 85, 470
Sanzovo, G. C., Singh, P. D., & Huebner, W. F. 1996, *A&AS* 120, 301

Sanzovo, G. C., de Almeida, A. A., Misra, A. *et al.* 2001, *MNRAS* 326, 852

Schleicher, D. G., Millis, R. L., Osip, D. J. *et al.* 1991, *Icarus* 94, 511.

Tancredi, G., Fernandez, J. A., Rickman, H. *et al.* 2000, *A&AS* 146, 73

Trevisan Sanzovo, D. T. 2006, *MSc Thesis*, State University of Londrina, Londrina, PR, Brazil

Weaver, H. A., Feldman, P. D., A'Hearn, M. F. *et al.* 1997, *Science* 275, 1900

Icy Bodies of the Solar System
Proceedings IAU Symposium No. 263, 2009
J.A. Fernández, D. Lazzaro, D. Prialnik & R. Schulz, eds.

© International Astronomical Union 2010
doi:10.1017/S1743921310001924

Spectroscopic Studies of Comets 9P/Tempel 1, 37P/Forbes and C/2004 Q2 (Machholz)

Enos Picazzio[1], Klim I. Churyumov[2], Larissa S. Chubko[3], Igor V. Lukyanyk[2], Valery V. Kleshchonok[2], Amaury A. de Almeida[1], and Roberto D. D. Costa[1]

[1]Department of Astronomy, Institute of Astronomy, Geophysics and Atmospheric Sciences,
University of São Paulo,
Rua do Matão 1226, Cidade Universitária 05508-900 São Paulo SP, BRAZIL
email:picazzio@astro.iag.usp.br

[2]Astronomical Observatory, Kyiv Shevchenko National University,
Box 04053, Observatorna str., 3, Kyiv, Ukraine
email: klim.churyumov@observ.univ.kiev.ua

[3]National Aviation University,
Box 03680, Kosmonavta Komarova ave. 1, Kiev, Ukraine

Abstract. The results of the analysis of the spectra of comets 9P/Tempel 1, 37P/Forbes and C/2004 Q2 (Machholz) observed in 2004-2005 at Observatório do Pico dos Dias (Brazil), and at Mount Pastukhov (SAO, Russia) are presented.

Keywords. comets: general, comets: individual (9P/Tempel 1, 37P/Forbes, C/2004 Q2 (Machholz))

1. Observations and processing of cometary spectra

High and middle-resolution optical spectra of comets obtained with long-slit spectroscopy allow to (1) calculate some physical parameters of the cometary neutral atmosphere (escape velocities of the gas in the coma, lifetime of particles, etc.) (2) search for new cometary emission lines, (3) estimate parameters of gas and dust activity of the comet nucleus, (4) detect the cometary luminescence continuum of non-solar nature (Churyumov *et al.* 1994; Churyumov *et al.* 1999; Lukyanyk & Churyumov, 2002; Churyumov *et al.*, 2002; Lukyanyk *et al.*, 2002; Picazzio *et al.*, 2002; Picazzio *et al.*, 2006; Chubko *et al.*, 2009).

The spectra of comets 9P/Tempel 1 and 37P/Forbes were observed at Observatório do Pico dos Dias (LNA - Laboratório Nacional de Astrofísica), Brasópolis (Brazil) during 3-5 July 2005 with the Cassegrain spectrograph using a 900/500 grating, attached to the Perkin & Elmer 1.6-m telescope of LNA. In addition spectra of comet 9P/Tempel 1 and C/2004 Q2 (Machholz) were obtained with SCORPIO (Spectra Camera with Optical Reducer for Photometrical and Interferometrical Observations) installed in the prime focus of the 6-m telescope, and with the 1-m Zeiss reflector equipped with the long-slit spectrograph of the Special Astrophysical Observatory of the RAS on 14-16 March and 3-4 July 2005 (Mount Pastukhov, Russia). Another 2 spectra of comet C/2004 Q2 (Machholz) were obtained during the night on 17-18 Dec. 2004 also with the 6-m telescope and the MPFS spectrograph of the SAO of RAS.

2. Physical parameters of comets 9P/Tempel 1, C/2004 Q2 (Machholz) and 37P/Forbes

In order to determine some physical parameters of the gaseous components of the neutral cometary atmosphere (the gas component expansion u and the lifetime of the particles τ) we constructed a photometric profile for the C_2, C_3, and CN emission lines along the slit for comets 9P/Tempel 1 and 37P/Forbes (Fig. 1). Then the obtained monochromatic profiles were processed by Shul'man's model. Within this model the surface brightness was determined through the following formulas (Shul'man, 1970):

$$lg\frac{I(\rho, \varphi + \pi)}{I(\rho, \varphi)} = 1.72\frac{\rho}{r_{0C}}\sin\Theta_0\cos\varphi \qquad (2.1)$$

$$\frac{1}{2}lg\left[I(\rho, \varphi + \pi)I(\rho, \varphi)\right] = const + lg\left[\frac{r_{0k}}{\rho}\int\limits_{\frac{\rho}{r_{0k}}}^{\infty}K_0(y)dy\right] \qquad (2.2)$$

where $I(\rho, \varphi + \pi)$ and $I(\rho, \varphi)$ represent the brightness surface of emission line along slit, ρ, φ are polar coordinates on the picture plane with the polar axis directed to the Sun, $r_{0c} = \frac{2u^2}{g}$ is the characteristic scale of the spherical symmetry region, u is the expansion velocity, g is the acceleration of molecules in the gravity field of the Sun, φ is an angle between the z axis and g-vector, $K_0(y)$ is the modified Bessel function of the second kind.

The physical parameters of the neutral gaseous molecules C_2 (5165 Å), C_3 (4050 Å), CN (4200 Å) (velocity of expansion, lifetime and scale length of parent and daughter molecules) are given in the Table 1.

From Table 1 we see that the measured expansion velocities of the C_2, C_3 and CN molecules in the coma of the three comets diverge noticeably from the gas expansion velocity determined by Delsemme's formula

$$v = \frac{0.58}{\sqrt{r}} \qquad (2.3)$$

which gives equal velocities for all molecules at the same heliocentric distance.

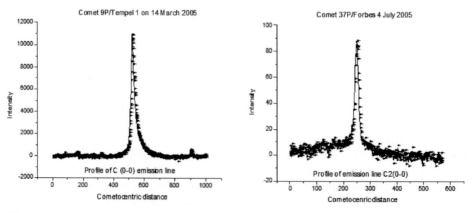

Figure 1. Profiles of brightness in the emission line $C_2(0-0)$ in the spectra of comets 9P/Tempel 1 and 37P/Forbes.

Table 1. Physical parameters of neutral gaseous cometary components of C_2, C_3 and CN (Shul'man's model).

Species	Velocity, $\left[\frac{m}{s}\right]$	Lifetime, $(10^6 s)$	Comet	Date
C_2 (5165 Å)	222	1.36	9P/Tempel 1	14/03/2005
C_3 (4050 Å)	102	0.38	9P/Tempel 1	14/03/2005
CN (4200 Å)	67	1.35	9P/Tempel 1	14/03/2005
CN (4200 Å)	> 300	> 50	Machholz C/2004 Q2	18/12/2004
C_3 (4050 Å)	363	5.13	Machholz C/2004 Q2	18/12/2004
C_2 (5165 Å)	535	39.9	Machholz C/2004 Q2	18/12/2004
C_2 (5165 Å)	855	73.4	Machholz C/2004 Q2	15/03/2005
C_2 (5165 Å)	209	25.6	37P/Forbes	04/07/2005

3. Search and detection of cometary luminescence continuum in spectra of comets C/2004 Q2 (Machholz), 9P/Tempel 1 and 37P/Forbes

A luminescence continuum was detected for the first time in the spectrum of comet 1P/Halley by G. Nazarchuk, who found two broad features with a maximum of intensity near 3950 Å and 5100 Å (Nazarchuk, 1987; Nazarchuk, 1987). They were part of the scattered solar continuum. Such a phenomenon is connected with the presence of an additional component of a continuous spectrum in cometary radiation. The source of this additional radiation could be the luminescence of organic cometary particles. Hence the spatial distribution of this source should have a very strong concentration close to the comet nucleus. To summarize: a cometary spectrum I_{com} consists of three components:

$$I_{com}(\lambda) = I_e(\lambda) + I_s(\lambda) + I_l(\lambda) \tag{3.1}$$

where I_e is the cometary emission spectrum, I_s is the solar spectrum reflected by cometary dust, I_l is the cometary luminescent continuum. Spectral regions with no strong emission lines ($I_e = 0$) are selected for the determination of the level of a luminescent continuum. In these regions of the spectrum it is accepted, that the level of a luminescent continuum does not vary. Thus, for the selected regions of a cometary spectrum it is possible to accept

$$I_{com}(\lambda) = k \cdot I_f(\lambda) + l \tag{3.2}$$

where $I_{com}(\lambda)$ is the known solar spectrum which is calculated taking into account the spectral resolution of cometary spectrum, its discontinuity, k, is the factor which characterizes the reflective ability of the cometary dust, and l is the intensity of the luminescent continuum. In practice the parameters k and l are selected such that the best agreement to a region of the cometary continuum is obtained. The given technique was used for studying a luminescent continuum in the spectra of comets C/2004 Q2 Machholz, 9P/Tempel 1 and 37P/Forbes (Fig.2).

The maximum of a luminescent continuum for comet C/2004 Q2 Machholz is close to 6300 Å (Churyumov, 1999; Lukyanyk & Churyumov, 2002; Lukyanyk *et al.*, 2002). In this region its intensity reaches 46% of the total cometary continuum. For comet 9P/Tempel 1 the level of the luminescent cometary continuum is 30% of the level of the total cometary continuum with the maximum near wavelength 5250 Å. For comet 37P/Forbes the level of the luminescent cometary continuum is 20% of the level of the total cometary continuum with the maximum near 4500 Å. Comparison of the spectra of the three comets shows that new comet C/2004 Q2 (in Oort's sense) has a higher level luminescent continuum, which may indicate a larger number of organic particles.

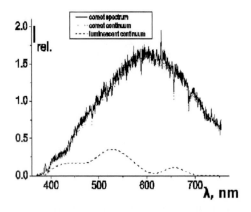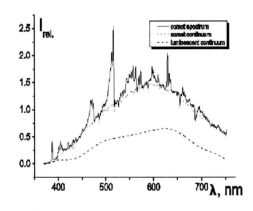

Figure 2. Spectra of comets 9P/Tempel 1 and C/2004 Q2 Machholz with the dedicated cometary continuums and the luminescence cometary dust levels of the total cometary continua.

4. Conclusion

The gas component expansion u and the lifetime of the particles in the comae of comets 9P/Tempel 1, C/2004 Q2 (Machholz) and 37P/Forbes are calculated. The spectra of these comets show evidence for a luminescent cometary continuum which may be connected to the luminescence of organic species in cometary dust particles (e.g. *CHON*-particles).

References

Chubko, L. S., Churyumov, K. I., Afanasiev, V. L., Lukyanyk, I. V., & Kleshchonok, V. V. 2009, *Deep Impact as a World Observatory Event: Synergies in Space, Time, and Wavelength*, ESO Astrophysics Symposia., p. 197

Churyumov, K. I., Kleshchenok, V. V., & Vlassyuk, V. V. 1994, *Pisma v Astronomicheskij Journal*, 20, 9

Churyumov, K. I., Kleshchenok, V. V., & Mussaev, F. A. 1999, *Earth, Moon and Planets*, 78, 1

Churyumov, K. I., Lukyanyk, I. V., Afanasiev, V. L. *et al.* 2002, *Proceedings of Asteroids, Comets, Meteors (ACM 2002)*, p. 657

Lukyanyk, I. V., Churyumov, K. I., Afanasiev, V. L. *et al.* 2002, *Proceedings of Asteroids, Comets, Meteors (ACM 2002)*, p. 717

Lukyanyk, I. V. & Churyumov, K. I. 2002, *Earth, Moon and Planets*, 90, 1

Nazarchuk, H. K. 1987, *Kometnyj Tsirkulyar*, 372, 2

Nazarchuk, H. K. 1987, *Kometnyj Tsirkulyar*, 377, 2

Picazzio, E., de Almeida, A. A., Churyumov, K. I., Andrievskii, S. M., & Lukyanyk, I. V. 2002, *Earth, Moon and Planets*, 90, 23

Picazzio, E., de Almeida, A. A., Churyumov, K. I., Andrievskii, S. M., Lukyanyk, I. V. 2006, *Advances in Space Research*, 10, 312

Shulaman, L. M. 1970, *Astrometry and Astrophysics*, 11, 26

Icy Bodies of the Solar System
Proceedings IAU Symposium No. 263, 2009
J.A. Fernández, D. Lazzaro, D. Prialnik & R. Schulz, eds.

Cometary gas relations 1P/Halley

Marcos R. Voelzke

Cruzeiro do Sul University
Regente Feijó Avenue 1295, 03342-000, São Paulo, SP, Brazil
email: mrvoelzke@hotmail.com

Abstract. Photographic and photoelectric observations of comet 1P/Halley's ionised gas coma from CO^+ at 4,250 Å and neutral gas coma from CN at 3,880 Å were part of the Bochum Halley Monitoring Program, conducted at the European Southern Observatory, La Silla, Chile, from February 17 to April 17, 1986.

In this spectral range it is possible to see the continuum formation, motion and expansion of plasma and neutral gas structures. To observe the morphology of these structures, 32 CO^+ photos (glass plates) from comet 1P/Halley obtained by means of an interference filter have been analysed. They have a field of view of 28.6×28.6 degrees and were obtained from March 29 to April 17, 1986 with exposure times between 20 and 120 minutes.

All photos were digitised with a PDS 2020 GM microdensitometer. After digitisation, the data were reduced to relative intensities, and those with proper calibrations were also converted to absolute intensities, expressed in terms of column densities. The CO^+ absolute intensity values still contain the continuum intensity. To calculate the CO^+ column density it is necessary to subtract this continuum intensity.

The relations between CO^+ and CN in average column density values ($\overline{N_{CO^+}/N_{CN}}$) are 11.6 for a circular diaphragm with average diameter ($\overline{\Phi}$) of 6.1 arcminutes which corresponds to a distance from the nucleus (ρ) equal to 6.3×10^4 km; 20.0 for $\overline{\Phi} = 7.1$ arcminutes and $\rho = 7.3 \times 10^4$ km; 8.1 for $\overline{\Phi} = 8.5$ arcminutes and $\rho = 8.7 \times 10^4$ km; 35.6 for $\overline{\Phi} = 11.9$ arcminutes and $\rho = 1.2 \times 10^5$ km; and 31.3 for $\overline{\Phi} = 16.7$ arcminutes and $\rho = 1.7 \times 10^5$ km. These values are in perfect agreement with the data for short distances (ρ from 3.9×10^3 to 1.2×10^4 km) and small slit diameters ($\overline{\Phi}$ from 0.4 to 1.2 arcminutes).

With the use of diaphragms with large diameters it is possible to get some information about the outer coma of the comet (in this paper, from 60,000 until 170,000 km away from the nucleus). At these distances, the CO^+ column density changes only due to the geometrical dilution, because the CO^+ parent molecules are already photoionised or photodissociated.

Keywords. comets: 1P/Halley - comets: general

---

1. Introduction

Comet 1P/Halley (1P/1982 U1) was observed photographically at the European Southern Observatory (ESO), La Silla, from February 17 to April 17, 1986 by a group of the *Astronomisches Institut der Ruhr-Universität Bochum*. Altogether 1,216 images were taken in 57 of consecutive 60 nights with exposure times between one second and 170 minutes. Photoelectric photometry of the cometary coma was obtained from February 24 to April 17, 1986 using the 61 cm-Bochum telescope. The observations aimed the study of structure, dynamics, and physical properties of the coma, dust and plasma tail in full spatial extent, and looked for correlations in different parts of the comet with the solar wind. The reader should refer to Celnik *et al.* (1988) for a complete list of all obtained cometary images and associated technical data.

Table 1. CO$^+$ Plasma images absolute intensities

Date	Image	Decimal Day in April 1986	Absolute Intensity in 10^{-9} erg s^{-1} cm^{-2}
01.04.1986	5530	1.340	1.405
02.04.1986	5546	2.252	1.269
02.04.1986	5556	2.304	1.108
04.04.1986	5602	4.239	0.868
04.04.1986	5619	4.335	0.672
06.04.1986	5675	6.243	1.473
06.04.1986	5685	6.339	1.316
07.04.1986	5693	7.090	1.103
07.04.1986	5716	7.255	1.158
08.04.1986	5732	8.094	0.691
08.04.1986	5751	8.248	1.233
13.04.1986	5881	13.234	2.229
14.04.1986	5913	14.057	0.606
14.04.1986	5949	14.223	0.621

2. Data Reductions

Thirty two CO$^+$-filter Wide-Field-Camera cometary images and their spot calibrations were digitised with a PDS 2020 GM microdensitometer at the *Astronomisches Institut der Westfälischen Wilhelms-Universität* in Münster. The step of the scan was Δx = Δy = 25 μm so that one pixel is 25 × 25 μm and corresponds approximately to 46.9 × 46.9 arcseconds.

These 32 cometary images from March 29 to April 17, 1986 are the best series of images available because at this time the comet 1P/Halley was nearest to Earth. So 1P/Halley's coma is shown in maximal resolution (approximately 30.0 arcseconds, about 10,210 km at a distance from 0.52 AU (Celnik & Schmidt-Kaler, 1987) and the comet was visible all night at this time, thus two long plasma images could be taken per night (t ⩾ 20 minutes).

Just twenty of 32 relative calibrated pictures could theoretically be calibrated absolutely since photometric measurements were not available in all the nights. In fact, only fourteen of these twenty pictures could be calibrated absolutely because their duration of exposure was that long that in six cases the nucleus region of the comet was wider than the largest diaphragm. These fourteen remaining plasma-images were analysed, i.e., they were effectively absolutely calibrated (Voelzke, 1996; Voelzke *et al.*, 1997) making use of MIDAS image processing system. The images and their absolute intensities are listed in Table 1.

3. Results of the absolute calibration of the plasma images

The values of the absolute intensities were calculated by subtracting the values obtained with the largest diaphragm from those values obtained with the second largest, i.e., the difference between the measured values of diaphragms 25 and 26 was determined.

The values of the entire brightness and the continuum background in the CO$^+$ emission filter were converted into relative units of intensity I'(CO$^+$). For the evaluation of the CO$^+$ radiation flow, A'Hearn & Vanysek (1986) equation was used. The thusly calculated CO$^+$ radiation flow has the dimension [erg s^{-1} cm^{-2}]. After the relative calibration of all the images, the background of the image was subtracted, then the cometocentrical coordinates were included.

These calculated absolute intensity values still contain the continuum intensity. To obtain a column density of CO$^+$, the intensity of the continuum must be subtracted, i.e., the values for continuum background in the CO$^+$ emission filter must be subtracted

Table 2. CO^+ radiation flow, CO^+ column density and proportion between CO^+ and CN column density for the diaphragms 25 and 26

Date	Picture	$F(CO^+)$ $(10^{-9}$ erg s^{-1} cm$^{-2})$	$N(CO^+)$ $(10^{13}$ molecules cm$^{-2})$	N_{CO^+}/N_{CN}
01.04.1986	5530	0.643	1.288	8.53
02.04.1986	5546	0.507	1.040	7.88
02.04.1986	5556	0.346	0.710	5.38
04.04.1986	5602	0.106	0.229	5.45
04.04.1986	5619	−0.09	—	—
06.04.1986	5675	0.711	1.612	15.35
06.04.1986	5685	0.554	1.259	11.99
07.04.1986	5693	0.341	0.789	17.15
07.04.1986	5716	0.396	0.920	20.00
08.04.1986	5732	−0.071	—	—
08.04.1986	5751	0.471	1.121	12.88
13.04.1986	5881	1.467	3.918	—
14.04.1986	5913	−0.156	—	—
14.04.1986	5949	−0.141	—	—

from the values of the total brightness (see Table 1). For this, first the CO^+ radiation flow equation of A'Hearn & Vanysek (1986) was used. Following to this equation, the estimated continuum in the CO^+ filter is partly stronger than the total flux observed in the filter because the emission band is weak compared to the continuum background. Consequently, different negative values for the flow of the band are obtained. This effect can also be observed at other comets (Wolf & Vanysek, 1987).

Table 2 shows the CO^+ radiation flow and column density as well as the proportion between the CO^+ and the CN column density.

4. Conclusions

(*a*) Regarding Figs. 1 and 2 it becomes clear that the results calculated for the large diaphragms apply well to the data determined for the smaller diaphragms. This suggests that the utilised method to avoid negative values for the CO^+ radiation flow provides correct results.

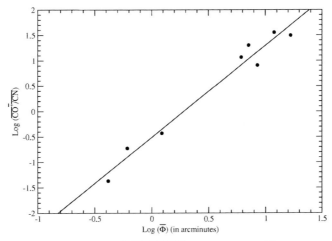

Figure 1. $\overline{N_{CO^+}/N_{CN}}$ as function of $\overline{\Phi}$

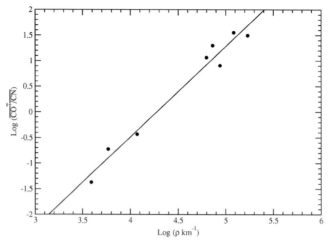

Figure 2. $\overline{N_{CO^+}/N_{CN}}$ as function of the distance ρ

(*b*) A possible explication for the increase of the curve in Fig. 2 is that the CN column density decreases the bigger the distance is, i.e., the lifetime of CN is shorter than the one of CO$^+$ (Schlosser *et al.*, 1992). The CN molecule photodissociates in C and N with a rate of 3.2×10^{-6} s^{-1} with a minimal sun activity (Huebner *et al.*, 1992). On the other hand, the lifetime of the CO$^+$ ion increases along with a larger distance to the comet since the electronical recombination and also the reaction with neutral water is reduced (Häberli *et al.*, 1995).

(*c*) The data of 1P/Halley comet make certainly a contribution to the question of resemblances and diversities of comets (for ex. Hyakutake (C/1996 B2) and Hale-Bopp (C/1995 O1)), which is of a basic interest. The optical data also help to make a connection betweeen data measured *in situ* (observations of spacecrafts) and earthbound data. The data of this work give a spacious resolution (column density as function of the cometary distance) which can be compared to data of the Giotto spacecraft measured *in situ*, or to results of comet missions - as for example Rosetta.

References

A'Hearn, M. F. & Vanysek, V. 1986, *IHW Photometry and Polarimetry New Letter*, 1986

Celnik, W. E. & Schmidt-Kaler, Th. 1987, *A&A*, 187, 233

Celnik, W. E., Koczet, P., Schlosser, W., Schulz, R., Svedja, P., & Weißbauer, K. 1988, *A&AS*, 72, 89

Haeberli, R. M., Altwegg, K., Balsinger, H., & Geiss, J. 1995, *A&A*, 297, 881

Huebner, W. F., Keady, J. J., & Lyon, S. P. 1992, *Ap&SS*, 195, 1

Schlosser, W., Schulz, R., Klawuhn, L., & Stüwe, J. A. 1992, *BAAS*, 24, 1025

Voelzke, M. R. 1996, *PASP*, 108, 1063

Voelzke, M. R., Schlosser, W., & Schmidt-Kaler, Th. 1997, *Ap&SS*, 250, 35

Wolf, M., & Vanysek, V. 1987, *Bull. Astr. Inst. Czechosl.*, 38, 136

Icy Bodies of the Solar System
Proceedings IAU Symposium No. 263, 2009
J.A. Fernández, D. Lazzaro, D. Prialnik & R. Schulz, eds.
© International Astronomical Union 2010
doi:10.1017/S1743921310001948

Near infrared photometry of comet C/2005 E2 (McNaught)

Enos Picazzio[1], Elysandra Figueredo[1], Amaury Augusto de Almeida[1], Claudia Mendes de Oliveira[1], and Klim Ivanovich Churyumov[2]

[1]Department of Astronomy, Institute of Astronomy, Geophysics and Atmospheric Sciences, University of São Paulo,
Rua do Matão 1226, Cidade Universitária 05508-900 São Paulo SP, BRAZIL
email:picazzio@astro.iag.usp.br

[2]Astronomical Observatory, Kyiv Shevchenko National University,
Box 04053, Observatorna str., 3, Kyiv, Ukraine
email: klim.churyumov@observ.univ.kiev.ua

Abstract. We present results of JHK photometry for comet C/2005 E2 (McNaught). Observations were made with the OSIRIS imager at the SOAR telescope, on Sept. 26-27, 2005. The dependence of the main parameters on the angular distance from the cometary photometric nucleus is discussed. We considered concentric rings around the photometric nucleus of the comet, with the radius determined by the aperture. Integrated flux in the ring, surface brightness, J-H and H-K color indices, and mean flux decay were obtained for each ring. A phometric radial profile was also obtained by tracing average values of the ace brightness for increasing values of radii in isophotes which allowed us to compute the azimuthally averaged surface brightness of the coma. The images centered on the photometric nucleus show three jets from active areas on the nucleus.

Keywords. comets: general, comets: individual C/2005 E2 (McNaught), techniques: image processing

1. Observations and processing

Comet C/2005 E2 (McNaught) was discovered on Mar. 12, 2005, when it was 15.5 mag., with a uniform 7 arcsec coma and 13 arcsec tail in phase angle 250°. It is a non-periodic comet (eccentricity = 1.0001388), with an orbital plane inclination of 17°, perihelion distance = 1.5196 AU, and perihelion on Feb. 23.4729229, 2006.

We observed C/2005 E2 on Sept. 26-27, 2005, at a geocentric distance of $\Delta = 2,492$ AU, a heliocentric distance of $r = 2.050$ AU, and a phase angle of 22.9^o, with a magnitude $m_V = 11.5$. Images were taken with OSIRIS (Ohio State InfraRed Imager/Spectrometer, F/7, FOV = 80 arcsec, and plate scale = 0.139 arcsec/pixel) at the 4.1-m SOAR (Southern Astrophysical Research telescope, Cerro Pachón, Chile). FWHM in arcsec of individual images in the wavebands are: 0.88 (J), 0.84 (H) and 0.83 (K). All images were corrected in terms of OSIRIS linearity coefficients and flatfield. Bad pixel mask and sky subtraction were accomplished using IRAF and Cirred (CTIO IR Reduction Package). The calibration star is P9181 (S234-E) from Persson *et al.* (1998), with the following magnitudes: J = 12.464 ± 0.011, H= 12.127 ± 0.008, and K= 12.095 ± 0.007.

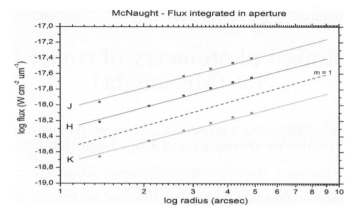

Figure 1. Surface brightness profile of Comet C/2005 E2 (McNaught). Also, plotted for reference is the case $m = 1$ line for steady state model.

2. Discussion

2.1. *Surface brightness profile*

Standard models invoking spherically symmetric constant dust production and optically thin coma predict that the surface brightness will fall with nucleocentric distance as r^{-1} [Gehrz & Ney (1992)]. Integrating the surface brightness over a circular aperture of the angular radius centred at the comet nucleus we obtain the brightness with a logarithmic slope m ($m = 1$ for a steady-state isotropic model and $m = 1.5$ for a radiation pressure dominated case).

As it can be seen from Fig. 1, the surface brightness of comet McNaught falls approximately as r^{-1}. For apertures smaller than 1 arcsec the surface brightness falls steeper, but there the fitting is not good.

2.2. *Integrated magnitude*

The dependence of the surface brightness on the aperture for comets McNaught and 1P/Halley is shown in Fig. 2. Comet McNaught presents similar behaviour in the J, H and K bands, but for comet Halley the similarity on J and K bands appears only on March 10 [Woodward *et al.* (1996)]. The surface brightness varies more steeply for comet Halley.

The magnitudes integrated in 10 arcsec (m) and the absolute magnitudes ($H_0 = m - 5 \log \Delta - 2.5n \log r$) in each waveband are: J ($m = 11.75$, $H_0 = 6.33$), H ($m = 11.29$, $H_0 = 5.84$) and K ($m = 11.28$, $H_0 = 5.74$). The photometric index (n) for the pre-perihelion phase was taken from Yoshida (2008).

The color of comet McNaught seems to be bluer than the Sun (Fig.3). It is known that fine grains scatter blue light much better than red light. Comet Halley, for example, showed fine dust grains near the nucleus.

2.3. *Color indices*

Color indices for comets McNaught and Halley, and Kuiper Belt Objects (KBOs) are shown in Table 1. Comet McNaught is very similar to some KBOs (Classical, Scattered and Resonant).

2.4. *Morphology of comet McNaught (C/2005 E2)*

Broadband images in J, H and K are shown in Fig.5. Three jets are projected on the plane of the sky. They seem to be very similar in the different bands. The K band image

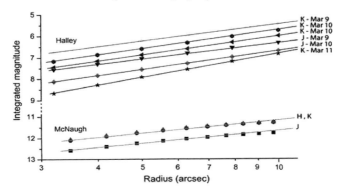

Figure 2. Integrated magnitude for comets 1P/Halley and C/2005 E2 (McNaught).

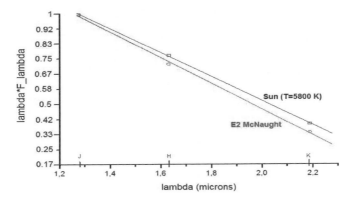

Figure 3. Color of comet McNaught and for the Sun.

Table 1. Color index for comets C/2005 E2 and 1P, and for KBOs.

Object	J - H	H - K	J - K	Nature	Ref
McNaugth	0.48 ± 0.01	0.01 ± 0.01	0.49 ± 0.01	Oort Cloud	A
Halley	0.48 ± 0.02	0.17 ± 0.03	0.65 ± 0.05	Jupiter's family	B
Halley	0.69	0.25	0.94	Jupiter's family	G
1999 CD$_{158}$	0.44 ± 0.10	0.03 ± 0.10	0.47 ± 0.09	Classical KBO	C,D
2000 OK$_{67}$	0.46 ± 0.08	0.04 ± 0.07	0.50 ± 0.09)	Classical KBO	C
1996 GQ$_{21}$	0.48 ± 0.07	0.05 ± 0.08	0.68 ± 0.11	KBO-SDO*	D,E,F
1998 VG$_{44}$	0.41 ± 0.06	0.01 ± 0.08	0.42 ± 0.08	Resonant KBO	E
1999 RZ$_{253}$	0.48 ± 0.09	0.10 ± 0.09	0.58 ± 0.06	Classical KBO	E,F

Notes: * SDO - Scattered Disk Object; Ref: A - this paper; B - Matsuura *et al.* (1987); C - Delsanti *et al.* (2004); D - Delsanti *et al.* (2006); E - McBride *et al.* (2003); F - Deressoundiram *et al.* (2003); G - Woodward *et al.* (1996).

suggests that the nucleus is rotating counter clockwise. Jets of gas and dust from comet nuclei are common features among comets. They come from active regions on the nucleus surface. Gas jets were observed in comets C/1996 B2 Hyakutake, C/1995 O1 Hale-Bopp, 109P/Swiff-Tuttle (found in long-slit spectra) and for the first time in comet 1P/Halley (A'Hearn *et al.* (1986)).

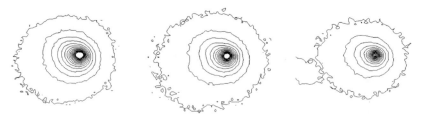

Figure 4. Isophotes of comet McNaught. From left to right, respectively, in J, H and K bands. The isophotes, in logarithmic scale, are asymmetric about the nucleus and show extension along anti-solar direction.

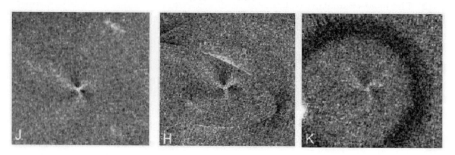

Figure 5. Morphology of comet McNaught; from left to right in J, H and K bands. North is up and East is right. Field of view in arcsec: J (32×29), H (27×28), K (27×28). Plate scale (in km) is indicated.

References

A'Hearn, M. F., Hoban, S., Birch, P. V., Bowers, C., Martin, Ralph., & Klinglesmith, D. A. 1986, *Nature*, 324, 649

Delsanti, A., Peixinho, N., Doressoundiram, A., Boehnhardt, H., Barucci, M. A., & Large Program. 2004, *Bulletin of the American Astronomical Society*, 36, 1102

Delsanti, A., Peixinho, N., Boehnhardt, H., Barucci, A., Merlin, F., Doressoundiram, A., & Davies, J.K. 2006, *AJ*, 131, 1851

Doressoundiram, A., Tozzi., G. P., Barucci, M. A., Boehnhardt, H., Fornasier, S., & Romon, J. 2003, *AJ*, 125, 2721

Gehrz, R. D., Ney, E. P. 1992, *Icarus*, 100, 162

Matsuura, O. T., Picazzio, E., & Kawara, K. 1987, *PASP*, 99, 154

McBride, N., Green, S. F., Davies, J. K., Tholen, D. J., Sheppard, S. S., Whiteley, R. J., & Hillier, J. K. 2003, *Icarus*, 161, 501

Persson, S. E., Murphy, D. C., Krzeminski, W., Roth, M., & Rieke, M. J. 1998, *AJ*, 116, 2475

Yoshida, S. 2008, *http://www.aerith.net/comet/catalog/2005E2/2005E2.html*, Updated on March 12

Icy Bodies of the Solar System
Proceedings IAU Symposium No. 263, 2009
J.A. Fernández, D. Lazzaro, D. Prialnik & R. Schulz, eds.
© International Astronomical Union 2010
doi:10.1017/S174392131000195X

Some active processes in comet icy nuclei: nucleus splitting and anti tail formation

Kh. I. Ibadinov, A. M. Buriev, and A. G. Safarov

Institute of Astrophysics, Academy of Sciences of Tajikistan, Tajikistan
email: Ibadinov@mail.ru

Abstract. The most dramatic display of variable activity of a comet is splitting of the nucleus. For the purpose of revealing the trends of splitting of comet nuclei and of formation of abnormal cometary tails, we have created two catalogues of comets: a catalogue of split nuclei, containing 99 comets, and a catalogue of comets with abnormal tails, including 60 objects. Statistical investigation reveals some general trends of these phenomena. The greatest number of recorded cases of nucleus splitting and abnormal tail (60%) occurs within an interval of heliocentric distance ranging from 0.6 AU to 1.6 AU (maximum at 1.1 AU) and geocentric distance ranging from 0.6 AU to 1.8 AU (maximum at 1.15 AU). Splitting of nuclei and abnormal tails are more often (75%) recorded close to the perihelia of the cometary orbits. Only 16% of splitting comets also exhibit abnormal tails. Some cases of nuclear splitting and large velocity (some km/s) eruptions of dust from a nucleus, as well as cases of abnormal tails developed at large heliocentric distances, may indicate collisions of comet nuclei with other bodies. Our results are of interest for the physics of comets, and for the distribution of meteoroids in solar system.

Keywords. Comets, Comet: nucleus, Come: splitting

1. Introduction

Comets are extremely variable objects, exhibiting brightness flashes, gas-dust jets, halos, heterogeneity in plasma tailis, abnormal tails and splitting of the nucleus. The most dramatic display of variable activity of a comet is nucleus splitting. As a result of splitting, observable daughter comets and meteoroid stream are formed; sometimes the comet disappears completely. Another kind of unusual activity of comet nuclei is the abnormal direction of the tail: towards the Sun rather than in the antisolar direction. It was suggested that such abnormal tails result from synchronous emission of meteoroid particles from the nucleus towards the Sun (Bredikhin 1934; Vsekhsvytskiy 1958). Splitting of the nucleus and formation of an abnormal tail attest to active processes that may take place in cometary nuclei. It is possible that both phenomena are the consequence of one process. The reasons and mechanisms of splitting of the nucleus and anti-tail formation are not always known. Very often it is not possible to define the exact time of onset of these phenomena.

The purpose of the present work is to reveal from observations the general trends of nucleus splitting and abnormal tail formation. To achieve this purpose, we we have created two catalogues. The first includes 99 comets for which nuclear splitting, or obvious signs of splitting have been recorded; it is a continuation of similar catalogues of Konopleva (1967), Golubev (1975), Dobrovolsky & Gerasimenko (1987) and Ibadinov (1998). The second catalogue includes 60 comets for which an abnormal tail was observed and provides a considerable supplement to Demenko's (1965) catalogue, which included only 16 comets. These catalogues will be published in a separate paper. On the basis of these catalogues, we investigate the conditions of nucleus splitting and formation of an abnormal tail in comets. We consider the dependence of these phenomena on the heliocentric (r) and geocentric (Δ) distances of the comet at the moment of their detection, on the inclination of the orbital plane of the comet with respect to the ecliptic (i), and on the perihelion distance (q) of the orbit. In the present paper we focus on statistical results, presenting the number of splitting comets N and the number of comets with an abnormal tail N_t as functions of each of the parameters r, Δ, i and q.

In Fig.1 (a-b-c) histograms are presented, showing the dependence of the number of splitting comets N on heliocentric distance r and geocentric distance Δ at the moment of registration of the splitting and the dependence of N on perihelion distance of the comet's orbit. Similar histograms are presented in Fig. 2(a–b–c) for comets with abnormal tails. It is noticeable that the maximum number of splittings and abnormal tails occurs within an interval of heliocentric and geocentric distances between 0.6–1.6 AU, which most likely reflects the visibility conditions of comets, i.e., near the orbit of the Earth, nucleus splitting and abnormal tails are more easily observed than far from it. The same effect is exhibited in Fig. 1(a) and 2(c), showing the dependence of N and N_t on q, where maximum is obtained again in the range 0.6–1.6 AU. In our opinion, this effect also is caused by observational bias (near Earth's orbit).

By contrast, the results presented in Fig. 1(d) and Fig. 2(d) can shed light on the general trends and the most probable mechanisms responsible for splitting of comet nuclei and formation of abnormal tails. In Fig. 1(d) the dependence of the number of comets N on the difference between the heliocentric distance r where splitting was recorded and perihelion distance q, i.e. $(r-q)$, is presented. We note that the maximum is obtained near $(r-q) \approx 0$. This result confirms Dobrovolsky & Gerasimenko (1987) and Ibadinov (1998) conclusions, that the maximum number of splitting of comets occurs near perihelion. We obtain similar results for the abnormal tails of comets: in Fig. 2(d) the dependence of the number of comets with an abnormal tail on $(r - q)$ is presented and the maximum

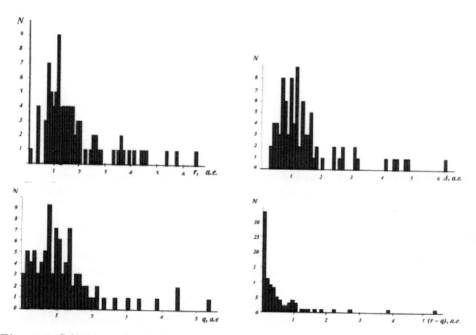

Figure 1. Splitting nuclei of comets. Dependence of number of comets N on: (a) heliocentric distance r; (b) geocentric distance Δ; (d) perihelion distance q; (e) $(r - q)$.

of the distribution $N(r - q)$ occurs again near $(r - q) \approx 0$, that is, in the vicinity of the perihelion.

In the distribution $N(r, \Delta, q, (r-q))$ separate peaks at large distances are observed, in the interval 3–5 AU. Here, the influence of Jupiter is considerable. This group of comets includes comet Hailey-Bopp, which has collapsed under the influence of Jupiter. It may be also possible that in this region collisions of comets with asteroids and meteoroids take place as well.

In summary, from the statistical analysis of comet nucleus splitting it is established with sufficient confidence that the highest probability for splitting is obtained near orbital perihelion. The probability of detecting nucleius splitting is highest when the comet is within heliocentric and geocentric distances 0.6–1.6 AU, due to observational bias (best observing conditions).

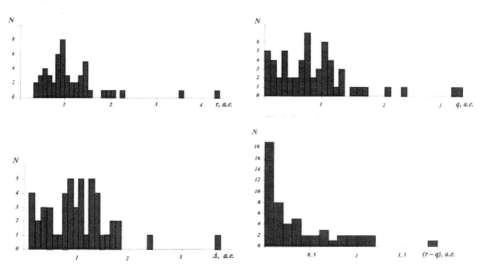

Figure 2. Anti-tail of comets. Dependence of number of comets N with anti-tail on: (a) heliocentric distance r; (b) perihelion distance q; (d) geocentric distance Δ; (e) $(r - q)$.

2. Conclusions

We have created catalogues of recorded comet nucleus splittings and abnormal cometary tails. Based on statistics, it is established that the probability of detecting the splitting of comet nuclei and the formation of abnormal tails is high, if these phenomena have occurred within heliocentric and geocentric distances in the range 0.6–1.6 AU. This is probably connected with observing conditions. The greatest number of splittings and abnormal tails is recorded near the perihelia of the cometary orbits. A plausible reason for splitting may be the tidal influence of the Sun and of Jupiter. We find that abnormal cometary tails are also formed at large distances from the Sun, between the orbits of Mars and Jupiter. Only for 16% of splitting comets were abnormal tails also detected.

References

Bredikhin, F. A. 1934, *About tails of comets*, Rus., M: GTTI

Golemubev, V. A. 1975, *Problems of Space Physics*, 10, 23

Demenko, A. A. 1965, *About tails of comets*, (dissertation), Kiev

Dobrovolsky, O. V., & Gerasimenko, S. I. 1987, *Bull. of Institute of Astrophysics of AS RT* 77, 3

Ibadinov, K. I. 1998, *Space Research Institute of Russian Academy of Sciences*, p. 296

Konopleva, V. P. 1967, in: *Active processes in comets*, (Kiev: Naukova Dumka), p. 57

Vsekhsvytskiy, S. K. 1958, *Physical properties of comets*, M. Prosveshenie

Icy Bodies of the Solar System
Proceedings IAU Symposium No. 263, 2009
J.A. Fernández, D. Lazzaro, D. Prialnik & R. Schulz, eds.

© International Astronomical Union 2010
doi:10.1017/S1743921310001961

A Simple Model for the Secular Light Curve of Comet C/1996 B2 Hyakutake

Eduardo Rondón and Ignacio Ferrín

Centro de Física Fundamental, Facultad de Ciencias, Universidad de Los Andes,
Mérida 5101, Venezuela
email: erondon@ula.ve

Abstract. The secular light curves of comets (Ferrín, 2005) give a large amount of physical information on the cometary nucleus. We have developed a model that allows the prediction of a secular light curve, from which we derive parameters like the orientation of the rotational axis (I, ϕ) and optical thickness of the cometary coma. The model is based on the paper published by (Cowan & A'Hearn, 1979). To do the calculation we found a correlation between the water production rate and the reduced magnitude. We obtain probable orientations of the nucleus pole for several combinations of parameters for comet C/1996 B2 Hyakutake.

Keywords. Hyakutake, Secular Light Curve, Water Production rate, Sublimation

1. Introduction

The secular light curve of a comet is a plot of the visual reduced magnitude vs logarithm of Sun-Comet distance. These light curves have a large amount of information on the cometary coma and nucleus: composition, age, size, structure, activity and cometary opacity (Ferrín 2005, Rondón 2009). These parameters are:

1)R_{ON} (AU): The turn on distance of the coma. Physically R_{ON} corresponds to the onset of steady activity. 2)R_{OFF} (AU): The turn off distance of the coma. It measures the end of activity of the nucleus. 3)V_{ON}: The magnitude at which the nucleus turns on. 4)V_{OFF}: The magnitude at which the nucleus turns off. 5)R_{OFF}/R_{ON}: An asymmetry parameter for the secular light curve. 6) m(1, 1): The absolute magnitude of the coma. For the nucleus the following parameters are listed:

8)V (1, 1, 0) = V_{NUC}: Absolute nuclear magnitude. 9)$A_{SEC} = V_{NUC} - m(1,1)$: Amplitude of the secular light curve. 10)D_{EFFE}: The effective diameter of the comet. 11)P-AGE = Photometric Age, it is related to the loss of volatiles. P-AGE = $\frac{1440}{A_{SEC}}$*(R_{ON}+R_{OFF}) comet years (cy).

2. Factors Affecting the Sublimation of a Comet

It is known that the main factor that affects the sublimation of a comet is the orientation of the axis of rotation (Cowan & A'Hearn, 1979). In our model we consider this parameter of the first order. We have assumed a chemical composition of water to study the vapor pressure and the heat latent function. The equation that describes the vaporization rate of a comet is the energy conservation equation, given by:

$$F_0(1 - A_v)r_H^{-2}\overline{cos(\theta)} = (1 - A_{ir})\sigma T^4 + Z(T)L(T) + K\frac{\partial(T)}{\partial(z)} \qquad (2.1)$$

A_{ir} is the Infrared albedo, r_H Sun-Comet Distance, $\overline{cos(\theta)}$ is the projection factor, σ is the Steffan Bolztmann constant, T is the temperature, Z(T) is the sublimation function, L(T) latent heat function, K thermal conductivity constant, z layer depth.
The projection factor is given by the piecewise function:
Where i is the angle between the sun-comet direction and the rotation axis, and b is the latitude of the sublimation point. The projection factor is equal to 1 when the angle between the normal to the surface and the incident solar radiation is equal to 0 degree, and the projection factor is equal to 0 when the angle between the normal to the surface and the incident radiation is equal to 90 degree.
For pure ice the vaporization of the comet is given by:

$$Z(T) = \frac{P(T)m}{2(\pi)kT} \tag{2.2}$$

where : $P(T)$ is the vapor pressure function, m is the molecular weight, k is the ideal gas constant. The vapor pressure function is given by:

$$\log(P) = \frac{-2445.5646}{T} + 82312\log(T) - 0.01677006T + 120514x10^{-5}T^2 - 6.757169 \tag{2.3}$$

The latent heat function is given by:

$$L(T) = 12420 - 4.8T \tag{2.4}$$

If we know the vapor pressure function and the latent heat function we can solve Z(i,b) at a variaty of values of i and b using the energy conservation equation(1) through $\overline{cos\theta}$. Then, for estimate the total sublimation we evaluate the integral using (Eq. 2.5)

$$Z_{total} = \overline{Z(i)} = \frac{1}{2}\int (Z(i,b)cos(b)db) \tag{2.5}$$

3. Modeling the Light Curve of Comet C/1996B2 (Hyakutake)

The first step for modeling the light curve is to calculate the sublimation rate of water, using our model. Then, we can find the correlation equation between the reduce visual magnitude (Fig. 1a) and the water production rate, (Fig. 1b), plotting the reduce visual magnitude vs Log(r/q), (Fig. 2), where r is the heliocentric distance and q is the perihelion distance.

The correlation law between the reduce visual magnitude and the water production rate for comet Hyakutake is (Fig. 2) .

$$m = 139.9 - 4.59logQ \tag{3.1}$$

If we solve this equation for Q we find the same relation obtained by Jorda (2008) (Eq. 3.2) but with slightly different coefficients:

$$logQ = 30.48 - 0.22m \tag{3.2}$$

With equation (3.1) we calculate the reduce visual magnitude as a function of the heliocentric distance, then we can plot the light curve considering an obliquity of $I = 90^o$ and varying the pole orbital longitude (Fig. 3) as a first approximation. In these plot we can see the effects produced by changing the orientation of the axis of rotation, and is able to explain the asimmetry of the light curve. We have calculated the standard deviation for our model with the observer data, and found that the (Schleicher 2003)

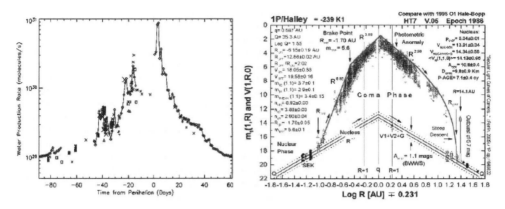

Figure 1. Figure 1a. Water production rate vs time from perihelion using observations by SOHO, SWAN and other authors (Combi, M., Makinen, T., *et al.*) (http://www-personal.umich.edu/~mcombi/SWANhya/index.html). Figure 1b. Secular light curve of Hyakutake Comet (Ferrín. 2009).

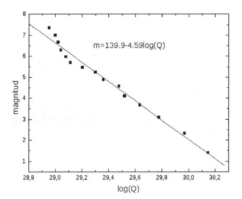

Figure 2. Calibration curve of the light curve, showing the equation of the magnitude as a function of the logarithm of the sublimation rate for comet C/1996 B2 Hyakutake, ajusted by least squares.

solution by the orientation of the axis of rotation $I = 108^o, \phi = 228^o$ has the smallest standar deviation (Fig. 4). We can also see the simmetry in the solution.

For small heliocentric distance, the light curve is affected for the increment of the dust rate produced for the comet, we are mesurement the effect produced by the dust and we have found the correlation beetwen the optical thickness and the heliocentric distance (Eq.3.3)(Fig. 5). Once found the correlation equation we can correct the reduce magnitude through equation. (Eq. 3.4).

$$\tau = 1.92 \mp 3.31 log(r/q) \qquad (3.3)$$

$$m_\Delta = m_{model} + 2.5 log(e^\tau) \qquad (3.4)$$

We can see (Fig. 6) that including the dust producion rate in the calculation of the optical thikness gives a better aproximation to the envelope of the observations.

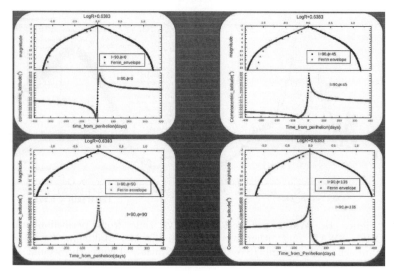

Figure 3. Seculars light curves calculate for the Hyakutake Comet considering $I = 90^o$ and varying the longitude of the orbital pole.

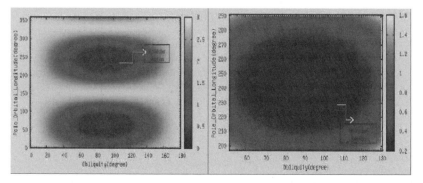

Figure 4. Contorn map made for our model for comet C/1996 B2 Hyakutake. The plot show the obliquity, I, versus the pole orbital longitude, ϕ. In this plot see the standard deviation for each of the orientations of the axis of rotation with a color profile of one degree.

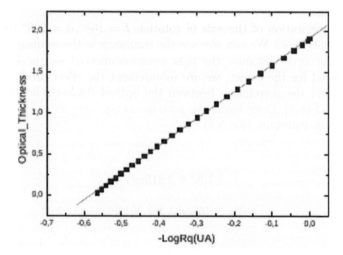

Figure 5. Optical thickness of the coma vs heliocentric distance.

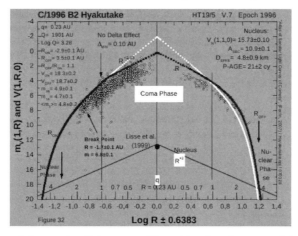

Figure 6. Modeling of the secular light curve. The white dots is our model without considering the effect produced by dust, and the blue dots is our model considering the effect produced by dust.

4. Conclusions

We have developed a theoretical model capable of reproducing the observational data of the secular light curve of comet C/1996 B2 Hyakutake. Since our model calculates the sublimation rate of water, we have found a correlation equation between the reduce visual magnitude and the sublimation rate, as a result yielding the same expression found by (Jorda 2008). The modeling of the envelope of the secular light curve is dependent on the orientation of the rotation axis and the dust production rate for small heliocentric distances. If we have the pole orientation we can calculate the water production rate as a function of time and the envelope of the secular light curve. We found a envelope that reproduces the same results obtained by (Schleicher 2003) with minimal standard deviation.

References

Cowan, J. & A'Hearn. 1979, *Icarus*, 178, 493
Ferrín, I. 2005, *Icarus*, 178, 493
Ferrín, I. 2009, *Planetary and Space Science*, in press
Jorda, L., Crovisier, J., & Green, D. W. E. 2003, *ACM 2008*, 8046
Rondón, E. 2009, *Master Thesis, University of Los Andes*
Schleicher, D. & Woodney, L. 2003, *Icarus*, 162, 190

Icy Bodies of the Solar System
Proceedings IAU Symposium No. 263, 2009
J.A. Fernández, D. Lazzaro, D. Prialnik & R. Schulz, eds.

© International Astronomical Union 2010
doi:10.1017/S1743921310001973

Observations of comets and minor planets at Kiev comets station (585)

Alexander R. Baransky, Klim I. Churuymov, and Vasyl A. Ponomarenko

Astronomical Observatory, Kyiv Shevchenko National University,
Box 04053, Observatorna str., 3, Kyiv, Ukraine
email: `klim.churyumov@observ.univ.kiev.ua`

Abstract. We present the results of astrometric and photometric observations of comets and minor planets obtained at the Kiev comet station (Code MPC 585) of the Astronomical Observatory of Kyiv Shevchenko National University in 2006-2009. The 2318 position observations of 176 comets, 302 observations of 57 numbered minor planets, and 220 observations of 30 unnumbered minor planets were obtained. The accuracy of the astrometric observations of the comets is analyzed.

Keywords. Astrometry, ephemeris, comets: general, minor planets

1. Introduction

CCD astrometric monitoring of new and short period comets and new asteroids is very important for the determination and improvement of their orbits and the study of the orbital evolution of new small bodies of the Solar System (Steel & Marsden, 1996). In 2006 a programme was started at observation station of the Astronomical observatory of the Kiev National University in Lisnyky ("Kiev Comet Station, code MPC 585 performing astrometric and photometric observations of comets and minor planets of Solar System. Observations are obtained according to the technique of the Minor Planet Center (MPC) (Holmes, 1995) with of the telescope reflector AZT-8 (D = 0.7 m), and CCD ST-8E which is accomodated in the primary focus of the telescope (focus of system $F = 2.8$ m, focal ratio $f/4$). The above noted equipment gives the chance to detect in integrated light the images of asteroids to a brightness of 21^m, and comets to 19.5^m.

2. The program of monitoring

The programme of monitoring minor Solar System bodies is directed to the objectives: 1) Observations of just discovered objects for the purpose of confirming their actually existence and receiving first astrometric and photometric observations. The list of objects which require confirmation is daily renewed by the Minor Planet Centre internet page – The NEO Confirmation Page. 2) Astrometric observations of known long- and short-period comets for the purpose of determining changes in the orbital characteristics which are connected to non-gravitational and gravitational perturbations. 3) Photometric observations of comets (integral and nucleus comet magnitude, size of the coma, degree of a central condensation of in the coma, length and position angle of the tail).

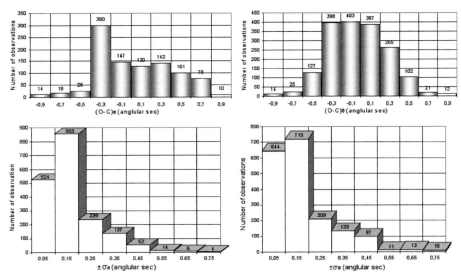

Figure 1. Distributions of the residuals $(O - C)_\alpha \pm \sigma_\alpha$ & $(O - C)_\delta \pm \alpha_\delta$.

4) Observations of unique comets during rare outbursts, split comets, appearance of peculiar tails. For example, we obtained a long series of observations of the unique comet 73P/Schwassmann-Wachmann, which nucleus had split into a considerable quantity of fragments (we identified 16 separate fragments of the comet). 5) Astrometric and photometric observations of asteroids which approach to the Earth (NEOs) and asteroids of the Main Belt.

3. Result and analyses of observations

The astrometric observations of comets and their accuracy are shown in the table 4

Figure (Fig. 1) shows diagrams of the accuray distribution of the observed (O-C) in angular sec for Right Accession $(O - C)_\alpha \pm \sigma_\alpha$ and for Declination $(O - C)_\delta \pm \sigma_\delta$. It is seen, that the basic quantity of observations is received with an accuracy of 0 - 0.4" for $(O - C)_\alpha$ and $(O - C)_\delta$. The diagram $(O - C)_\alpha$ is a little asymmetric, which is probably connected with restriction of accuracy of the telescopes clock mechanism. On the diagrams for $\pm\sigma_\alpha$ and $\pm\sigma_\delta$ the error of the majority of observations does not exceed 0,2".

4. Conclusion

Between 2.04.2006 - 9.05.2009 (140 nights) a total of 2318 observations were obtained of 176 comets, and 522 observations of 87 asteroids. The maximum quantity of observations has been obtained with an accuracy from 0,1" to 0,2". Results of astrometric and photometric observations were contributed to the database of the Minor Planet Centre (MPC), the International Comet Quarterly (ICQ), and to the database of the British Astronomical Association (BAA).

Table 1. Accuracy of astrometric observations of comets. Column 1: comet name, column 2: period of observations, column 3: number of observations, column 4: residual (O-C) for R.A., column 5: (O-C) for Declination.

Comets	Period observ.	N	$(O-C)_\alpha \pm \sigma_\alpha$	$(O-C)_\delta \pm \sigma_\delta$
4P/Faye	2006	30	0.22 ± 0.11	0.03 ± 0.09
6P/d'Árrest	2008	17	0.10 ± 0.17	-0.04 ± 0.15
8P/Tuttle	2008	8	-0.14 ± 0.22	0.12 ± 0.26
17P/Holmes	2007 − 2008	39	0.31 ± 0.11	0.29 ± 0.08
19P/Borrelly	2008 − 2009	4	-0.67 ± 0.11	0.50 ± 0.22
22P/Kopff	2009	18	-0.67 ± 0.19	-0.19 ± 0.16
29P/Schwassmann-Wachmann	2006, 2008 − 2009	33	0.15 ± 0.16	-0.07 ± 0.18
41P/Tuttle-Giacobini-Kresak	2006	2	0.30 ± 1.20	0.05 ± 0.65
44P/Reinmuth	2007 − 2008	10	0.20 ± 0.23	0.25 ± 0.20
46P/Wirtanen	2007 − 2009	16	-0.20 ± 0.21	-0.11 ± 0.25
50P/Arend	2007 − 2008	14	-0.08 ± 0.18	-0.15 ± 0.07
51P/Harrington	2008	7	-0.23 ± 0.28	0.24 ± 0.14
59P/Kearns-Kwee	2008	10	-0.21 ± 0.16	-0.45 ± 0.13
65P/Gunn	2008	4	0.11 ± 0.08	-0.71 ± 0.14
67P/Churyumov-Gerasimenko	2009	9	0.49 ± 0.26	0.19 ± 0.10
68P/Klemola	2008	4	0.35 ± 0.77	-0.05 ± 0.53
74P/Smirnova-Chernykh	2007 − 2009	23	-0.04 ± 0.16	0.19 ± 0.09
77P/Longmore	2009	23	-0.05 ± 0.11	-0.21 ± 0.13
79P/du Toit-Hartley	2008	3	0.17 ± 0.18	-0.77 ± 0.37
84P/Giclas	2006	9	-0.33 ± 0.20	-0.84 ± 0.30
87P/Bus	2007	3	0.77 ± 0.35	0.43 ± 0.64
88P/Howell	2009	5	0.52 ± 0.07	0.05 ± 0.23
98P/Takamizawa	2006	3	0.73 ± 0.18	0.17 ± 0.52
110P/Hartley	2007 − 2008	3	-0.33 ± 0.18	0.30 ± 0.36
112P/Urata-Niijima	2006	11	-0.20 ± 0.26	-0.44 ± 0.29
114P/Wiseman-Skiff	2006	3	-0.20 ± 0.87	0.33 ± 0.62
116P/Wild	2008 − 2009	8	0.15 ± 0.19	0.10 ± 0.18
124P/Mrkos	2008	10	-0.480 ± 0.06	0.02 ± 0.14
128P/Shoemaker-Holt	2006 − 2008	2	0.10 ± 0.50	0.15 ± 0.25
143P/Kowal-Mrkos	2009	10	-0.18 ± 0.09	-0.03 ± 0.15
144P/Kushida	2008 − 2009	13	0.64 ± 0.29	0.02 ± 0.16
173P/Mueller	2008	5	0.75 ± 0.35	-0.10 ± 0.19
177P/Barnard	2006	50	-0.32 ± 0.06	-0.41 ± 0.10
179P/Jedicke	2007 − 2008	11	0.27 ± 0.44	-0.28 ± 0.18
180P/NEAT	2008	10	-0.22 ± 0.23	0.08 ± 0.26
182P/LONEOS	2007	4	-0.07 ± 0.44	0.15 ± 0.72
183P/Korlevic-Juric	2008	7	-0.65 ± 0.13	-0.62 ± 0.03
188P/LINEAR-Mueller	2008	7	-0.30 ± 0.48	0.10 ± 0.23
189P/NEAT = P/2007 N2	2007	3	0.00 ± 0.40	-0.17 ± 0.87
191P/McNaught = P/2007 N1	2007	11	-0.21 ± 0.42	-0.47 ± 0.21
192P/Shoemaker-Levy	2008	5	0.20 ± 0.31	-0.79 ± 0.26
194P/LINEAR	2008	3	0.18 ± 0.05	0.07 ± 0.20
197P/LINEAR	2008	6	0.36 ± 0.40	0.14 ± 0.41
200P/Larsen	2008	5	-0.03 ± 0.39	0.25 ± 0.14
202P/Scotti	2008	5	-0.25 ± 0.54	-0.33 ± 0.08
205P/Giacobini	2008	28	-0.94 ± 0.13	-0.34 ± 0.07
206P/ Barnard-Boattini = P/2008T3	2008	16	-0.31 ± 0.17	0.01 ± 0.14
210P/Christensen	2009	10	-0.23 ± 0.11	-0.27 ± 0.07
C/2002 VQ94 (LINEAR)	2006, 2008	4	0.22 ± 0.36	-0.16 ± 0.35
C/2003 WT42 (LINEAR)	2006 − 2007	13	-0.19 ± 0.17	0.10 ± 0.17
C/2004 B1 (LINEAR)	2006	45	-0.01 ± 0.09	0.44 ± 0.49
P/2004 VR 8 (LONEOS)	2006	3	0.18 ± 0.22	1.02 ± 0.13
P/2005 JY 126 (Catalina)	2006	4	0.53 ± 0.10	0.15 ± 0.23
C/2005 L3 (McNaught)	2007 − 2008	60	0.38 ± 0.07	0.40 ± 0.09
C/2005 S4 (McNaught)	2006 − 2008	10	-0.09 ± 0.17	0.47 ± 0.36

Comets	Period observ.	N	$(O-C)_\alpha \pm \sigma_\alpha$	$(O-C)_\delta \pm \sigma_\delta$
P/2005 SB 216 (LONEOS)	2006	5	-0.24 ± 0.20	0.29 ± 0.14
C/2006 CK10 (Catalina)	2006	5	0.46 ± 0.20	-0.4 ± 0.80
P/2006 F1 (Kowalski)	2006	7	0.88 ± 0.26	-0.00 ± 0.31
P/2006 H1 (McNaught)	2006	8	0.64 ± 0.30	0.73 ± 0.17
P/2006 HR 30 (Siding Spring)	$2006 - 2007$	21	0.06 ± 0.19	-0.14 ± 0.10
P/2006 K2 (McNaught)	2006	3	0.89 ± 0.35	0.06 ± 0.90
C/2006 K4 (NEAT)	2006	4	0.43 ± 0.35	-0.44 ± 0.44
C/2006 M1 (LINEAR)	2006	9	0.02 ± 0.14	-0.07 ± 0.21
C/2006 M4 (SWAN)	2006	4	-0.43 ± 0.27	-0.35 ± 0.11
C/2006 O2 (Garradd)	2006	5	-0.20 ± 0.31	-0.23 ± 0.50
P/2006 R2 (Christensen)	2006	6	0.10 ± 0.41	-0.64 ± 0.37
C/2006 OF2 (Broughton)	$2007 - 2009$	72	-0.32 ± 0.05	-0.22 ± 0.04
C/2006 Q1 (McNaught)	2009	11	-0.68 ± 0.24	0.53 ± 0.06
P/2006 S1 (Christensen)	2006	13	-0.18 ± 0.21	0.48 ± 0.39
C/2006 S2 (LINEAR)	2006	8	-0.55 ± 0.15	0.69 ± 0.24
C/2006 S3 (LONEOS)	2006	3	0.88 ± 0.90	0.02 ± 0.28
P/2006 S4 (Christensen)	2006	14	0.38 ± 0.27	-0.10 ± 0.33
C/2006 S5 (Hill)	2006, 2008	34	0.39 ± 0.13	-0.10 ± 0.08
P/2006 S6 (Hill)	2006	19	0.28 ± 0.12	0.29 ± 0.11
C/2006 V1 (Catalina)	2007	4	0.33 ± 0.34	0.22 ± 0.24
C/2006 VZ13 (LINEAR)	2007	8	0.54 ± 0.30	0.09 ± 0.36
C/2006 W3 (Christensen)	$2008 - 2009$	83	0.62 ± 0.04	1.38 ± 0.05
P/2006 U1 (LINEAR)	2006	9	0.60 ± 0.25	-0.14 ± 0.12
P/2006 U5 (Christensen)	2006	3	-0.17 ± 0.05	0.05 ± 0.14
C/2007 B2 (Skiff)	$2007 - 2008$	29	-0.24 ± 0.09	-0.20 ± 0.09
C/2007 D1 (LINEAR)	2007	7	0.22 ± 0.10	0.02 ± 0.15
C/2007 E1 (Garradd)	2007	35	0.48 ± 0.13	-0.14 ± 0.13
C/2007 E2 (Lovejoy)	2007	64	0.46 ± 0.14	-0.28 ± 0.14
C/2007 F1 (LONEOS)	2007	6	0.05 ± 0.32	-0.27 ± 0.36
C/2007 G1 (LINEAR)	$2007 - 2008$	15	0.15 ± 0.18	0.12 ± 0.10
P/2007 H1 (McNaught)	2007	10	0.15 ± 0.17	0.06 ± 0.19
C/2007 M1 (McNaught)	2007	10	-0.51 ± 0.29	-0.21 ± 0.34
C/2007 JA21	2007	4	0.12 ± 0.54	0.14 ± 0.59
C/2007 M3 (LINEAR)	2007	27	-0.40 ± 0.17	0.02 ± 0.07
C/2007 N3 (Lulin)	$2007 - 2009$	44	0.01 ± 0.11	0.32 ± 0.07
C/2007 O1 (LINEAR)	2007	5	-0.05 ± 0.18	-0.26 ± 0.42
P/2007 Q2 (Gilmore)	2007	9	-0.41 ± 0.34	-0.31 ± 0.32
P/2007 R1 (Larson)	2007	8	-0.02 ± 0.34	-0.05 ± 0.44
P/2007 R2 (Gibbs)	2007	10	0.34 ± 0.22	-0.22 ± 0.29
P/2007 S1 (Zhao)	2007	5	1.37 ± 0.33	-1.15 ± 0.49
C/2007 S2 (Lemmon)	2007	9	0.33 ± 0.17	-0.19 ± 0.39
C/2007 T1 (McNaught)	2007	5	-0.50 ± 0.20	0.62 ± 0.23
P/2007 T2 (Kowalski)	2007	11	0.60 ± 0.22	0.59 ± 0.21
C/2007 T5 (Gibbs)	$2007 - 2008$	19	-0.28 ± 0.37	-0.41 ± 0.45
P/2007 T6 (Catalina)	2007	9	-0.34 ± 0.32	-0.42 ± 0.19
C/2007 U1 (LINEAR)	2008	10	0.12 ± 0.16	0.30 ± 0.14
P/2007 V1 (Larson)	$2007 - 2008$	9	0.37 ± 0.32	0.31 ± 0.33
C/2007 W1 (Boattini)	$2007 - 2008$	30	-0.30 ± 0.13	-0.37 ± 0.15
C/2007 W3 (LINEAR)	2008	15	0.65 ± 0.32	-0.23 ± 0.36
C/2007 Y1 (LINEAR)	2008	28	-0.28 ± 0.18	0.29 ± 0.19
P/2008 A2 (LINEAR)	2008	18	-0.30 ± 0.23	-0.18 ± 0.22
C/2008 C1 (Chen-Gao)	2008	46	-0.23 ± 0.12	-0.31 ± 0.12
C/2008 E1 (Catalina)	2008	6	0.94 ± 0.15	-0.28 ± 0.27
C/2008 FK75 (Lemmon-Siding Spring)	2009	5	0.04 ± 0.09	$1.01 \pm .096$
C/2008 G1 (Gibbs)	2008	5	0.12 ± 0.09	0.56 ± 0.22
C/2008 H1 (LINEAR)	2008	16	-0.15 ± 0.24	0.40 ± 0.21
C/2008 J1 (Boattini)	2008	33	-0.14 ± 0.07	-0.00 ± 0.13
P/2008 J2 (Beshore)	2008	27	$+0.57 \pm 0.10$	$+0.19 \pm 0.13$

Comets	Period observ.	N	$(O-C)_\alpha \pm \sigma_\alpha$	$(O-C)_\delta \pm \sigma_\delta$
C/2008 J5 (Garradd)	2008	5	-0.48 ± 0.51	0.20 ± 0.28
C/2008 J6 (Hill)	2008	22	0.30 ± 0.17	-0.18 ± 0.20
P/2008 L2 (Hill)	2008	13	0.12 ± 0.07	-0.02 ± 0.07
C/2008 L3 (Hill)	2008	4	1.31 ± 0.06	0.83 ± 0.20
C/2008 N1 (Holmes)	2008	9	-0.10 ± 0.14	0.18 ± 0.12
P/2008 O2 (McNaught)	2008	20	0.411 ± 0.17	0.11 ± 0.17
C/2008 Q1 (Maticic)	$2008 - 2009$	37	-0.29 ± 0.10	0.11 ± 0.11
P/2008 Q2 (Ory)	2008	15	0.41 ± 0.07	-0.03 ± 0.04
P/2008 QP20 (LINEAR-Hill)	2008	8	-0.37 ± 0.15	-0.57 ± 0.17
C/2008 R3 (LINEAR)	2008	6	0.41 ± 0.18	-0.19 ± 0.14

References

Steel, D. I. & Marsden, B. G. 1996, *Earth, Moon, and Planets*, 74, 2
Holmes, A. 1995, *CCD Astronomy*, 2, 1

Part VIII

Space Missions to Icy Bodies: Past, Present and Future

Icy Bodies of the Solar System
Proceedings IAU Symposium No. 263, 2009
J.A. Fernández, D. Lazzaro, D. Prialnik & R. Schulz, eds.

© International Astronomical Union 2010
doi:10.1017/S1743921310001985

New Horizons: Encountering Pluto and KBOs

Leslie A. Young[1] and S. Alan Stern[2]

[1]Southwest Research Institute,
1050 Walnut St., Suite 300,
Boulder CO 80302
email: `layoung@boulder.swri.edu`

[2]Southwest Research Institute,
1050 Walnut St., Suite 300,
Boulder CO 80302
email: `alan@boulder.swri.edu`

Abstract. New Horizons is a NASA mission to explore the Pluto system and the Kuiper Belt. The spacecraft was launched on 19 January 2006 and will begin its encounter studies of Pluto in early 2015, culminating on 14 July 2015 with a close approach just 12,500 km from Pluto. The spacecraft carries panchromatic and color images, IR and UV mapping spectrometers, a radio science package, two in situ plasma instruments, and a dust counter. We describe the capabilities of this instrument suite and the spacecraft, the observations planned for Pluto and its system of satellites, and our plans for KBO flybys to take place late in the 2010s.

Keywords. space vehicles, Pluto, Kuiper Belt

1. Introduction

New Horizons is a NASA mission to explore the Pluto system and the Kuiper Belt. Details of the New Horizons mission has recently been described in a series of papers in a special issue of Space Science Reviews (issue 140, 2008). This paper will recap the New Horizons mission, the mission design, its payload, and the planned observations, with a strong emphasis on the changes since the Space Science Review articles were resubmitted. This update is particularly timely for the planned observations, since the observation plans for the nine days around closest approach are well defined, with sequences built and tested on the spacecraft simulator.

2. Mission Overview

As described in Stern (2008) and Young *et al.* (2008), NASA defined three categories of science goals for a mission to Pluto: Group 1 (required), Group 2 (strongly desired), and Group 3 (desired). Young *et al.* (2008) presents the scientific motivation behind these goals, based on our knowledge of the Pluto system and its role in comparative planetology. Since the discovery of Pluto's moons Nix and Hydra in 2005, the New Horizons science team defined additional goals (Table 1). In general, Nix and Hydra goals paralleled the non-atmospheric goals for Pluto and Charon, but at one level lower in priority. We also took this opportunity to formalize additional science goals for Pluto and Charon that were not originally defined in the AO.

To address these goals, New Horizons carries four remote sensing and three in situ instruments, reviewed in Weaver *et al.* (2008). Alice (Stern *et al.* 2008), an ultraviolet spectrometer covering 465–1880 Å with a spatial resolution of 5 mrad/pixel, will

Table 1. Pluto-system Science Goals

Group 1 Objectives: REQUIRED	
Specified by NASA	Added by New Horizons Science Team
Global geology and morphology of Pluto & Charon Surface composition of Pluto & Charon Pluto's neutral atmosphere and its escape rate	None

Group 2 Objectives: STRONGLY DESIRED	
Specified by NASA	Added by New Horizons Science Team
Time variability	Composition of dark surfaces on Pluto
Stereo of Pluto & Charon	Far-side imaging of Pluto & Charon
High resolution terminator images of Pluto & Charon	Far-side color/comp. of Pluto & Charon
High resolution surface composition of Pluto & Charon	High resolution imaging of Nix & Hydra
Pluto's ionosphere and solar wind interaction	Color & Composition of Nix & Hydra
Search for minor neutral species in Pluto's atmosphere	Shapes of Nix & Hydra
Search for an atmosphere around Charon	
Bolometric Bond albedos for Pluto & Charon	
Map the surface temperatures of Pluto and Charon	

Group 3 Objectives: DESIRED	
Specified by NASA	Added by New Horizons Science Team
Energetic particle environment of Pluto & Charon	Surface microphysics of Pluto & Charon
Refine bulk parameters and orbits of Pluto & Charon	Measure temperatures of Nix & Hydra
Search for magnetic fields of Pluto & Charon	Measure the phase curve of Nix & Hydra
Search for additional satellites and rings	Image Nix and Hydra in stereo
	Education/Public Outreach

be primarily used for studying Pluto's atmosphere with airglow and UV occultations. The Ralph instrument (Reuter *et al.*, 2008) feeds two focal planes, the Multicolor Visible Imaging Camera (MVIC) and the Linear Etalon Imaging Spectral Array (LEISA). Ralph/MVIC, an imager with visible panchromatic arrays (400–975 nm) and color arrays (Blue, Red, methane, and near-IR) and a spatial resolution of 20 μrad/pixel, can observe large fields of view efficiently by scanning the spacecraft. Ralph/LEISA, an infrared imaging spectrometer spanning 1.25–2.5 μm at a resolution ($\lambda/\Delta\lambda$) of 240 to 550, is used to measure composition, temperatures, and mixing states of the surfaces in the Pluto system. The Long Range Reconnaissance Imager (LORRI, Cheng *et al.*, 2008) is a high-resolution (5 μrad/pixel) panchromatic imager (350–850 nm) that is used for our highest resolution images, as well as observations on approach and optical navigation. The Radio Experiment (REX, Tyler *et al.*, 2008) is used for an uplink occultation to probe Pluto's atmosphere and to provide radiometry at 4.2 cm. The Pluto Energetic Particle Spectrometer Science Investigation (PEPSSI; McNutt *et al.*, 2008) is an energetic particle detector with 12 energy channels spanning 1–1000 keV, designed to study pickup ions from Pluto's escaping atmosphere. The Solar Wind at Pluto (SWAP, McComas *et al.*, 2008), a solar wind analyzer with a resolution $\Delta E/E < 0.4$ for energies between 25 eV and 7.5 keV, is used to measure the interaction of Pluto's atmosphere with the solar wind. The Venetia Burney Student Dust Counter (VB-SDC. Horányi *et al.*, 2008) measures the size distribution and density of particles with masses greater than $10^{-12} g$ during cruise.

The spacecraft was launched on January 19, 2006, used a Jupiter gravity assist with a closest approach on February 28, 2007, and will fly past Pluto on July 14, 2015. After

Pluto, we expect to be able to encounter one or two Kuiper Belt Objects (KBOs) between 2016 and 2020. The mission design is described in Guo & Farquhar (2008). Updates to the trajectory presented in Guo & Farquhar (2008) were made by the science team in 2007, when they optimized the arrival date and distance. The encounter date in Guo & Farquhar (2008), chosen to allow solar and Earth occultations by both Pluto and Charon, was fortuitously the best for surface studies, giving good views of Pluto's bright terrain, dark terrain, the bright/dark transition, and the CO-rich longitudes. It was also near optimal for observing Nix and Hydra, and had Pluto/Charon separations that needed little slewing time. The flyby distance in Guo & Farquhar (2008) was acceptable for all Group 1 (required) goals, but allowed very little time for high-resolution observations at moderate or high-phase angles. The science team, recognizing the importance of getting a diversity of observations and a diversity of terrains for this first flyby of the Pluto system, chose a slightly more distant flyby, at the expense of slightly lower resolutions at closest approach. In the current trajectory, closest approach to Pluto is at 2015 July 14 11:50:00 UT, and New Horizons flies 13,695 km, 29,432 km, 22,012 km, and 77,572 km from the centers of Pluto, Charon, Nix, and Hydra.

3. Planned Observations at the Pluto System

Young *et al.* (2008) presents the strawman observing plan as formed pre-launch, between 2001 and 2005. These plans have been refined beginning in 2008. In particular, the time span covering seven days before to two days after closest approach have been completely defined by the science team and sequenced by the science operations team. At the time of writing, this nine-day sequence is being delivered by Science Operations to be run on our spacecraft simulator. Here, we describe updates to the planned observations at Pluto.

Approach Phase 1 runs from 180 to 100 days before closest approach (P-180 to P-100 days). During this time, the plasma instruments, SWAP and PEPSSI, will be measuring the ambient plasma to characterize the differences near Pluto. All four objects in the Pluto system (Pluto, Charon, Nix, and Hydra) are well separated from one another and detectable by LORRI, with Pluto being barely resolved (about 2.4–3 LORRI pixels in diameter). From 170 to 130 days before closest approach, LORRI will observe the system to improve the orbits of Pluto, Charon, Nix and Hydra, and look for variability in albedo patterns.

Approach Phase 2 runs from P-100 to P-21 days. Observations include the plasma, orbital, and panchromatic variability observations of Approach Phase 1. In addition, we begin looking for color variability, and search for satellites and rings. The start of this phase is chosen to roughly coincide with the time when LORRI resolution is better than that obtainable by HST.

Approach Phase 3 runs from P-21 to P-1 days, of which P-7 to P-1 days has been sequenced. Since Pluto rotates once every 6.4 days, this includes Pluto's best, second best, and third-best rotation before closest approach. We continue the observations from Approach Phase 2. During this period, PEPSSI and SWAP may detect pickup ions and the bow shock, contributing to the Group 1 goal of measuring atmospheric escape. In this period, Ralph/LEISA and Alice can begin looking for evidence of variability in the IR or UV. Near the end of this period, we can search for clouds or hazes, and track winds if discrete clouds exist. Several particularly photogenic shots for Education/Public Outreach happen in this phase, such as shots of the entire system just filling a LORRI field of view. Perhaps most importantly, this phase allows global maps of Pluto and Charon at longitudes other than those seen near closest approach.

Near Encounter Phase runs from P-1 to P+1 days, and has been entirely sequenced. The start of this phase continues the plasma, global mapping and cloud tracking observations from the end of Approach Phase 3. This phase also contains observations designed to study the shape and topography of Pluto and Charon (such as observations when the target fills a LORRI frame), UV observations for surface reflectance, high-resolution geologic, color, and compositional observations, and observations to measure temperatures with IR spectra and radiometry. This phase includes most of the Group 1 observations.

Departure Phase 1 runs from P+1 to P+21 days, of which P+1 to P+2 days has been sequenced. Plasma observations of pickup ions, the magnetotail (if present), and the ambient plasma environment, relating to the Group 1 goal of atmospheric escape, are taken throughout this phase. Remote sensing observations of Pluto and Charon are taken only during the best departure rotation, ending at P+6.4 days. These relate to the Group 1 goal of measuring Pluto's phase function. In this phase, we also measure nightside temperatures of Pluto with REX at two different longitudes, observe Nix and Hydra at high phase, and search for rings in forward scattering.

Departure Phase 2 runs from P+21 to P+100 days. Plans for Departure Phase 2 and 3 are less well developed than the other phases. SWAP and PEPSSI operate in this phase, as in all the phases. There will be a remote sensing campaign near P+30 days, including additional searches for rings or material in the Pluto-Charon L4/L5 points.

Departure Phase 3 runs from P+100 to P+180 days. The spacecraft is very quiet in this phase, with no remote sensing observations planned. SWAP and PEPSSI will continue to monitor the ambient plasma environment, and the VB-SDC will measure the dust environment near Pluto.

The plan described here differs from the pre-launch strawman timeline in Young *et al.* (2008) in five significant ways. First, the observations from P-7 to P+2 days have been sequenced, and include realistic overhead times (time to slew and settle, allocate the solid-state recorder, power cycle instruments, etc.), and realistic error ellipses (including time-of-flight knowledge and the uncertainty of the orbits of Nix and Hydra around the system barycenter). Second, it addresses the additional goals added by the Science Team (Table 1). Third, the encounter is much more robust. All Group 1 goals are addressed by one or more main observation, backup observations in case a single observation fails, and an observation with alternate instrument in case of a failure of a single instrument. For example, the Group 1 global maps of Pluto are taken with Ralph/MVIC in panchromatic at 0.46 km/pix, with Ralph/MVIC in color at 0.64 km/pix as a backup, and with LORRI at 0.85 km/pix as an alternate. Most Group 2 and 3 observations are similarly robust. Most goals are addressed with complementary measurement techniques, such as using both IR spectroscopy and 4.2-cm radiometry to measure surface temperatures. Fourth, all requested observations have been defined and ranked by the Pluto Encounter Planning team, a team that includes Leslie Young (Deputy Project Scientist and Pluto Encounter Planning lead), Alan Stern (New Horizons Principal Investigator), Jeff Moore (Geology, Geophysics, and Imaging lead), Will Grundy (Surface Composition lead), Randy Gladstone (Atmospheres lead) and Fran Bagenal (Plasma lead). This ranking has allowed the planning team to resolve all demands on resources (time, recorder volume, thruster usage) based on overall science return. Fifth, all observations from a May 4, 2009 version of the sequence have been audited by the Pluto Encounter Planning team for pointing, exposure time, roll angle, etc.; the October 2009 version of the sequence includes changes identified by the science team.

Nearly all the Group 1 observations are addressed in the period from 2.5 hours before to 1 hour after Pluto closest approach, from the 8.4 km/pixel Charon IR maps to the Pluto solar and Earth occultations (Table 2). This includes the Group 1 Pluto and Charon

Table 2. Observation plans near Closest Approach

Time[1]	Target	Instrument	Description	Resolution (km)	Solar phase angle (°)
−02:30	Charon	LEISA	C_LEISA_LORRI_1	8.4	27.2
		LORRI		0.68	
		Alice		238	
−02:15	Pluto	LEISA	P_LEISA_Alice_2a	6.7	22.2
		Alice		190	
−02:01	Pluto	LEISA	P_LEISA_Alice_2b	6.0	23.1
		Alice		169	
−01:45	Nix	LEISA	N_LEISA_LORRI_BEST	3.6	10.5
		LORRI		0.29	
−01:34	Pluto	LORRI	P_LORRI_STEREO_MOSAIC	0.37	25.4
		MVIC/Pan		1.56	
−01:14	Charon	LEISA	C_LEISA_HIRES	4.7	37.2
		LORRI		0.38	
		Alice		133	
−01:06	Charon	MVIC/Color	C_COLOR_2	1.4	39.4
−00:55	Pluto	LEISA	P_LEISA_HIRES	2.7	32.7
		LORRI		0.22	
		Alice		77	
−00:39	Pluto	MVIC/Color	P_COLOR_2	0.64	39.6
		Alice		56	
−00:32	Nix	MVIC/Pan	N_PAN_CA	0.46	92.2
−00:27	Pluto	MVIC/Pan	P_MPAN_1	0.46	49.8
		LORRI		0.11	
−00:23	Pluto	Alice	P_ALICE_AIRGLOW_HELD_1	121	55.2
−00:13	Pluto	MVIC/Pan	P_MVIC_LORRI_CA	0.29	75.8
		LORRI		0.07	
−00:07	Pluto	Alice	P_ALICE_AIRGLOW_HELD_2	78	91.8
−00:01	Charon	MVIC/Pan	C_MVIC_LORRI_CA	0.61	85.9
		LORRI		0.15	
+00:02	Pluto	MVIC/Pan	P_PHOTSCAN	0.26	121.4
+00:05	Pluto	REX	P_REX_THERMSCAN	250	132.4
+00:15	Pluto	MVIC/Pan	P_HIPHASE_HIRES	0.38	151.7
		LORRI		0.09	
+00:22	Pluto	MVIC/Pan	P_CHARONLIGHT	0.47	161.2
+01:02	Pluto/Sun	Alice	P_OCC	–	180.0
	Pluto/Earth	REX		–	

[1] Time from Pluto Closest Approach, 2015 July 15 11:50:00 UT, in hours:minutes

panchromatic maps (P_MPAN_1 and C_LEISA_LORRI_1), color maps (P_COLOR_2, C_COLOR_2), IR maps (P_LEISA_Alice_2a/2b, C_LEISA_LORRI_1), and the solar and Earth Pluto occultation (P_OCC).

Geology and panchromatic imaging. In addition to the Group 1 hemispheric observations of Pluto and Charon at 0.5 and 0.6 km/pixel, New Horizons will take higher resolution regional images. One pair – LORRI at 0.4 km/pix, 25° phase and Ralph/MVIC at 0.3 km/pix, 76° phase – will be used for stereo imaging on Pluto, to derive digital elevation models to 110 m accuracy over about a quarter of Pluto's visible disk. LORRI is used simultaneously with Ralph scans to observe three high-resolution stripes across Pluto's disk at 0.07, 0.11, and 0.22 km/pixel, and one across Charon at 0.15 km/pixel. The philosophy of the Pluto Encounter Planning team is to observe both Nix and Hydra well, but to emphasize Nix in order to well characterize one of Pluto's small moons. Therefore, we plan to observe Hydra at 1.14 km/pixel (40–146 pixels across Hydra's disk), and Nix at 0.46 km/pixel (87–307 pixels across Nix). A LORRI observation taken simultaeously with Ralph/LEISA has the potential for 0.29 km/pixel on Nix, depending where Nix falls in its error ellipse. Measuring the albedo of Pluto's dark pole will be important for constraining volatile transport models, and so we attempt this high-phase measurement twice, once with Ralph/MVIC just after closest approach (P_CHARONLIGHT), and once with LORRI 14 hours later. During Pluto's final approach rotation, we make global maps of the sunlit portions of Pluto and Charon at resolutions better than 30 km/pixel with LORRI, while on departure, observations taken every 3 to 6 hours allow us to build up maps of Pluto and Charon at high phase.

Color, composition, and temperature. The requirement for color images is 1.5–10 km/pix. New Horizons will exceed this, with Pluto and Charon color images at 0.7 km/pix and 1.4 km/pix, respectively. Nix and Hydra are also observed in color, at 2.0 and 4.6 km/pix respectively, with Ralph/MVIC. These color observations include a filter matched to methane's 890 nm absorption feature, so that, at least on Pluto, we will be able to derive crude compositional maps at geologic scales. For compositions, temperatures, and mixing states from infrared spectroscopy, we observe the entire hemispheres of Pluto, Charon, Nix, and Hydra at 6.0, 8.4, 3.6, and 14.6 km/pixel respectively with Ralph/LEISA. Hydra is expected to be barely resolved, but, as previously mentioned, the goal is to characterize Nix well. Regional IR maps of Pluto and Charon are at higher resolutions, 2.7 and 4.7 km/pixel, respectively. During Pluto's final approach rotation, we make global maps of the sunlit portions of Pluto and Charon at resolutions better than 120 km/pixel in color with Ralph/MVIC, and better than 370 km/pixel with Ralph/LEISA. On departure, observations are taken at specific longitudes to characterize the phase function and IR spectra of different types of terrain, and break some grain-size/temperature/mixing-ratio ambiguities. In addition to the disk-integrated brightness at 0.1 K sensitivity of the day and nightsides of Pluto and Charon, REX will observe Pluto with resolved thermal radiometry along two tracks at 0.3 K sensitivity per 325 km footprint, where one track crosses the specular point and the other crosses Pluto's nightside pole.

Atmosphere and Plasma. We observe both solar and Earth occultations of both Pluto (extending well above any expected ionoshere) and Charon (extending out to Charon's Hill sphere). In addition, we observe a stellar occultation by Pluto as a backup and to probe different areas of Pluto's atmosphere, and two Pluto appulses to further characterize Pluto's upper atmosphere. In the Near Encounter Phase, Alice observes 6.5 hours of airglow on approach and 3.6 hours on departure, with two opportunities near closest approach to resolve airglow at 78–121 km resolution (P_ALICE_AIRGLOW_HELD_1 and _2), and airglow of Charon and Nix to search for signs of an atmosphere. Images at high phase angle will be taken to look for hazes (P_HIPHASE_HIRES and others). Starting at P-1.5 days, we observe Pluto at a variety of time cadences to search for clouds, plumes, and wind motions. To search for a cloud of atomic H around Pluto, Alice will look for Lyman-α absorption near Pluto. The plasma instruments will be in nearly continuous operation in 2015, with continuous operation in the Near Encounter Phase. Regular rolls of the spacecraft will insure that the instruments will observe the planetary pickup ions.

4. Kuiper Belt Object flyby

Shortly after Pluto closest approach, we plan for New Horizons to execute a trajectory correction maneuver to direct the spacecraft to its next target, a Kuiper Belt Object (KBO), The KBO has not yet been identified or selected. Most likely, the encounter will be in 2018 or 2019, near 42 AU, due to the intrinsic peak in KBO distribution, the narrower cone of accessibility at smaller distances, and the faintness of more distant KBOs (Spencer *et al.* 2003).

The search area is in the Milky Way, making the search for faint KBOs (down to $V = 27$) difficult. The search area shrinks with time as it converges on the spacecraft trajectory, defined by KBO velocity dispersion, not available delta-V. We plan to select and fund KBO search teams in 2010 for searches in 2011 and 2012. Interested teams should contact Alan Stern (alan@boulder.swri.edu).

References

Cheng, A. F. *et al.* 2008, *Space Sci. Revs*, 140, 215
Guo, Y. & Farquhar, R. W. 2008, *Space Sci. Revs*, 140, 49
Horányi, M. *et al.* 2008, *Space Sci. Revs*, 140, 387
McComas, D. *et al.* 2008, *Space Sci. Revs*, 140, 261
McNutt, R. L. *et al.* 2008, *Space Sci. Revs*, 140, 315
Reuter, D. C. *et al.* 2008, *Space Sci. Revs*, 140, 129
Stern, S. A. 2008, *Space Sci. Revs*, 140, 3
Stern, S. A. *et al.* 2008, *Space Sci. Revs*, 140, 155
Spencer, J. S. *et al.* 2003, *Earth, Moon and Planets*, 92, 483
Tyler, G. L. *et al.* 2008, *Space Sci. Revs*, 140, 217
Weaver, H. A. 2008, *Space Sci. Revs*, 140, 75
Young, L. A. *et al.* 2008, *Space Sci. Revs*, 140, 93

Icy Bodies of the Solar System
Proceedings IAU Symposium No. 263, 2009
J.A. Fernández, D. Lazzaro, D. Prialnik & R. Schulz, eds.

© International Astronomical Union 2010
doi:10.1017/S1743921310001997

The Rosetta Mission: Comet and Asteroid Exploration

Rita Schulz

ESA Research and Scientific Support Department, ESTEC,
Keplerlaan 1, NL-2001 AZ Noordwijk, The Netherlands
email: rschulz@rssd.esa.int

Abstract. In March 2004 the European Space Agency launched its Planetary Cornerstone Mission Rosetta to rendezvous with Jupiter-family comet 67P/Churyumov-Gerasimenko. The Rosetta mission represents the next step into the improvement of our understanding of comet nuclei naturally following the four successful comet nucleus fly-by missions carried out in the past. It will however not perform a simple fly-by at its target comet, but combines an Orbiter and a Lander Mission. The Rosetta spacecraft will go in orbit around the comet nucleus when it is still far away from the Sun, and escort the comet for more than a year along its pre- and post-perihelion orbit while monitoring the evolution of the nucleus and the coma as a function of increasing and decreasing solar flux input. Different instrumentations will be used in parallel, from multi-wavelength spectrometry to in-situ measurements of coma and nucleus composition and physical properties. In addition the Rosetta Lander Philae will land on the nucleus surface, before the comet is too active to permit such a landing (i.e. at around r = 3 AU) and examine the surface and subsurface composition as well as its physical properties. Two fly-bys at main belt asteroids have been scheduled for the Rosetta spacecraft during its journey to the comet. The first fly-by at E-type asteroid (2867) Steins was already successfully executed in September 2008. The second and main fly-by at asteroid (21) Lutetia is scheduled for July 2010.

Keywords. comets: general, individual (67P/Churyumov-Gerasimenko), minor planets, asteroids

1. Introduction

Cometary matter is believed to constitute a unique repository of information on the sources that contributed to the proto-solar nebula, as well as on the condensation processes that resulted in the formation of planetesimals, which later formed larger planetary bodies. Their high content of frozen volatiles and organics make comets particularly interesting in view to understanding solar system ices. However, it is these characteristics which make comets so difficult to study. Direct evidence on the volatiles in a comet nucleus is particularly difficult to obtain, as remote observations, even during fly-by missions, only cover species in the coma, which have been altered by physico-chemical processes such as sublimation and interactions with solar radiation and the solar wind. The Rosetta mission represents one possible solution to the problem by combining two strategies of characterizing the properties of a comet nucleus (Schulz, 2009). The approach of a rendezvous mission with a probe staying close to the nucleus along a major part of the orbit and performing comprehensive remote-sensing and analytical investigations of material from the nucleus and the coma, guarantees by design minimal perturbations of the comet material as analyses are performed in situ, at low temperatures, and in a microgravity environment. Furthermore, nucleus material will be analysed in-situ by the instruments on board the Philae Lander at a time when the nucleus is still at a low state of activity.

2. Mission Overview

On 2 March 2004 Rosetta started its 10-year journey to rendezvous with Jupiter-family comet 67P/Churyumov-Gerasimenko with an Ariane-5 launch from Kourou in French Guiana. To acquire sufficient orbital energy for being able to rendezvous with its target comet and go in orbit around the nucleus the spacecraft has performed four planetary gravity assist manoeuvres (Earth 4 Mar. 2005, Mars 25 Feb. 2007, Earth 13 Nov. 2007, Earth 13 Nov. 2009). After the second and third Earth gravity assist there is a close fly-by of the spacecraft at a main-belt asteroid. The first of these close asteroid encounters took place on 5 September 2008, when Rosetta passed E-type asteroid (2867) Steins at a distance of 802.6 km (Schulz, 2009b). On 10 July 2010 the spacecraft will fly-by (21) Lutetia, a large asteroid with an estimated diameter of 95 km which has been classified to be either of C-type or of M-type. After this second and last asteroid fly-by Rosetta will move into the outer solar system to rendezvous with its target comet at r = 4.5 AU in May 2014 and go into close orbit in September 2014, when the comet is at r = 3.4 AU and Δ = 2.8 AU. Rosetta will start the global observation and mapping of the nucleus during which the spacecraft will fly down to distances of a few kilometres from the surface. The landing of Philae on the surface of the comet nucleus is planned for 10 November 2014 when the comet is about 3 AU away from the Sun. After a five day prime Lander mission, both, the Orbiter and the Lander will enter the routine scientific phase, escorting the comet to perihelion (12 August 2015) and beyond. The nominal mission will end on 30 December 2015.

3. Spacecraft and Payload

The Rosetta spacecraft is a 3-axis stabilised aluminium box with dimensions of 2.8 × 2.1 × 2.0 metres. The scientific payload is on the top panel (Payload Support Module) either body mounted or attached to one of the deployable booms, while the subsystems are on the base panel (Bus Support Module). The steerable 2.2-m high-gain antenna and the Lander Philae are attached to two opposite sides of the spacecraft and the solar panel wings extend from the other two sides. In the vicinity of comet 67P/Churyumov-Gerasimenko the scientific instruments on the orbiting spacecraft point almost always towards the comet, while the antenna and solar arrays point towards the Earth and the Sun. Each solar panel has a length of 14 m which results in a total area of 64 m^2 and a total span of 32 m from tip to tip. The total launch mass of Rosetta was estimated to be about 2900 kg including 1720 kg of propellant, 165 kg of scientific payload on the orbiter plus 110 kg for the lander Philae.

The payload of the Rosetta orbiter consists of 11 instruments (Table 1) which con-jointly have unprecedented capabilities to study the volatile and the refractory material in the coma and the composition and physical properties of the comet nucleus. The re-mote sensing suite of instruments characterizes the nucleus surface in a wide wavelength range from the far-UV (70 nm) to millimetre (1.3 mm) with high spatial resolutions. On top of determining the size, shape, rotational state, and detailed surface topography the OSIRIS camera system also characterizes sublimation and erosion processes on the nucleus surface. The infrared imaging spectrometer VIRTIS focuses on the detection and characterization of specific signatures such as typical spectral bands of minerals and molecules, to identify and quantify different constituents of comet material. Together with the MIRO instrument it also studies the thermal evolution of the comet nucleus as a function of solar radiation input. MIRO determines at micro-wavelength the surface and near-surface temperatures as well as the temperature gradient in the nucleus. It also

Table 1. Rosetta Orbiter Payload

Short Name	Objective	Principal Investigator
	Remote Sensing	
OSIRIS	Multi-Colour Imaging	H. U. Keller, MPS
	(Narrow and Wide Angle Camera)	Katlenburg-Lindau, Germany
ALICE	UV-Spectroscopy	A. Stern, SRI
	(70 nm - 205 nm)	Boulder, CO, USA
VIRTIS	VIS and IR Mapping Spectroscopy	A. Coradini
	(0.25 μm - 5 μm)	IAS-CNR, Rome, Italy
MIRO	Microwave Spectroscopy	S. Gulkis, NASA-JPL
	(0.5 mm and 1.3 mm)	Pasadena, CA, USA
	Mass Spectrometres	
ROSINA	Neutral Gas and Ion Mass Spectroscopy	H. Balsiger
	DFMS: 12-150 AMU, M/ΔM $\sim$ 3000	Univ. Bern, Switzerland
	RTOF: 1-350 AMU, M/ΔM $>$ 500	
	COPS gas density and velocity	
COSIMA	Dust Mass Spectrometer	M. Hilchenbach, MPS
	(SIMS, m/Δm $\sim$ 2000)	Katlenburg-Lindau, Germany
	Dust Flux and Physical Properties	
GIADA	Grain Impact Analyser	L. Colangeli, Oss. Astro.
	Dust Accumulator	Capodimonte, Italy
MIDAS	Grain Morphology, nm resolution	W. Riedler, IWF
	(Atomic Force Microscope)	Graz, Austria
	Radio Experiments	
CONSERT	Radio Sounding	W. Kofman, CEPHAH
	Nucleus Tomography	Grenoble, France
RSI	Radio Science Experiment	M. Pätzold
		Univ. Köln, Germany
	Comet Plasma Environment, Solar Wind Interaction	
RPC	Langmuir Probe (LP)	A. Eriksson, IRF
		Uppsala, Sweden
	Ion and Electron Sensor (IES)	J. Burch, SRI
		San Antonio, TX, USA
	Flux Gate Magnetometer (MAG)	K.-H. Glassmeier, IGEP
		Braunschweig, Germany
	Ion Composition Analyser (ICA)	R. Lundin, IRF
		Kiruna, Sweden
	Mutual Impedance Probe (MIP)	J.G. Trotignon, LPCE/CNRS
		Orleans, France
	Plasma Interface Unit (PIU)	C. Carr, Imperial College
		London, UK
SREM	Radiation Environment Monitor	

measures outgassing rates and isotopic ratios of certain major volatile species by measuring their molecular transitions. The ultraviolet imaging spectrograph ALICE determines production rates of numerous atoms and gas molecules, studies small dust grains and ions in the coma, and characterizes the surface of the nucleus at UV wavelengths.

Four instruments collect and analyse gas and dust samples from the inner coma focussing on different aspects. The ROSINA instrument measures the global molecular, elemental, and isotopic composition of neutral gas and ions by in-situ mass spectroscopy. The in-depth analysis of the dust coma is shared by three instruments. The Grain Impact Analyser and Dust Accumulator, GIADA explores the dust flux evolution and grain dynamic properties with position and time, while the secondary ion mass spectrometer COSIMA performs the in-situ compositional analysis of the coma dust including chemical

Table 2. The Payload of Lander Philae.

Short Name	Objective	Principal Investigator
	Imaging	
ÇIVA	Panoramic imaging	J.P. Bibring, IAS
	Vis-IR microscopic imaging	Orsay, France
	of samples	
ROLIS	Descent and	S. Mottola, DLR
	Down-Looking Camera	Berlin, Germany
	Composition	
APX	α-X-ray Spectrometer	G. Klingelhöfer
	Elemental surface composition	Univ. Mainz, Germany
COSAC	Evolved Gas Analyser	F. Goesmann, MPS
	Molecular composition and	Katlenburg-Lindau
	chirality of samples	Germany
Ptolemy	Evolved Gas Analyser	I. P. Wright
	Isotopic composition	Open University, UK
	of light elements in sample	
	Physical Properties	
MUPUS	Multi-Purpose Sensor for	T. Spohn, DLR
	(Sub-)Surface Science	Berlin, Germany
SESAME	Comet Acoustic Surface Sounding	K.J. Seidensticker, DLR
	Experiment (CASSE)	Köln, Germany
	Dust Impact Monitor (DIM)	I. Apathy, KFKI, Hungary
	Permittivity Probe (PP)	W. Schmidt, FMI, Finland
SD2	Drilling and sampling and	A. Ercoli-Finzi
	Distribution Device	Politechnico, Milano, Italy
ROMAP	Magnetic and Plasma Monitoring	U. Auster, IGEP
		Braunschweig, Germany
CONSERT	Radio Sounding	W. Kofman, CEPHAG
	Nucleus Tomography	Grenoble, France

characterisation of main organic components, present homologous and functional groups, as well as the mineralogical and petrographical classification of inorganic phases. The dimensions and microstructure of individual dust grains are determined by the MIDAS instrument, the first atomic force microscope on a space mission.

The nucleus mass, bulk density, gravity coefficients, and moments of inertia are determined by radio science techniques (RSI) and the deep interior of the comet nucleus is investigated by radio sounding with the CONSERT instrument, which transmits long wavelengths radio waves through the nucleus. The comet plasma environment and its interaction with the solar wind is monitored in-situ by five sensors, which have been combined into one instrument suite sharing common subsystems and run by the Rosetta Plasma Consortium (RPC). The radiation environment is measured by a Standard Radiation Monitor (SREM).

The Rosetta lander Philae carries another 10 scientific experiments (Table 2) and will perform the first in situ analysis of comet nucleus material. Philae investigates in-situ the composition and physical properties of the comet nucleus by measuring the elemental, molecular, mineralogical, and isotopic composition of surface and subsurface material down to a depth of about 20 cm, and determining the nucleus mechanical, electrical, and thermal properties. Also the nucleus close environment, its large scale structure, interior, and activity will be studied.

Detailed descriptions of the Rosetta payload can be obtained in Schulz *et al.* (2009).

4. Concluding Remarks

The Rosetta rendezvous mission to comet 67P/Churyumov-Gerasimenko determines the composition and physical properties of a comet nucleus through in-situ measurements and remote sensing observations. The evolution of the comet along the orbit around the Sun is investigated by monitoring the nucleus and the near-nucleus environment from a pre-perihelion distance of about 3.4 AU (orbit insertion) through perihelion passage at 1.24 AU and back out to about 2 AU post-perihelion. Rosetta studies how a comet nucleus develops its activity at large heliocentric distance and how such a process functions for high solar radiation close to the Sun. The chemical, mineralogical and isotopic composition of both, the comet nucleus (near the surface) and the inner coma is determined allowing identification and quantification of the chemical reaction chains by which the observed coma gas species are produced from the original icy material on the nucleus. The mission will provide information on "how a comet works" and lead us to a better understanding of the early solar nebula and the evolution of our planetary system. The study of the chemical and isotopic composition of the ices and the dust grains in the comet provides information on the processes through which these compounds formed and about pre-biotic chemistry. Rosetta help answering the question what role comets played in the evolution of life on Earth.

References

Schulz, R. 2009a, *Solar System Research*, 43/4, 343
Schulz, R. 2009b, *Bull. Am. Astron. Soc.*, 41, 563
Schulz, R., Alexander, C., Boehnhardt, H., Glassmeier, K.-H. 2009, *Rosetta ESAs Mission to the Origin of the Solar System* (Springer Science+Business Media), ISBN 978038777517-3.

Icy Bodies of the Solar System
Proceedings IAU Symposium No. 263, 2009
J.A. Fernández, D. Lazzaro, D. Prialnik & R. Schulz, eds.

Deep Impact ejection from Comet 9P/Tempel 1 as a triggered outburst

Sergei I. Ipatov[1] and Michael F. A'Hearn[2]

[1] Catholic University of America
Washington DC, USA
email: siipatov@hotmail.com

[2] Dept. of Astronomy, University of Maryland,
College Park, MD, USA
email: ma@astro.umd.edu

Abstract. Ejection of material after the Deep Impact collision with Comet Tempel 1 was studied based on analysis of the images made by the Deep Impact cameras during the first 13 minutes after impact. Analysis of the images shows that there was a local maximum of the rate of ejection at time of ejection ∼10 s with typical velocities ∼100 m/s. At the same time, a considerable excessive ejection in a few directions began, the direction to the brightest pixel changed by ∼50°, and there was a local increase of brightness of the brightest pixel. The ejection can be considered as a superposition of the normal ejection and the longer triggered outburst.

Keywords. comets: general; solar system: general

1. Analysis of images

In 2005 the Deep Impact (DI) impactor collided with Comet 9P/Tempel 1 (A'Hearn *et al.* 2005). Our studies (Ipatov & A'Hearn 2008, 2009) of the time variations in the projections v_p of characteristic velocities of ejected material onto the plane perpendicular to the line of sight and of the relative rate r_{te} of ejection were based on analysis of the images made by the DI cameras during the first 13 min after the impact. We studied velocities of the particles that were the main contributors to the brightness of the cloud of ejected material, that is, mainly particles with diameter $d < 3$ μm. Below we present a short description of analysis of the images and the conclusions based on our studies. Details of the studies and more figures and references are presented in Ipatov & A'Hearn (2008). More complicated models of ejection will be studied in future.

Several series of images taken through a clear filter were analyzed. In each series, the total integration time and the number of pixels in an image were the same. As in other DI papers, original images were rotated by 90° in anti-clockwise direction. In DI images, calibrated physical surface brightness (hereafter CPSB, always in W m^{-2} sterad^{-1} micron^{-1}) is presented. For several series, we considered the differences in brightness between images made after impact and a corresponding image made just before impact. Overlapping of considered time intervals for different series of images allowed us to calculate the relative brightness at different times t after impact, though because of non-ideal calibration, the values of peak brightness at almost the same time could be different (up to the factor of 1.6) for different series of images. Variation in brightness of the brightest pixel and the direction from the place of ejection to the brightest pixel were studied. At $t > 100$ s, some DI images do not allow one to find accurately the relative brightness of the brightest pixel and the direction from the place of ejection to this pixel. First, there are large regions of saturated pixels on DI images, which may not allow one to calculate accurately the peak brightness. Secondly, at $t \sim 600 - -800$ s, coordinates of the brightest pixel were exactly the same for several images. It can mean that some pixels became 'hot' when the distance between the spacecraft and the nucleus of the comet became small.

We analyzed the time dependencies of the distances L from the place of ejection to contours of CPSB = const for several levels of brightness and different series of images (Fig. 1). Based on the supposition that the same moving particles corresponded to different local maxima (or minima) of $L(t)$ (e.g., to values L_1 and L_2 on images made at times t_1 and t_2), we calculated the characteristic velocities $v = (L_1 - L_2)/(t_1 - t_2)$ at time of ejection $t_e = t_1 - L_1/v$. In this case, we use results of studies of series of images in order to obtain one pair of v_p and t_e. Such approach to calculations of velocities doesn't take into account that particles ejected at the same time could have different velocities. According to theoretical studies presented by Housen et al. (1983), velocities of ejecta are proportional to $t_e^{-\alpha}$ (with α between 0.6 and 0.75), and the rate r_{te} of ejection is proportional to $t_e^{0.2}$ at $\alpha = 0.6$ and to $t_e^{-0.25}$ at $\alpha = 0.75$. Our estimates of the pairs of v_p and t_e were compared with the plots of $v_{expt} = v_p = c(t/0.26)^{-\alpha}$ at several pairs of α and c (0.26 was considered because the second ejection began mainly at $t_e \approx 0.26$ s). The comparison testifies in favor of mean values of $\alpha \sim 0.7$. Destruction and sublimation of particles and variation in their temperature could affect on the brightness of the DI cloud, but, in our opinion, don't affect considerably on our estimates of velocities and slightly change estimates of α.

For the edge of bright region (usually at CPSB=3), the values of $L=L_b$ in km at approximately the same time can be different for different series of DI images considered. Therefore, we cannot simply compare L_b for different images, but need to calculate the relative characteristic size L_r of the bright region, which compensates non-ideal calibration of images and characterizes the size of the bright region. Considering that the time needed for particles to travel a distance L_r is equal to $dt=L_r/v_{expt}$, we find the time $t_e=t-dt$ of ejection of material of the contour of the bright region considered at time t.

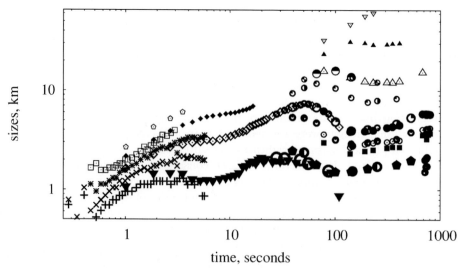

Figure 1. Time variations in sizes L (in km) of regions inside contours of CPSB=C. Different signs correspond to different series of images at different C. The curves have local minima and maxima that were used for analysis of time variations in velocities.

The volume V_{ol} of a spherical shell of radius L_r and width h is proportional to $L_r^2 h$, and the number of particles per unit of volume is proportional to $r_{te} \cdot (V_{ol} \cdot v)^{-1}$, where v is the velocity of the material. Here r_{te} corresponds to the material that was ejected at t_e and reached the shell with L_r at time t. The number of particles on a line of sight, and so the brightness B_r, are approximately proportional to the number of particles per unit of volume multiplied by the length of the segment of the line of sight inside the DI cloud, which is proportional to L_r. Actually, the line of sight crosses many shells characterized by different r_{te}, but as a first approximation we supposed that $B_r \propto r_{te}(v \cdot L_r)^{-1}$. For the edge of the bright region, $B_r \approx$ const. Considering $v = v_{expt}$, we calculated the relative rate of ejection as $r_{te}=L_r t^{-\alpha}$. Based on this dependence of r_{te} on time t and on the obtained relationship between t and t_e, we constructed the plots of dependences of r_{te} on t_e (Fig. 2). Because of high temperature and brightness of ejecta, the real values of r_{te} at $t_e < 1$ s are smaller than those in Fig. 2. As typical sizes of ejected particles increased with t_e, the real rate of ejection decreased more slowly than the plot in this figure. If, due to the outburst, typical velocities of observed ejected particles did not decrease much at $t_e > 100$ s, then the values of r_{te} could be greater than those in Fig. 2.

Excessive ejection in several directions ('rays' of ejection) was studied based on analysis of the form of contours of constant brightness. Bumps of the contours considered by Ipatov & A'Hearn (2008) include the upper-right bump ($\psi \sim 60 - 80°$, still seen at $t \sim 13$ min), the right bump (ψ increased from $90°$ at $t \sim 4 - 8$ s to 110-$120°$ at $t \sim 25 - 400$ s), the left bump ($\psi \sim 245 - 260°$), which transformed with time into the down-left bump ($\psi \sim 210 - 235°$), the upper bump (backward ejection, ψ varied from 0 to -25°, the bump consisted mainly of particles ejected after 80 s), where ψ is the angle between the upper direction and the direction to a bump measured in a clockwise direction. Together with hydrodynamics of the explosion, icy conglomerates of different sizes at different places of the ejected part of the comet could affect on the formation of the rays.

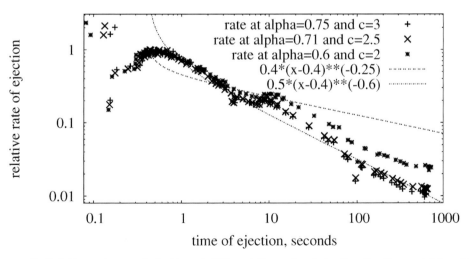

Figure 2. Relative rate r_{te} of ejection at different times t_e of ejection for the model in which the characteristic velocities of the edge of the observed bright region at time t are equal to $v_{expt}=c \cdot (t/0.26)^{-\alpha}$ (in km/s), for three pairs of α and c. The maximum rate at $t_e > 0.3$ s is considered to be equal to 1. Two curves of the type $y=c_r \cdot (x - c_t)^\beta$ are also presented for comparison.

2. Conclusions

Our studies showed that there was a local maximum of the rate of ejection at time of ejection $t_e \sim 10$ s (Fig. 2) with typical projections (onto the plane perpendicular to the line of sight) of velocities $v_p \sim 100 - -200$ m/s. At the same time, the considerable excessive ejection in a few directions (rays of ejecta) began, the direction to the brightest pixel quickly changed by about $50°$, and there was a local increase of brightness of the brightest pixel. On images made during the first 10–12 s, the direction was mainly close to the direction of impact; after the jump, it gradually came closer to the direction of impact. These features at $t_e \sim 10$ s were not predicted by theoretical models of ejection and could be caused by the triggered outburst. At the outburst, the ejection could be from entire surface of the crater, while the normal ejection was mainly from its edges. Starting from 10 s, the ejection of more icy material could begin. The increase in the fraction of icy material caused an increase in the observed ejection rate and the initial velocities (compared to the normal ejection).

At $1 < t_e < 3$ s and $8 < t_e < 60$ s, the plot of time variation in estimated rate r_{te} of ejection of observed material was essentially greater than the exponential line connecting the values of r_{te} at 1 and 300 s. The difference could be mainly caused by that fact that the impact was a trigger of an outburst. The sharp decrease of the rate of ejection at $t_e \sim 60$ s could be caused by the decrease of the outburst and/or of the normal ejection. The contribution of the outburst to the brightness of the cloud could be considerable, but its contribution to the ejected mass could be relatively small. Duration of the outburst (up to 30–60 min) could be longer than that of the normal ejection (a few minutes). The studies testify in favor of a model close to gravity-dominated cratering.

Projections v_p of the velocities of most of the observed material ejected at $t_e \sim 0.2$ s were about 7 km/s. As the first approximation, the time variations in characteristic velocity at $1 < t_e < 100$ s can be considered to be proportional to $t_e^{-0.75}$ or $t_e^{-0.7}$, but they could differ from this exponential dependence. The fractions of observed small particles ejected (at $t_e \leqslant 6$ s and $t_e \leqslant 14$ s) with $v_p > 200$ m/s and $v_p > 100$ m/s were estimated to be about 0.13–0.15 and 0.22–0.25, respectively, if we consider only material observed during the first 13 min and $\alpha \sim 0.7 - -0.75$. These estimates are in accordance with the previous estimates (100–200 m/s) of the projection of the velocity of the leading edge of the DI dust cloud, based on various ground-based observations and observations made by space telescopes. The fraction of observed material ejected with velocities greater than 100 m/s was greater than the estimates based on experiments and theoretical models. Holsapple & Housen (2007) concluded that the increase of velocities was caused by vaporization of ice in the plume and by fast moving gas. In our opinion, the greater role in the increase of high-velocity ejecta could be played by the outburst (by the increase of ejection of bright particles), and it may be possible to consider the ejection as a superposition of the normal ejection and the longer triggered outburst. Time variations in velocities could be smaller (especially, at $t_e > 100$ s) for the outburst ejecta than for the normal ejecta.

The excess ejection of material in a few directions (rays of ejected material) was considerable during the first 100 s and was still observed in images at $t \sim 500 - -770$ s. The sharpest rays were caused by material ejected at ~ 20 s. In particular, there were excessive ejections, especially in images at $t \sim 25 - -50$ s, in directions perpendicular to the direction of impact. Directions of excessive ejection could vary with time.

Acknowledgements

The work was supported by NASA DDAP grant NNX08AG25G.

References

A'Hearn, M. F. *et al.* 2005, *Science*, 310, 258

Holsapple, K. A. & Housen, K. R. 2007, *Icarus*, 187, 345

Housen, K. R., Schmidt, R. M., & Holsapple, K. A. 1983, *J. Geophys. Res.*, 88, 2485

Ipatov, S. I. & A'Hearn, M. F. 2008, *http://arxiv.org/abs/0810.1294*

Ipatov, S. I. & A'Hearn, M. F. 2009, *Lunar. Planet. Sci.* XL, 1022 (abstract)

Author Index

Object Index

Subject Index